Introducing C++ for Scientists, Engineers and Mathematicians

Springer
London
Berlin
Heidelberg
New York
Barcelona
Hong Kong
Milan
Paris
Singapore
Tokyo

Derek Capper

Introducing C++ for Scientists, Engineers and Mathematicians

Second Edition

 Springer

Derek Capper

British Library Cataloguing in Publication Data
Capper, D.M. (Derek M.)
   Introducing C++ for Scientists, engineers and mathematicians. – 2nd ed.
   1. C++ (Computer program language)
   I. Title II. C++ for scientists, engineers and mathematicians
   005.1'33
ISBN 1852334886

Library of Congress Cataloging-in-Publication Data
Capper, D.M. (Derek M.), 1947
   Introducing C++ for scientists, engineers and mathematicians / Derek Capper.—2nd ed.
      p. cm.
   Rev. ed. of: The C++ programming language for scientists, engineers and
mathematicians. 1994.
   Includes bibliographical references and index.
   ISBN 1-85233-488-6 (alk. paper)
   1. C++ (Computer program language) I. Capper, D.M. (Derek M.). 1947-. C++
programming language for scientists, engineers and mathematicians. II. Title
   QA76.73.C153 C36 2001
   005.13'3—dc21
                                                        2001034207

ISBN 1-85233-488-6 2nd edition Springer-Verlag London Berlin Heidelberg
ISBN 3-540-76042-3 1st edition Springer-Verlag Berlin Heidelberg New York
A member of BertelsmannSpringer Science+Business Media GmbH
http://www.springer.co.uk

First published 1994
Second edition 2001

Typeset by the author using LaTeX
Printed and bound at TJ International, Padstow, Cornwall, UK
34/3830-543210 Printed on acid-free paper SPIN 10834582

Contents

Preface to the Second Edition

Since the first edition of *C++ for Scientists, Engineers and Mathematicians*, many things have changed. Perhaps the most important is that the ANSI C++ Standard was approved in 1998. The first edition of this book was incompatible with some parts of the Standard and every effort has been made to rectify this. The facilities offered by the C++ language have also grown enormously since the first edition. It is not just the language itself that has grown, but the ANSI Standard also defines an extensive range of library facilities.

I have also made two changes in emphasis since the first edition. Firstly, since there is so much important material to cover, some less important features of the language have been removed. The second change is to try to illustrate the language by means of complete programs, rather than code fragments. Hopefully, this will give you more confidence in applying new techniques.

Finally, although C++ is a bigger language than it was, it is even more fun to use. So, enjoy it!

Preface to the First Edition

Computers are being used to solve problems of ever increasing complexity in science and engineering. Although some problems can be solved by means of "off the shelf" packages, many applications would benefit greatly from using a sophisticated modern programming language. Without such a language, it is difficult to develop and maintain complex software and the difficulties are considerably worse for multi-processor systems. Of the many possible languages, C is being increasingly used in scientific applications, but it has some dangers and disadvantages owing to its system programming origins. These drawbacks have largely been overcome by the introduction of C++, an object-oriented version of C, which was developed at AT & T Bell Laboratories by Bjarne Stroustrup.

C++ has proved to be an enormously popular language, but most of the available books assume a prior knowledge of C. Unfortunately, most books on C (in common with those on C++) are not specifically intended for scientists or engineers and, for instance, often use manipulation of character strings as the basis for examples. To the scientist struggling to solve a non-linear differential equation, rather than writing a text editor, such examples may not be very illuminating. C programmers also seem to revel in writing very terse code, which can be difficult for the non-expert to penetrate.

This book is specifically intended for scientists, engineers and mathematicians who want to learn C++ *from scratch* by means of examples and detailed applications, rather than a formal language definition. The examples used are deliberately numerical in character and many other topics of possible interest (such as databases, computer algebra and graph algorithms) are left untouched. For someone with little or no programming experience there is much to be gained by learning C++ rather than C. Instead of becoming a good C programmer and then converting to an object-oriented language, the reader can progressively use those aspects of C++ that are both relevant and accessible. For instance inline functions instead of macros, together with function name overloading and data hiding by means of classes, are all useful programming techniques that can be introduced at an early stage. Another advantage is that in many ways C++ is a safer language than C and encourages better programming practices.

No prior knowledge of any programming language is assumed and the approach is to learn by means of relevant examples. The mathematical techniques involved in these examples have been kept deliberately simple and are fundamental to any serious numerical application of computers. For any reader not familiar with a particular technique, the explanations given in the text should be sufficient and additional background material can be found in the bibliography. This bibliography is annotated to provide the reader with a guide to parallel reading whilst studying this book and to suggest pathways through the literature for more advanced study. In particular, we do not dwell on those aspects of C++ that only provide backward compatibility with C. Such features of the language may well be important to readers wishing to integrate existing C code into a C++ program and help can be obtained from the bibliography. Since our approach is to learn by example, rather than by formal definition, the bibliography (in particular [10] and [1]) should also be consulted if a precise definition of the language is required.

Structure of this Book

It is very unlikely that any application will need every C++ feature described in this book. However, with the exception of Chapter 18, the order in which material is presented means that you can progress as far as necessary in order to meet your particular needs, without working through the entire book. Even if a fairly complete understanding of C++ is required, it is worth studying the language in deliberate stages, interspersed with projects from your own field of interest. The following stages are suggested:

1. Chapters 1 to 5 introduce the basics of C++, including control structures and functions. These features are common to most languages that have found nu-

merical application and are essential for writing any significant program. Most applications would also need an understanding of Chapter 6, which introduces arrays. This stage is roughly equivalent to what you would have in a language such as FORTRAN. Most C++ programs make considerable use of pointers and, although this topic is introduced in Chapter 6, it is developed further in the following chapter. Here you will also find a description of strings and dynamic memory management.

2. Classes, objects, data hiding and member functions are introduced in Chapter 8, and Chapters 9 and 10 are devoted to operator overloading, constructors and destructors. Although by themselves such techniques are not sufficient to constitute object-oriented programming, they should enable you to write safer code for fairly large applications. After working through these chapters you should also be able to use C++ class libraries, such as matrix algebra, complex arithmetic, strings etc. However, you will not be able to use inheritance to extend existing classes to meet any special needs of your own specific applications.

3. Bitwise operations are described in Chapter 11. Such operations may not be central to your interests and you could decide to omit this chapter. However, an understanding of how numbers are stored in memory is very useful and you would miss the neat application to the Sieve of Eratosthenes.

4. Chapter 12 describes single inheritance. At this stage you can truly claim to be doing object-oriented programming with the ability to construct your own classes and to extend existing classes.

5. Multiple inheritance is introduced in Chapter 13. Although many experts would claim that multiple inheritance is an essential part of object-oriented programming, you will probably be developing quite large applications before using such techniques.

6. Chapters 14, 16 and 15 are about namespaces, templates and exception handling. Namespaces are useful for avoiding potential name clashes in large projects and templates provide a way of reusing code. Exception handling provides techniques for dealing with errors, and is particularly important for errors that are detected by library facilities. All three chapters are prerequisites for the penultimate chapter.

7. The ANSI C++ Standard defines an important and extensive set of library facilities. Chapter 17 provides an overview of this Standard Library. Chapter 18 describes input and output, which is actually part of the Standard Library. This means that many details of input and output are not covered until this final chapter. For example, file I/O is not introduced until this stage, although you may have applications that lead you to consult this chapter before working through the preceding chapters. The reason for leaving details of input and output until Chapter 18 is that it involves classes, operator overloading, constructors, destructors, inheritance, templates, exception handling etc. In fact the I/O facilities offered by the Standard Library employ just about every idea of the preceding chapters. However, don't despair! If you do need facilities such as file I/O at an

early stage, then you should be able to extract the necessary information without working through the entire book!

If used as an undergraduate text, it is likely that the material contained in all eighteen chapters would be too much for a single course of lectures. Rather it is envisaged that Chapters 1 to 9 would be used in a first course, with the following chapters being suitable for a subsequent course.

This book introduces some advanced features of C++ that could be omitted if a complete knowledge of the language is not required. The following techniques are used to identify such features:

1. Items that fit naturally into a particular chapter, but could reasonably be postponed until later, are marked by a single dagger (†). You would probably have to return to such sections in order to understand later chapters and you may have to postpone attempting some of the exercises. A typical example is Section 4.13.3 on Conditional Compilation; you will probably need to use this information sometime, but skipping the section on a first reading will not do any harm.

2. There are two kinds of section which are sufficiently self-contained that they can be omitted altogether without any loss of continuity and these are marked by double daggers (††). Some sections are only intended for reference. Other sections will only be of importance to particular applications; for example, bitwise operations (Chapter 11) could be crucial for a number theory application but of no relevance to solving a differential equation.

Lack of space prevents the inclusion of a chapter on object-oriented design in the context of numerical applications. However, this is a topic which deserves an entire book to itself and some recompense may be obtained by consulting the bibliography ([2], [7] and [10]).

Acknowledgements

I would like to thank Helen Arthan for conscientiously working through the various drafts of the first edition of this book and Mark Burton for removing many technical errors. Peter Dew and Ian Robson both provided inspiration for some of the applications.

Above all I wish to thank my wife, Janet, for her constant support, without which neither edition of this book would have been completed.

Chapter 1

Introduction

1.1 Getting Started

Probably the most frustrating stage in learning any language is getting started. Our first steps need to be modest and we start by simply sending a one-line message to the output device. Very few assumptions are made in this book, but it is assumed that you have access to a C++ compiler and that you know how to use an editor.[1] Suppose the following lines of code are entered into a file by means of an editor:

```
// A simple C++ program:

#include <iostream>
using namespace std;

int main()
{
    cout << "This is our first C++ program.\n";
    return(0);
}
```

This complete C++ program must go through a two-stage process before it can be run. First of all the code needs to be compiled in order to generate instructions that can be understood by our computer and then this collection of instructions must be linked with libraries.[2] The exact way in which the compilation and linking is carried out will depend on the details of the particular system. Typically, if this first program is in the file example.cxx, then both stages may be accomplished by entering:

```
g++ -o example example.cxx
```

---

[1]Word processors put undesirable characters in their output files. However, those that can be persuaded to output plain text files could also be used to edit C++ code. Alternatively, you may have access to an IDE (*Integrated Development Environment*) that has an integrated editor and C++ compiler.

[2]A library consists of a collection of compiled code that performs tasks that are not intrinsic to the language. Typical examples are input, output and the mathematical functions (square root, logarithm etc.).

This command is consistent with the GNU C++ compiler running under Linux.[3] However, it may be conventional on your system to use a different extension for source code (such as .cc, .cpp, .C or .c++) and the compile and link command may be different (perhaps CC or gxx). You may also have to use <iostream.h> in place of <iostream>. The -o example part of the compile and link command results in the executable code being placed in a file called example. You should then be able to execute this file by entering the command:

    example

If you omit -o example from the compile and link command, then the GNU compiler will put the executable code in the file a.out. However, with other compilers the file in which the executable code is placed may be different; for example, on an MS-DOS or Microsoft Windows system it may be example.exe.

    If you have overcome all of the system-dependent hurdles, the following output should appear:

    This is our first C++ program.

    The lines of code that produce this output require a few words of explanation. The first line is a comment since any code following // on a line is ignored. A larger program would normally be introduced by a more significant comment. The third non-blank line is the start of a function definition. Every complete C++ program must contain a function called main and in our example, main is the only function. The function name is preceded by the type specifier, int, since main should return an integer. Parentheses, (), follow the name and are used to indicate that main is a function. In some circumstances these parentheses may contain variables, known as arguments.

    The function, main(), contains two *statements*; the end of each statement is denoted by a semicolon, sometimes known as a *statement terminator*. The first statement outputs a message and the second statement effectively returns control to the operating system. The standard output stream (usually the screen) is known as cout and the << operator is used to send a string of characters to this stream, the group of characters between the double quotes, " ", being known as a string. In our example, the string ends with \n, which is used to denote a new line. This technique of sending character strings to the output stream is not strictly part of the C++ language, but is part of a library and is one of many standard functions that are available with any C++ compiler.

    Since our program is compiled before it is linked to code for the output stream, we need to supply information about the interface to cout. This information is contained in the file, iostream, which is made available to the compiler by the instruction:

    #include <iostream>

The consequence of this instruction is that the iostream file is copied (that is *included*) at this point. Files such as <iostream> are usually known as *header files* and are described in more detail in Section 4.13.1. The < > notation is not part of the file

---

[3]This is the environment that was used to test the code given in this book, although the code should work with little modification in other environments.

name, but rather an indication that the file exists in a special place known to the compiler.

The statement:

```
using namespace std;
```

makes certain names that belong to what is known as the Standard Library visible to the program. The Standard Library is introduced in Chapter 17 and **namespaces** are described in Chapter 14. Until Section 14.8, the above statement will appear in every file that includes header files with the < > notation. In Section 14.8 we explain how to avoid using the statement.

1.2 Solving a Quadratic Equation

No doubt our interests are more numerical than simply outputting lines of text, so our second example consists of solving a quadratic equation:

$$ax^2 + bx + c = 0$$

The solutions are, of course, given by:

$$x = \frac{-b \pm \sqrt{b^2 - 4ac}}{2a}$$

which can be implemented by the following (not very robust) program:

```cpp
// Solves the quadratic equation:
// a * x * x + b * x + c = 0

#include <iostream>
#include <cmath>          // For sqrt() function.
using namespace std;

int main()
{
    double root1, root2, a, b, c, root;

    cout << "Enter the coefficients a, b, c: ";
    cin >> a >> b >> c;
    root = sqrt(b * b - 4.0 * a * c);
    root1 = 0.5 * (root - b) / a;
    root2 = - 0.5 * (root + b) / a;
    cout << "The solutions are " << root1 << " and " << root2 <<
        "\n";

    return(0);
}
```

This program introduces floating point variables, which can store numbers that include a decimal point, such as 0.5. Notice that we use 0.5 in the above code, rather than divide by 2.0, since multiplication is usually faster then division. The variables `root1`, `root2`, `a`, `b` and `c` are defined to be floating point numbers by virtue of the type specifier, `double`.

The program first uses the `cout` stream to send text to the output device, instructing the user to input three coefficients. Values for `a`, `b`, and `c` are read from the input stream, `cin`, which is usually the keyboard. For example, you could try typing:

```
1  -3  2
```

and then hit the enter key. Having obtained the values for `a`, `b`, and `c`, the code makes use of the assignment, `=`, multiplication, `*`, addition, `+`, and subtraction, `-`, operators in self-evident ways.

In general, solving a quadratic equation involves finding the square root of a number. This is achieved by invoking the `sqrt()` function,[4] which is defined for us in a system-supplied mathematics library. Since our program is first compiled and then linked with compiled code for the libraries, we must also include the header file, `<cmath>`, which contains a *function declaration* for the `sqrt()` function, specifying the argument and return types (in this case `double`). Notice that we really do mean `<cmath>`, rather than `<math>`, since the `sqrt()` function was originally part of the C, rather than C++, libraries.[5]

On some systems it may be necessary to specify that this example requires the mathematics library for linking. A typical form of the compile and link command in such cases is:

```
g++ -o example example.cxx -lm
```

but you may need to consult the compiler reference manual for your particular system.

1.3   An Object-oriented Example

The techniques illustrated by our examples so far have not differed much from other languages, such as FORTRAN, BASIC or Pascal. Now consider (but don't try to compile) the following program, which multiplies two matrices and lists the result:

```
// Program to test matrix multiplication:

#include "matrix.h"
using namespace std;

int main()
{
    int i, j;
    matrix A(6, 4), B(4, 6), C(6, 6);
```

---

[4]Within the text we write a function with an empty pair of parentheses following the name. This is to emphasize that we really do mean a function, but does not necessarily imply that the function takes no arguments. Within a segment of code, any necessary arguments are, of course, included.

[5]C++ is an extension of C, a language which was invented by Kernighan and Ritchie [5] and then adopted by ANSI (the American National Standards Institute). C++ has also been adopted by an ANSI committee. (See [1].)

```
// Assign some numbers to the elements of A:
for (i = 1; i <= 6; ++i)
    for (j = 1; j <= 4; ++j)
        A(i, j) = i * j;

// Assign some numbers to the elements of B:
for (i = 1; i <= 4; ++i)
    for (j = 1; j <= 6; ++j)
        B(i, j) = i * j;

// Multiply A and B:
C = A * B;

// Output the result:
cout << C;

return(0);
}
```

This program uses some features of C++ to perform operations on matrices. The statement:

```
matrix A(6, 4), B(4, 6), C(6, 6);
```

defines matrices, A, B, C, which are of order 6×4, 4×6 and 6×6 respectively. The important point to notice about the next seven lines of code is that A(i, j) and B(i, j) are used to access the ij elements of the matrices A and B. Notice that the included file matrix.h has a .h extension and also double quotes are used rather than the < > notation. This is the convention when such files are not provided by the system and is covered in more detail in Section 4.13.1.

The program first assigns integers to the elements of two non-square matrices. The statement that then multiplies these two matrices is wonderfully simple:

```
C = A * B;
```

Notice how the code closely parallels the mathematics and is not obscured by function calls. Moreover, if we make a mistake and write:

```
C = B * A;
```

which incorrectly attempts to assign a 4×4 matrix to a 6×6 matrix, then an error message results.

Finally, the statement:

```
cout << C;
```

is all that is required to list the matrix, C:

30	60	90	120	150	180
60	120	180	240	300	360
90	180	270	360	450	540
120	240	360	480	600	720
150	300	450	600	750	900
180	360	540	720	900	1080

Matrices are not part of the C++ language, but are an example of a user-defined type, or class, which can be tailored to meet the exact needs of the application.[6] For instance, some applications may be restricted to real matrix elements, whilst others may be complex. In some circumstances bounds-checking may be an intolerable overhead, whereas in others it may be essential. An application may even require that matrix elements are themselves matrices. We will learn how to construct such classes. We will also learn how to build on existing classes in order to create new classes, thus enabling us to reuse software *without* understanding its detailed design.

The three matrices, A, B and C, are instances of the `matrix` class and are known as *objects*. This example only skims the surface of object-oriented techniques, but the essential idea is that a matrix is considered as an entity on which we perform operations. A matrix object encapsulates details of how the data is stored, the value of each element, the numbers of rows and columns, etc. Moreover, the `matrix` class also shields the application programmer from details of how matrix operations, such as multiplication, are implemented.

1.4   Why Object-oriented?

Solving realistic problems can require programs of ten or even a hundred thousand lines. Such programs may be in use for many years and have been written by many different programmers. Three key requirements are that programs are *maintainable, reusable* and *efficient*.

A program must be maintainable because it may need modifying when hardware, software or requirements change, or when errors in the code are inevitably discovered. Developers of a program are frequently not involved in maintenance and, in any case, the changes may be made long after the program was written. Throughout this book you will therefore come across techniques that encourage maintainability, such as *readability, modularity, type-safe, self-documenting* etc. These techniques are in turn facilitated by object-oriented programming.

The code for a complicated application may involve many years of effort, so it makes sense to be able to reuse as much of this work as possible. Functions can aid re-usability to a limited extent; instead of implementing a cosine every time it is

---

[6]The phrase "user-defined" refers to a user of the C++ language as distinct from a user of a program. Unfortunately, this potentially confusing terminology is well-entrenched in the literature.

required, we can use a function. But a function to perform other operations may be less straightforward. For instance, a function to invert a matrix needs information on how the matrix is stored, its size and type. It would be much easier to reuse a matrix inversion function if it simply operated on matrices as objects, which encapsulated details of storage, size etc. Also, rather than simply reusing the matrix class, we may want to modify it (perhaps to introduce bounds-checking) and ideally we should be able to make changes without tampering with the original code. Object-oriented techniques enable us to do all of these things.

Scientists and engineers are often working at the leading edge of what is computationally possible and are also well aware that inefficiency costs money. It might seem that efficiency is incompatible with the requirements of maintainable and reusable code. It is also true that object-oriented techniques usually introduce a slight run-time overhead when compared with traditional methods. However, the key to efficiency is identifying the *appropriate* objects. For example, if an application manipulates matrices, then it should be written in terms of matrix objects, rather than matrix elements. The details of matrix multiplication, addition etc. are then hidden from the application programmer. Moreover, since the details are hidden, the implementation can be made as efficient as possible. Such increases in efficiency far outweigh any run-time overhead due to using object-oriented techniques.

1.5 Summary

- Every program must have one (and only one) function called `main()`.

- Compiling and linking a program is system dependent. In this book we use the GNU C++ compiler and the command line

    ```
    g++ -o example example.cxx
    ```

  gives `example` as the executable file. Using

    ```
    g++ example.cxx
    ```

  gives `a.out` as the executable file.

- `cout << x` sends the value of `x` to the output stream.

- `cin >> i` reads the value of `i` from the input stream.

- The object-oriented features of C++ enable us to write reusable code in a way that is natural for the particular application.

1.6 Exercises

1. Our program to solve a quadratic equation is not very robust. Try to make the program fail in as many different ways as possible. What would you *like* to do to improve the program? (We haven't covered the techniques required to actually carry out the necessary improvements yet.)

2. Using the quadratic equation program as a guide, write a short program that reads a, b and c from the keyboard and lists the results for a+b, b*c and a/c. Again, you should be able to input data that causes your program to fail.

Chapter 2

The Lexical Basis of C++

The most basic study of any language is lexical; without knowing the rules for constructing words, we cannot begin to write books, or even to construct a single sentence. Communication is impossible. Likewise, before we can write a meaningful C++ program, we must learn the rules for constructing "words", or more correctly *tokens*.

2.1 Characters and Tokens

A C++ program consists of one or more files, which contain a series of characters. The details of how characters are represented internally are dependent on the C++ compiler, but one byte per character is normal and one of the most popular representations, the ASCII character set, is given in Appendix A.[1] The allowed characters are shown in Table 2.1. A compiler resolves (or parses) the C++ program into a series of tokens. There are five types of token: identifiers, keywords, constants, operators and other separators.

upper case letters:	A B C D ... X Y Z
lower case letters:	a b c d ... x y z
digits:	0 1 2 3 4 5 6 7 8 9
special characters:	! " % ^ & * ( ) [ ] { }_ - + = | \ < , > . ? / ~ # : ; '
non printing characters:	blank, carriage return, new line etc.

Table 2.1: Permissible characters.

---

[1]A byte is the smallest addressable element of memory and is usually eight bits. Indeed, throughout this book a byte is assumed to be synonymous with eight bits. A bit is the smallest element of memory and can only take the values 0 and 1. Bits and bytes are considered in more detail in Chapter 11.

2.2 Comments and White Space

Comments are an essential feature of good programming practice. In C++ there are two ways of denoting comments. Two forward slashes, //, indicate the start of a comment that continues until the end of a line:

```
// This is a comment.
x = 1;  // The first part of this line is not a comment.
```

An alternative technique is to use /* to indicate the start of a comment that continues until */ is encountered. Such comments can continue over many lines:

```
/* This is a comment */
/* This
is
a
longer
comment */
```

The /*   */ technique may also include one or more // style of comments:

```
/* This is a comment. // This is a second comment
   and this is a continuation. */
```

However, the /*   */ style of comments cannot themselves be nested:

```
/* /* This comment should not get past the compiler.
      If it does then the compiler does not conform
      with the ANSI C++ Standard.
      Such extensions are best avoided. */ */
```

This restriction is a minor nuisance, since temporarily commenting out sections of code by means of /*   */ is very convenient, but cannot be used if the code already contains /*   */ style comments. This is a good reason for only using the // style of comment in new code. However, an alternative technique for making the compiler temporarily ignore sections of code is given in Section 4.13.3.

Notice that blanks are not allowed within the tokens defining either style of comment:

```
/ *   This is not a valid comment. * /
/ /   Neither is this.
```

and that an incomplete /*   */ style of comment can lead to errors:

```
/* Get the coefficients of the equation:
   cout << "Enter the coefficients a, b, c: ";
   cin >> a >> b >> c;
/* Now find the roots: */
   root = sqrt(b * b - 4.0 * a * c);
```

In this example, everything between the first /* and */ is taken as a comment and, as a result, a, b and c have arbitrary values.

Comments, blanks, vertical and horizontal tabs, form feeds and new lines are collectively known as *white space*. White space is not allowed in any token, except in character or string constants. The compiler ignores any white space that occurs between tokens. Such white space is effectively a separator.

2.3 Identifiers

An identifier is a sequence of some combination of letters, digits and the underscore symbol, _. Both upper and lower case characters are valid and are distinct. However, it isn't a very good idea to rely on the case being the only distinguishing feature between identifiers. In both C and C++ there is a tradition of using mainly lower case, except where some special significance is being highlighted, as with a global constant. This tradition makes for easily readable code.

The ANSI C++ Standard doesn't impose any maximum limit on the number of characters in an identifier. Since real computers have finite resources your computer will have some limit, but this is very unlikely to prove a problem.

An important restriction on identifiers is that they must start with a character or underscore rather than a digit. Some examples of valid identifiers are:

```
main
cout
position_1
initial_velocity
```

whereas invalid identifiers include:

```
velocity$       // $ is not a digit, letter, or _.
1velocity       // Identifiers cannot start with a number.
v e l o c i t y // White space is not allowed.
initial-velocity // Don't confuse - with _.
0xygen_level    // Don't confuse 0 with o.
```

Other examples of valid identifiers include those with a leading underscore:

```
_velocity
```

or embedded double underscores:

```
initial__velocity
```

Both of these combinations are best avoided, since leading underscores and embedded double underscores are often generated internally by compilers or used in libraries. Likewise, it is worth avoiding trailing underscores:

```
velocity_
```

Apart from these minor restrictions, it is good programming practice to employ names that are meaningful in the context in which they are used, such as:

```
water_temperature
```

instead of:

```
T12
```

Meaningful names make the code self-documenting and avoid the need for excessive and distracting comments. While some programmers use an underscore to distinguish different parts of an identifier, others prefer to use an upper case letter:

```
waterTemperature
```

You will probably find that you like one particular style and loath the other.

2.4　Keywords

Keywords are special identifiers that have a significance defined by the language rather than the programmer. A complete but, at this stage, rather impenetrable list is given in Table 2.2. You are not expected to memorize this list, rather it is given in order to warn you to only use these keywords in ways that are defined by the language.[2]

asm	else	new	this
auto	enum	operator	throw
bool	explicit	private	true
break	export	protected	try
case	extern	public	typedef
catch	false	register	typeid
char	float	reinterpret_cast	typename
class	for	return	union
const	friend	short	unsigned
const_cast	goto	signed	using
continue	if	sizeof	virtual
default	inline	static	void
delete	int	static_cast	volatile
do	long	struct	wchar_t
double	mutable	switch	while
dynamic_cast	namespace	template	

Table 2.2: C++ keywords.

There are also some alternative representations of C++ operators. These are given in Table 2.3 and should only be used as defined by the language. For reference, the operators that they represent together with the section where they are introduced in this book are also given.

A few other keywords that may be reserved for some compilers are:

ada	fortran	pascal	overload
huge	near	far	

---

[2]Not all of these keywords are described in this book.

Representation	Operator	Section Defined
and	&&	4.2
and_eq	&=	11.1.6
bitand	&	11.1.2
bitor	|	11.1.4
compl	~	11.1.1
not	!	4.2
not_eq	!=	4.3
or	||	4.2
or_eq	|=	11.1.6
xor	^	11.1.3
xor_eq	^=	11.1.6

Table 2.3: Alternative representations of C++ operators.

and these are best avoided as user-defined identifiers in order to ensure portability. For some compilers such keywords may have a leading underscore, as in _near.

Examples of the valid use of keywords are:

```
int i;     // The keyword int declares i
           // to be an integer variable.
char c;    // The keyword char declares c
           // to be a character variable.
```

However, the following are invalid:

```
int switch; // Attempts to declare switch to be
            // an integer variable.
char class; // Attempts to declare class to be
            // a character variable.
i n t i;    // White space is not allowed.
INT i;      // Keywords must be lower case.
```

2.5 Constants

Constants, which are also known as literals, can be integer, Boolean, floating point, character or string. More details of these four types are given in the next chapter, but they are introduced here.

2.5.1 Integer Constants

Integer constants consist of a sequence of digits, such as:

```
14768
```

but not:

```
14,768        // An embedded comma is not allowed.
14 768        // Embedded white space is not allowed.
-14768        // This is an integer expression.
```

Notice that a negative integer is actually an integer constant expression rather than an integer constant. Integer constants are usually given in decimal (base ten) notation, but octal and hexadecimal constants are also possible. Octal (base eight) constants start with 0 and cannot include the digits 8 or 9. An example of an octal constant is:

```
024
```

Hexadecimal (base sixteen) integer constants start with 0x (or 0X) and may include the letters a to f (or A to F, the case is not significant) as, for example:

```
0x7a1f
```

2.5.2 Boolean Constants

There are two Boolean constants, namely **true** and **false** and these have their obvious meanings. Further details are given in Section 3.1.10.

2.5.3 Floating Point Constants

Floating point constants include a decimal point or an exponent, or both. Some examples are:

```
2500.1      2.5001e3      25001E-1
```

where the value following e (or E) is the exponent. Negative floating point numbers, such as

```
-2500.1      -2.5001e3    -25001e-1
```

are actually floating point constant expressions, rather than constants.

2.5.4 Character Constants

A character inside single quotes, such as:

```
'a'        '1'        ' '
```

is a character constant and is usually represented internally by one byte. Note that '1' is not equal to the integer 1; for instance in the ASCII character set, '1' is actually represented by 49. (See Appendix A for more detail.) Notice also that a character constant has the form 'a', rather than 'a'.

Certain hard to get at characters are represented by using an *escape sequence*, starting with a backslash; for instance '\n' signifies a new line. The complete list of all such sequences is given in Table 2.4. The generalized escape sequence consists of up to three octal or as many hexadecimal digits as are required. For instance, '\61' and '\x31' both represent the character, '1'.

new line	\n	backslash	\\
horizontal tab	\t	question mark	\?
vertical tab	\v	single quote	\'
backspace	\b	double quote	\"
carriage return	\r	octal number	\032
form feed	\f	hex. number	\0x32
alert (bell)	\a		

Table 2.4: Escape sequences.

Exercise[3]

To demonstrate that both '\61' and '\x31' represent the same character, try the following simple program:

```
#include <iostream>
using namespace std;

int main()
{
    char a, b;
    a = '\61';
    b = '\x31';
    cout << a << ' ' << b << '\n';
    return(0);
}
```

and then modify it to sound your terminal bell by means of a suitable hexadecimal escape sequence.

2.5.5   String Constants

String constants are sequences of characters between double quotes, as in:

```
"Hello world"
```

Notice that the double quote is a single character, ", rather than, '', and that white space is significant in string constants so that "Hello world" is not the same as "Helloworld". Internally a string is represented by an array of characters. This means that "Hello world" is actually twelve characters since the escape sequence, '\0', is used after the last character to denote the end of a string.[4] Consequently 'C' is *not* the same as "C"; whereas the first consists of one character, the second consists of two.

Adjacent string constants are concatenated, so that:

```
cout << "Hello" " World";
```

---

[3]Occasionally we include simple, un-numbered exercises in the text. You should do these as you come across them. You are also encouraged to attempt the exercises at the end of each chapter!

[4]'\0' is sometimes known as the null character or the end-of-string character.

produces the output:

```
Hello World
```

2.6 Operators

An operator is a language-defined token, consisting of one or more characters, that instructs the computer to perform some well-defined action. An example is the assignment operator, =, used in the following:

```
i = 1;        // Assign 1 to the memory location
              // holding the value of i.
```

A complete list of operators is given below:

```
()      .       []          ->      ::      typeid
->*     &       *           !       ~       dynamic_cast
++      --      -           /       %       static_cast
+       .*      <<          >>      <       reinterpret_cast
<=      >       >=          ==      !=      const_cast
^       |       &&          ||      ?:      throw
=       +=      -=          *=      /=      new
%=      <<=     >>=         &=      ^=      delete
|=      ,       sizeof
```

Notice that some keywords are also operators and that multi-character operators form a single token. Once again it should be emphasized that white space is not allowed within a token, meaning, for instance, that + = is not a valid operator, unlike +=. However, in contrast to some languages, two operators can follow in succession; for instance x * -1 is equivalent to x * (-1). More detail on most of these operators will be given later in this book.

2.7 Programming Style

There is no enforced programming style in C++. White space is ignored (except within tokens), there is no restriction to typing statements in certain columns (as in Fortran), there is no required indentation (as in occam) and statements can flow across many lines. Consequently it is possibly to produce code that is very difficult to read, as in:

```
#include <iostream>
#include <cmath>
using namespace std;int main(){double x1,x2,a,b,c,root;cout
<<"Enter the coefficients a, b, c: ";cin>>a>>b>>c;root=
sqrt(b*b-4.0*a*c);x1=0.5*(root-b)/a;x2=-0.5*(root+b)/a;cout
<<"The solutions are "<<x1<<" and "<<x2<<"\n";return(0);}
```

However, it is up to the programmer to adopt a clear style that can easily be understood by others. There is no unique *good style*, but it is worth adopting something similar to that used in this book. In particular it is a good idea to keep to one statement per line (except in special circumstances) and to indent by one tab to denote blocks of code. With this in mind, it is worth adjusting the tab on your editor to four (or perhaps three) characters per tab position. If you don't do this, your code will march across the page too quickly. Further hints on style will be given when more of the C++ language has been introduced.

There is one final point that should be made in this chapter. Very occasionally there may appear to be be an ambiguity as to how one or more lines of code are resolved by the compiler into tokens. For instance, given that both -- and - are valid C++ operators, how should

```
x = y---z;   // x = (y--) - z; or x = y - (--z);
```

be read? The compiler resolves this ambiguity by adopting a *maximal munch* strategy;[5] the parsing stage of the compilation process bites off the largest sequence of characters that form a valid token. So the example above really does mean:

```
x = (y--) - z;
```

It is a good idea to resolve such apparent ambiguities by using white space or parentheses.

2.8 Summary

- A C++ source file consists of a sequence of tokens that are made up of one or more characters.

- White space (including blanks, new lines, tabs etc.) is not allowed within tokens.

- There are five types of token: identifiers, keywords, constants, operators and other separators.

- Use meaningful names for user-defined identifiers, such as `flow_rate`.

- Identifiers with language-defined significance are known as keywords, examples of which are `int`, `class`, `if`, etc. The significance of keywords cannot be altered by the programmer.

- Constants can be integer, Boolean, floating point, character or string. Examples are:

```
10     true     2.01     'a'     "Hello"
```

- Operators are tokens representing operations, such as assignment, =, addition, +, multiplication, *.

---

[5]See [6].

2.9   Exercises

1. How many different types of token are there? Classify the following tokens:

    ```
    new         ->                -              >
    ()          ;                 i              friend
    int         pressure          14             *
    1.4         '\n'              public         []
    ```

2. What is wrong with the following statements:

    (a)   `speed$of$light = 186000;`

    (b)   `initial velocity = 0.0;`

    (c)   `#zero = 10;`

    (d)   `180_degrees = 3.1415926535897932;`

    (e)   `-x = 360.0;`

    (f)   `void = 0;`

    (g)   `pressure := 760.0;`   `// in pascals.`

    (h)   `/* Perhaps this fix will work:`
    `          volume = 100.0;          /* in litres. */`
    `    */`

    (i)   `case = 2;`

3. Which of the following are valid constants:

    (a)   `27`

    (b)   `11,111`

    (c)   `'Hello world'`

    (d)   `'a'`

    (e)   `-3.1415926535897932`

    (f)   `'\t'`

    (g)   `25L4`

    (h)   `'\o32'`

Chapter 3

Fundamental Types and Basic Operators

C++ is a typed language; that is the variables and constants that occur in a program each have a type and this controls how they may be used and what storage is required. For example, variables of the `int` type, which are used to represent integers, may require two bytes of memory, and variables of the `float` type, which are used to represent floating point numbers, may need four bytes. Furthermore, the `float` and `int` types are represented in memory in entirely different ways. The language-defined types that are used to represent integers, Booleans, characters and floating point numbers are known as *fundamental* types and are the concern of this chapter.[1] It is also possible to define what are known as *derived* types in order to manipulate objects such as matrices, complex numbers or 3-dimensional solids.[2] Techniques for creating derived types are described in later chapters.

The motivation for using a typed language is that the compiler can catch many programming errors by performing type-checking. For example, if an object is declared to be constant (by using the `const` keyword, which is explained later in this chapter) then an attempt to assign a value to the object is a compile-time error since it has the wrong type for such operations.

In this chapter we examine all of the fundamental types defined by the C++ language, together with some of the basic operators. This will enable us to carry out the usual arithmetic operations and to manipulate characters. Compared with some languages, the fundamental types are rather limited; for instance, unlike Fortran, complex numbers are not defined. Although this may seem a bit restrictive for many applications, derived types can effectively extend the language indefinitely. For example, complex arithmetic is part of the ANSI C++ Standard Library. (See Section 17.3.1.)

---

[1]The fundamental types are sometimes known as *built-in* or *standard types*.

[2]Derived types are sometimes known as *user-defined types*.

3.1   Integral Data Types

The integral data types consist of three groups: the integer, Boolean and character types. The integer group contains the **short**, **int** and **long** types, all of which can be either **signed** or **unsigned**.[3] The integer types are used to store and perform arithmetic operations on "whole" numbers (0, 1, −50 etc.). Boolean variables can only be used to store the two constants **true** and **false**. The character types are **char**, **signed char** and **unsigned char**. Since characters are actually represented internally by small integers, the character types can also be used for storing and operating on such integers.

3.1.1   Type int

The **int** data type is the most important of the integral types and can hold both positive and negative integers. For **int**s represented by two bytes, the maximum number that can be represented is 32767 and the minimum is −32768. (For some compilers these numbers may be 32768 and −32767 respectively, since sixteen bits can hold $2^{16} = 65536$ different numbers, including 0.) The exact limits for all integral data types are given in a header file, called <**climits**>. It is worth examining this file at some stage to see the limits for your particular system. On a UNIX or Linux system, the header files are usually in the directory **/usr/include/g++**, or something similar. You should be aware that the <**climits**> notation only implies that the compiler knows where to find the file. It does not necessarily imply that the file is actually called **climits**. The contents of the C++ header file <**climits**> are the same as the C header file <**limits.h**>, so you may find that <**climits**> merely includes <**limits.h**>. If this is the case, then you will have to search for <**limits.h**>. On a UNIX or Linux system this will probably be in the **/usr/include** directory.

If we wish to use the variables **i**, **j**, **k**, then they can be defined by:

```
int i;
int j;
int k;
```

It is also possible to define a list of variables:

```
int i, j, k;
```

We say "define" rather than "declare"; in C++ the distinction is both subtle and important. Defining an object, as above, not only specifies the type, but also reserves the appropriate amount of memory. As will be seen later, it is possible to make a declaration, such as

```
extern int i;
```

without reserving any memory (which must of course be done somewhere). In a complete program there must be one (and only one) definition of an object, even though there can be many declarations. It is worth emphasizing that, although an object may be defined in the sense of memory being allocated, its "value" can still be arbitrary.

---

[3]The integer types are **signed** by default. It is therefore very rare and completely redundant to have **signed int**, **signed long** or **signed short**.

An object must be defined or declared before it can be used in a program, otherwise a compile-time error results. This is more restrictive than some other languages, but helps to eliminate errors such as mistyped identifiers.

The operator, =, is used for integer assignment, so the following defines i, j, k and sets them to 1, 2, 3 respectively:

```
int i, j, k;
i = 1;              // Notice that every statement
j = 2;              // ends with a semicolon.
k = 3;
```

Addition and subtraction of integers are carried out using the usual + and - operators, as in:

```
int i, j, k;
i = 1;
j = 2;
k = 3;
i = i + j;
k = k - i;
```

Notice that "=" is the assignment rather than equality operator, so the statement:

```
i = i + j;
```

means add the current values of i and j (1 and 2) and assign the result (3) to i. The fact that "=" is the assignment operator means that expressions such as:

```
i + 1 = 7;        // WRONG
```

which look mathematically respectable, are not correct C++ statements. Conversely, some expressions that are valid in C++, would make no sense as mathematical equations. For example, in the above code fragment, the statement:

```
i = i + j;
```

does not imply that j is zero; in fact j is 3 in this context. It is also worth pointing out that := is *not* a valid assignment operator, nor even an operator:

```
i := 10;          // WRONG:   := is not an operator.
```

Our short sequence of assignment statements does not form a complete C++ program but, following the examples in Chapter 1, we can easily rectify this situation to produce the following code:

```
#include <iostream>     // For cout
using namespace std;
int main()
{
    int i, j, k;
    i = 1;
    j = 2;
```

```
    k = 3;
    i = i + j;
    k = k - i;
    cout << "i = " << i << "   k = " << k << "\n";
    return(0);
}
```

Notice that `main()` is a function that returns the `int` type and in this example we return the value 0. (More discussion of this `return` value is given later, but in some circumstances it may be important for the operating system to know whether or not the program has run successfully, as indicated by 0.) We also include a statement so that the program can actually display the result of the calculations by using the insertion operator to send the values of i and k to the output stream. The resulting output is:

```
    i = 3   k = 0
```

Exercise

Enter the above program into a file on your system, making sure the file has the appropriate extension, such as `.cxx`. Then compile, link and run the program. Check that you get the expected result. Try modifying the program to add different combinations of integers.

Since the `int` type occupies a fixed and finite amount of memory, only a limited range of integers can be stored. Consequently it is possible to perform operations which mathematically would give results above the maximum value that can be represented or below the minimum value. Such invalid operations are known respectively as *integer overflow* and *integer underflow*. These are demonstrated in the following program:

```
    #include <iostream>
    #include <climits>
    using namespace std;
    int main()
    {
        int i;
        i = INT_MAX + 10;          // Integer overflow.
        cout << "i = " << i << "\n";
        i = INT_MIN - 10;          // Integer underflow.
        cout << "i = " << i << "\n";
        return(0);
    }
```

`INT_MAX` and `INT_MIN` are the maximum and minimum values that can be stored by the type `int` and are defined in the header file, `<climits>`. These values depend on the particular compiler, as does the result of underflow or overflow.

Exercise

What values are given to i for these two statements on your system? Show that your results are consistent with the values given in `<climits>`.

3.1.2 Integer Multiplication

As might be expected, the token, *, is used to denote integer multiplication. The usual rules of arithmetic are followed, as in:

```
#include <iostream>
using namespace std;
int main()
{
    int i, j, k, m, n;
    i = 2;
    j = -3;
    k = -4;
    m = i * j;      // Assigns -6 to m.
    n = j * k;      // Assigns 12 to n.
    i = 3 * 6;      // Assigns 18 to i.
    cout << "m = " << m << "\nn = " << n << "\ni = " << i <<
        "\n";
    return(0);
}
```

Exercise

Check that the above program gives the results you would expect on your system.

3.1.3 Integer Division

The forward slash is used to denote the integer division operator, as in:

```
int k;
k = 3 / 2;
```

which means divide 3 by 2 and put the result in k. Since 2 does not exactly divide 3, there is a potential ambiguity as to whether k is set to 1 or 2. In fact, in common with most languages, the C++ result is 1; integer division with both numbers positive always truncates. The same is true when both numbers are negative, so that -3 / -2 yields 1 rather than 2. If only one number is negative, then the result depends on the compiler: -3 / 2 could give either -1 or -2. Relying on a feature provided by a particular compiler is not good programming practice and should be avoided if at all possible. (In general, if you do need to use a compiler-dependent feature, then you should have a clear comment on this in your program.)

In mathematics, dividing any non-zero number by zero gives infinity and dividing zero by zero is undefined; in C++, any integer division by zero is undefined.

3.1.4 Integer Modulus or Remainder Operator

The % token is the modulus or remainder operator. If i and j are both positive integers, then i % j (read as "i modulo j") gives the remainder obtained on dividing i by j.

For instance, 5 % 3 yields the value 2. If both i and j are negative then the result is negative; that is -5 % -3 yields -2. If only one of i and j is negative, then the result is again dependent on the C++ compiler, but should be consistent with the dependency for division. That is:

    (i / j) * j + i % j

should give the value of i. Again, if j is zero the result of i % j is not defined and may not be consistent with the (undefined) result of i / j. (Saying that a piece of code is "undefined" means that the C++ language definition doesn't say how it should be interpreted.) Notice that 0 % j gives 0 for any (non-zero) integer, j.

Exercise

Try out the following program for various values of i and j (positive, negative and zero) and explain the output. What happens if you enter a non-sense value, such as "accidentally" hitting the w key instead of 2? Why?

```
#include <iostream>
using namespace std;
int main()
{
    int i, j, m, n, a, b;
    cout << "enter an integer: ";
    cin >> i;
    cout << "enter an integer: ";
    cin >> j;
    m = i / j;
    n = i % j;
    a = m * j;   // (i / j ) * j
    b = a + n;   // (i / j) * j + i
    cout << "i = " << i << "    j = " << j << "\n\n" <<
        "i / j = " << m << "    i % j = " << n <<
        "\n\nThe following result should be " << i <<
        ":\n" << "(i / j) * j + i % j = " << b << "\n";
    return(0);
}
```

In this example we use the C++ input stream, cin, so that the values of i and j can be entered from the keyboard. The expression:

    cin >> i;

extracts the value of i from the input stream. The operator, >>, is known as an *extraction* operator. It is possible to concatenate such operators so that the first few lines can be replaced by:

    cout << "enter two integers: ";
    cin >> i >> j;

but notice that the following is *not* equivalent:

    cin >> i, j;   // WRONG!

The statement does compile, but no value is assigned to j when the program is run.

3.1.5 Increment and Decrement Operators

The value of a variable can be changed without using the assignment operator. Instead of:

```
i = i + 1;
```

we can use:

```
++i;          // Prefix increment operator.
```

or:

```
i++;          // Postfix increment operator.
```

In all three cases, i is increased by one. Whereas the prefix operator increments i by one before the value of i is used, the postfix operator increments i after its value has been used. Either version of the increment operator has the same effect in the above code fragments, but note that this is not true for the following program:

```
#include <iostream>
using namespace std;
int main()
{
    int i, j, k, m, n;
    i = 2;
    j = i;
    k = 3;
    m = i++ / k;     // Assigns 0 to m.
    n = ++j / k;     // Assigns 1 to n.
    cout << "m = " << m << "\nn = " << n << "\ni = " << i <<
        "\nj = " << j << "\n";
    return(0);
}
```

In this example, the expression for m is evaluated before i is incremented, but n is evaluated after j is incremented.

> **Exercise**
>
> Modify the above program so that it reads i and k from the keyboard. Try entering a few different values of i and k, and check that you get the results you expect for m and n.

As might be expected, there are also prefix and postfix decrement operators so that:

```
--i;
```

and:

```
i--;
```

are equivalent to:

```
    i = i - 1;
```

Whereas the prefix decrement operator decreases the value of i by one before it is used
in an expression, the postfix decrement operator decreases i after the value has been
used. The use of increment and decrement operators can lead to very compact code,
but it is worth exercising some restraint otherwise the code can get difficult to read.

Exercise

Replace the increment by decrement operators in your program for the
previous exercise.

It may seem obvious, but constants cannot be incremented or decremented:

```
++3.142;          // WRONG!
10--;             // WRONG!
```

In any case, it is difficult to imagine how such statements could be useful.

3.1.6   Associativity and Precedence of Integer Operators

Although we have only considered simple expressions so far, we have learned how to
use quite a few operators:

```
    =    +    -    *    /    %    ++    --
```

The =, *, /, % tokens are all binary operators; that is they are defined for an operand
on each side of the operator, such as in:

```
    k = i * j;        // Valid C++.
```

rather than:

```
    k = i *;          // WRONG: where is the second operand?
```

The ++ and -- operators are both unary operators, that is they require a single operand:

```
    ++x;              // Unary - operator takes only one operand.
```

The - operator can be both a binary operator, as in:

```
    k = i - j;        // Binary - operator.
```

or a unary operator, as in:

```
    k = -j;           // Unary - operator takes only one operand.
```

The unary + operator, as in:

```
    k = +j;           // Valid but useless unary + operator.
```

is not expressly forbidden. However, it doesn't serve any useful purpose and you won't
come across the unary + operator again in this book.

In complicated expressions, such as:

```
    k = i + j / m * n - k;
```

the two concepts of associativity and precedence are used to determine the order in which the evaluation is carried out. Operator precedence controls which operator in an expression is applied first; the associativity determines the order in which operators with the same precedence are applied. That is, the compiler first uses operator precedence to determine the order of evaluations in an expression and then any remaining ambiguities are removed by using operator associativity.

The properties of the operators introduced so far are given in Table 3.1. Notice

Operator		Associativity	Section First Defined
++	postfix increment	right to left	3.1.5
--	postfix decrement	right to left	3.1.5
++	prefix increment	right to left	3.1.5
--	prefix decrement	right to left	3.1.5
-	unary minus	right to left	3.1.6
*	multiplication	left to right	3.1.2
/	division	left to right	3.1.3
%	modulo	left to right	3.1.4
+	addition	left to right	3.1
-	subtraction	left to right	3.1
=	assignment	right to left	3.1
+=	add and assign	right to left	3.4.3
-=	subtract and assign	right to left	3.4.3
*=	multiply and assign	right to left	3.4.3
/=	divide and assign	right to left	3.4.3
%=	modulo and assign	right to left	3.4.3

Table 3.1: Common operators.

that the postfix increment and decrement operators have a higher precedence than the prefix increment and decrement operators. All operators shown in the same group have the same precedence and associativity, but a group higher up in the table has a higher precedence than one further down. For example, in the expression:[4]

```
i = 1 - 2 * 3 + 4;   // * has higher precedence than - or +.
```

the multiplication operator has the highest precedence and is applied first, yielding:

```
i = 1 - 6 + 4;       // -, + have higher precedence than =.
```

The subtraction and addition operators have equal precedence, greater than that of the assignment operator, so the right-hand side of the expression is evaluated left to right, giving the result of -1. Finally -1 is assigned to i. These rules, which are similar to those of many other languages, do not necessarily give the results that would be expected mathematically. For instance, due to the left to right associativity of the divide operator, the statement:

---

[4]Throughout this book short fragments of code, which are not complete programs, are frequently given. You should now be able to convert these code fragments into programs, by introducing **main()**, the appropriate **#include** directives and sending results to the output stream. Indeed, you are strongly encouraged to try out as many such programs as possible.

```
i = 3 / 4 / 2;
```

is evaluated as:

```
i = (3 / 4) / 2;
```

rather than:

```
i = (3 * 2) / 4;
```

Any expression inside parentheses is evaluated first, which means that:[5]

```
i = (1 - 2) * (3 + 4);
```

is reduced to:

```
i = -1 * 7;
```

which assigns -7 to i. If you are in doubt about the precedence of any operator, it is always possible to use parentheses.

Exercise

Write a program to evaluate the following expressions:

```
i = 2 / 3 + 3 / 2;
i = 2 / 3 / 3;
i = 3 / 2 % 7;
i = 7 % 2 * 3;
i = 7 / 2 * 3;
```

Are the results what you would expect?

3.1.7   Long Integers

Long integers have the same properties as integers, except that at least the same number of bytes must be used in their representation as are used for the int data type. Typically four bytes are used and integers between -2147483648 and 2147483647 can be represented, as in:

```
long i, j, k;
i = 2140001111;      // Too big for 2 bytes.
j = -2140001111;     // Too small for 2 bytes.
k = i - j;
```

However, it is fairly common for a compiler to define both the int and long data types to have the same four-byte representations. The limits for a particular compiler are again given in the <climits> header file.

Notice that the keyword, long, is used to define (and declare) long integers. The definition:

---

[5]When describing the four different bracket types, the common terminology employed is: parentheses for ( ), braces for { }, square brackets for [ ], and angle brackets for < >. Note that the various brackets must always occur in pairs.

```
long int i, j, k;
```

is equivalent to that given above, but is less widely used.

The suffix L (or l) is used to distinguish a `long` constant:

```
long i;
i = 1L;
```

3.1.8 Short Integers

The `short` type has the same properties as `int`, except that a particular C++ compiler should not use more bytes to represent `short` than are used for `int`. Typically the `short` type uses two bytes and has a range of -32767 to 32768. This type can be useful for saving memory as in:

```
short i, j, k;
i = 1;              // Can easily be represented by 2 bytes.
j = 2;              // Can easily be represented by 2 bytes.
k = i - j;
```

However, if the representation is less than the natural size suggested by the hardware, then there may be a performance penalty. It is best to avoid short integers unless there is a very good reason to do otherwise. In any case, a compiler will often use the same number of bytes for `short` as for `int`.

Notice that the keyword, `short`, is used to define short integers. The equivalent definition:

```
short int i, j, k;
```

is again less widely used.

3.1.9 Unsigned Integers

The `short`, `int` and `long` types all have corresponding `unsigned` types. These occupy the same amount of memory as the `signed` type and obey the laws of arithmetic modulo 2^n, where n is the number of bits in the representation. For instance, if n is 16 (two bytes), then numbers between 0 and 65535 (that is $2^{16} - 1$) can be represented. Unsigned addition does not overflow and subtraction does not underflow; they simply wrap round. This is demonstrated by the following program:

```
#include <iostream>
#include <climits>
using namespace std;
int main()
{
    unsigned int i;
    i = UINT_MAX;
    i = i + 1;
    cout << "UINT_MAX + 1 = " << i << "\n";
    i = i - 1;
```

```
        cout << "0 - 1 = " << i << "\n";
        return(0);
}
```

The program gets UINT_MAX (the largest unsigned appropriate for the particular compiler) from the header file <climits>. Adding one to UINT_MAX gives zero, and subtracting one from zero gives UINT_MAX.

It is not usually worth using an unsigned rather than signed integral type just to gain a higher upper limit on the positive integers that can be stored. An example of the resulting possible pitfalls is given in Exercise 2 of Chapter 4; an example where using an unsigned type *is* worthwhile appears in Section 11.4.1.

A U (or u) is used to denote an unsigned constant, as in:

```
unsigned int i;
unsigned long j;
i = 1U;
j = 1UL;
```

Any combination of U,u and L,l in any order can be used for an unsigned long constant.

3.1.10   Booleans

A Boolean can have one of two values, true or false. In C++, a Boolean is represented by the bool data type, as in:

```
bool b1, b2;
b1 = true;
b2 = false;
```

We can convert between the types bool and int. By definition, true takes the value 1 on conversion to type int, and false takes the value 0. The rule for the conversion of type int to bool is that 0 is converted to false and any non-zero integer corresponds to true. So the output from:

```
#include <iostream>
using namespace std;
int main()
{
    bool b1, b2;
    b1 = true;
    b2 = false;
    cout << "b1 is " << b1 << "\n" <<
        "b2 is " << b2 << "\n";
    return(0);
}
```

is:

```
b1 is 1
b2 is 0
```

The operators for Booleans are introduced in Section 4.2.

3.1.11 Character Types

The type, char, is represented by a sufficient number of bytes to hold any character belonging to the character set for the particular compiler. This representation is usually one byte. Rather confusingly C++ distinguishes three character types: char, unsigned char and signed char. The signed char and unsigned char data types obey the same rules of arithmetic as their integer counterparts. If the type is simply specified as char, then it is compiler-dependent as to whether or not the high order bit is treated as a sign bit. This makes no difference when storing the standard printing characters, but if other data is stored as type char it may appear to be negative on one computer and positive on another. If the sign of such data is significant, then the type should be specified as either signed or unsigned char, as appropriate. Such distinctions may seem a bit obscure to you at present, but may be relevant in the future when you are puzzled by why your code behaves differently on different computers.

Although the char types obey the corresponding rules of integer arithmetic, they are typically used to hold character data, as in:

```
#include <iostream>
using namespace std;
int main()
{
    char c1, c2;
    c1 = 'C';
    c2 = '+';
    cout << c1 << c2 << c2 << "\n";
    return(0);
}
```

Exercise

What does the above program output? Verify the answer on your system.

Since characters are represented by numbers, it is possible to make assignments directly rather than by using the character constant notation. For instance an obscure way of rewriting the above code, for the ASCII character set, is:

```
#include <iostream>
using namespace std;
int main()
{
    char c1, c2, c3;
    c1 = 67;
    c2 = 43;
    c3 = 10;
    cout << c1 << c2 << c2 << c3;
    // Does the output agree with the previous exercise?
    // See the ASCII character set in Appendix A.
    return(0);
}
```

but such manipulations are rarely useful in scientific programming.

3.2   Floating Point Data Types

Floating point numbers, such as 3.142 and 2.9979×10^8, are essential to almost every scientific calculation. In C++, there are three floating point types: `float`, `double` and `long double`. The type `float` is typically represented by 4 bytes, a `double` by 8 bytes and a `long double` by 10 or 12 bytes. However, all that is required is that a `double` uses at least the same number of bytes as a `float` and a `long double` uses at least as many as a `double`. We first consider the type `double`, since from many points of view it is the standard floating point type.

3.2.1   Type `double`

Identifiers of type `double` are defined by using the `double` keyword, as in:

```
double pi, c;
pi = 3.1415926535897932;
c   = 2.997925e8;
```

where the floating point assignment operator is denoted by the token, `=`. The expression for `c` is typical of a floating point number and may be split into three parts:

2	the integer part
997925	the fractional part
8	the exponent.

A decimal point separates the integer and fractional parts, while e (E is also valid) separates the fractional part from the exponent. A floating point constant must contain a decimal point or an exponent or both, since otherwise there would be nothing to distinguish it from an integer. If there is a decimal point, then either an integer or a fractional part must be present. If there is no decimal point, then there must be both an integer part and an exponent (but there can be no fractional part of course). Some examples of valid floating point assignments are:

```
double x, y, z;
x = 1.0;
y = 0.1;
z = 1e10;
x = 1.1e10;
y = 11e9;
z = .11e11;
```

However, the following are not valid `double` constants:

```
.e9          // WRONG: no integer or fractional part.
1,000.0      // WRONG: an embedded comma is not permitted.
1 000.0      // WRONG: an embedded space is not permitted.
1000         // WRONG: this is an integer constant.
e10          // WRONG: no integer part.
```

Apart from the use of e or E, all of this should be straightforward to anyone familiar with scientific notation.

We can perform all of the usual arithmetic operations of assignment, addition, subtraction, multiplication, division, increment, decrement and unary minus. In fact all of the operators we introduced for integers, except for the modulus operator, are also valid for `double` identifiers and constants. The following program demonstrates these operations:

```
#include <iostream>
using namespace std;
int main()
{
    double x, y, z;
    x = 3.1416;         // Assignment.
    y = -x;             // Unary minus.
    z = -3.1416;        // Unary minus.
    ++x;                // Increment (prefix version).
    y--;                // Decrement (postfix version).
    x = x - 1.0;        // Subtraction.
    y = y + 1.0;        // Addition.
    z = x * y;          // Multiplication.
    z = z / 3.1416;     // Division.
    cout << z << "\n";
    return(0);
}
```

All of these operators have the same precedence and associativity as for the corresponding integer versions. Notice that `-3.1416` is actually an expression involving a unary minus together with a `double` constant and that (as you might expect) floating point division does not truncate to an integer, unlike integer division.

Exercise

What is the final result for z in the above code? Verify your answer by running the program.

Only a finite subset of all floating point numbers can be represented by one of the defined types, such as `double`. There are many ways of representing a floating point number by a fixed number of bytes, but most C++ compilers conform to what is known as the IEEE standard. For such compilers a `double` of eight bytes gives a range of about 10^{-308} to 10^{308} and an accuracy of sixteen decimal places. More details of how floating point numbers are represented by a computer are given in Chapter 11. For the present it is worth pointing out that in numerical calculations it is very easy to attempt to generate a number whose absolute magnitude is too big to be represented (causing floating point overflow) or too small (causing floating point underflow). It is also possible to attempt to perform invalid operations, such as a divide by zero. For a C++ compiler conforming to the IEEE standard, all such meaningless results are flagged as NaNs (Not a Number). Once a NaN is created it propagates through the calculation, ensuring that any meaningless final answers are also suitably flagged. However, unless there is an appropriate compiler option, a NaN is not flagged when it is generated. Such an option is invaluable for large numerical applications.

Unlike some other languages, such as FORTRAN, there is no operator for raising a number (either integer or floating point) to a power. In C++, such operations are carried out by a function call. For small integer powers it is, in any case, usually much faster to use repeated multiplication, as in:

```
velocity = 0.5 * acceleration * time * time;
```

3.2.2 Type `float`

The type `float` should use at least four bytes (but not more than for `double`) to represent a floating point number. For compilers conforming to the IEEE standard, four bytes gives a range of about 10^{-38} to 10^{38} and an accuracy of seven decimal places. The type `float` has the advantage of using less memory than `double` and in fact four-byte floating point arithmetic is accurate enough for many scientific and engineering applications. Also, calculations using the `float` type may be significantly faster than `double`. However, this will depend on the processor used in a particular machine. Older versions of C++ only provided `double` implementations for the standard mathematical functions (sine, cosine etc.). Fortunately, the C++ Standard Library provides both `float` and `double` implementations and so it is easier for the programmer to make an appropriate choice of floating point type.

Variables of type `float` are defined by the keyword, `float`, as in:

```
float pi, c;
pi = 3.142f;
c = 2.9979e8f;
```

A `float` constant is distinguished from its `double` counterpart by means of the suffix `f` (or `F`). Omitting the `f` does not cause an error in the above examples, as the compiler inserts a conversion from `double` to `float`. Such omissions are quite common. Notice that the `f` is appended; it does not replace the `e` (or `E`):

```
c = 2.9979f8;        // WRONG: f is not legal.
```

The operators available for the types `float` and `double` are identical. The precedence and associativity of the operators are also identical.

3.2.3 Type `long double`

The type `long double` is defined by means of the `long double` keyword, as in:

```
long double pi, gamma;
pi = 3.14159265358979323846L;    // pi.
gamma = 0.57721566490153286061L; // Euler's constant.
```

A `long double` constant is distinguished from `double` and `float` by means of the suffix, `L`. The letter `l` is also valid, but is probably best avoided since `l` looks too similar to 1.[6] You *do* need to append the `L` (or `l`) or else the constant will be read as a

---

[6]For the same reason, it is best to avoid using a single `l` as a user-defined identifier.

double, even if it is then assigned to a long double. For an IEEE 64-bit long double, 18 decimal places are significant.

The available operators for the types float, double and long double are identical. The C++ Standard Library provides long double implementations for the standard mathematical functions (sine, cosine etc.), in addition to the float and double implementations. However, for some C++ compilers there may be no difference between the representations of double and long double (or even double and float).

3.3 Changing Types

The concept of type is needed in a language, such as C++, in order to be able to distinguish the different uses we make of different bytes of memory. Having different types can help to prevent us carrying out incompatible operations on this memory. However, it is sometimes necessary to convert a value having one type directly into another. This may either be done automatically (and often silently) by the compiler, or we may have to force the change. Since to some extent types exist in order to protect us from our own foolishness, we had better be certain of what we are doing if we force the compiler to do a type conversion!

3.3.1 Type Promotion and Conversion

In many situations arithmetic binary operators have operands of different types, as demonstrated by the following code fragment:

```
double x;
float y;
int i;
long j;
i = 2;
x = 1 + 3.7;      // + operator has integer and double operands.
y = 3.7;          // = operator has float and double operands.
j = i;            // = operator has long and int operands.
```

Such statements can often be avoided by careful programming, but they are valid C++ since automatic conversions take place for arithmetic expressions containing mixed types. The rules for conversions are such that binary operations involving mixed types are performed using the type best able to handle the operands. For instance in the expression, $1 + 3.142$, the right operand is of type double and the left is of type int, which therefore gets implicitly converted to a double. There is a hierarchy of conversions for the operands of binary operators, with the first match in the hierarchy being the one that is actually used. This hierarchy is complicated and you should consult [10] and [1] if you need more detail. However, numerical applications mainly use either the float or double types, with int typically used as an iteration counter. As a result, mixed operand types are mostly: double *and* int, float *and* int, and double *and* float.

As mentioned in Section 3.1.10, it is also possible to have a promotion from the type bool to the type int. In this case false is converted to zero and true is converted to one.

Demotions may also occur, as in:

```
int i, j, k;
i = 3.142;          // double truncated to int.
j = -3.142;         // double truncated to int.
k = 2.9979e40;      // Undefined.
```

In the first case the `double` constant is truncated and 3 is assigned to `i`. In the second case the right operand is negative and the direction of truncation depends on the C++ compiler; the result could be either -3 or -4. The result in the third case is undefined since an integer cannot hold such a large value.

Exercise

If `x` has type `double`, why do

```
x = (1.0 + 1) / 2;
```

and

```
x = 0.5 + 1 / 2;
```

give different results?

3.3.2   Casts

It is possible to perform an explicit conversion, known as a *cast*. This is a very risky thing to do since it throws away any type checking and in most cases explicit conversions can be avoided by careful programming. In order to provide at least a little safety, there are four types of cast in C++. These are the `static_cast`, `reinterpret_cast`, `dynamic_cast` and `const_cast`. In each case the target type is specified in angled brackets and the object that is being cast from is in parentheses. An example is:

```
double x;
int i = 4;
x = static_cast<double>(i);
```

Used in this way, `static_cast<double>()` is actually a unary operator. In mixed arithmetic expressions casts are unnecessary since they are automatically inserted by the compiler. The three remaining casts are not used in this book, but if you should need them, they are described in [10].

3.4   Some Basic Operations

In this section we introduce a few basic operations. We first consider the `sizeof` operator, which is an essential part of the language. Then we go on to learn various new assignment operators and how to make an initialization in a definition; these features are all useful rather than fundamental.

3.4.1 The sizeof Operator

The number of bytes used to represent the fundamental types is dependent on the C++
compiler being used. However, it is frequently necessary to know the size of these types,
for instance when copying an object between different memory locations. As an aid to
writing portable code, there is an unary `sizeof` operator which, as its name suggests,
gives the size of an object in bytes. To be precise, the result of the `sizeof` operator is
an unsigned integer of type `size_t`, which is specified in the header file, `<cstddef>`.
So, if it is necessary to declare an identifier to hold the size of an object, we should
really define it be of type `size_t`, as in the following program:

```
#include <iostream>
#include <cstddef>
using namespace std;

int main()
{
    size_t size;
    double x, y;
    int i;
    cout << "\nRelative storage sizes are:\n\n";
    size = sizeof(char);
    cout << "char: \t\t" << size << "\n";
    size = sizeof(int);
    cout << "int: \t\t" << size << "\n";
    size = sizeof(long int);
    cout << "long int: \t" << size << "\n";
    size = sizeof(bool);
    cout << "bool: \t\t" << size << "\n";
    size = sizeof(float);
    cout << "float: \t\t" << size << "\n";
    size = sizeof(double);
    cout << "double: \t" << size << "\n";
    size = sizeof(x * y);
    cout << "(x * y): \t" << size << "\n";
    size = sizeof(x * i);
    cout << "(x * i): \t" << size << "\n";
    return(0);
}
```

This subtlety is not important for the objects given here, since the compiler will insert
an implicit cast if necessary.

The `sizeof` operator actually gives the storage as a multiple of what is required
for the `char` type; that is the result of `sizeof(char)` is one by definition. However,
throughout this book we assume that the `char` type takes one byte of memory.

It is a good idea to use the `sizeof` operator since it is much more meaningful to see
`sizeof(double)` in a program, rather than 8. The `sizeof` operator helps to make code
portable and carries no performance overhead, since it is evaluated at compile-time.

Exercise

Run the above program on your machine and note the sizes of the various types.

There are actually two forms of the sizeof operator. If we want the size of a type, then we must include the brackets, as in

```
size_t size;
size = sizeof(char);     // O.K.
size = sizeof char;      // WRONG
```

If we want the size of an expression, then we don't need to include the brackets, so the following *is* valid:

```
size_t size;
double x;
size = sizeof x;         // O.K.
```

However, the sizeof operator has a higher precedence than many common operators, such as multiplication. This can sometimes lead to unexpected results. For example, in the following program the values assigned to size_1 and size_2 are different:

```
#include <iostream>
#include <cstddef>
using namespace std;
int main()
{
    size_t size_1, size_2;
    double x = 3.142;
    int i = 10;
    size_1 = sizeof x * i;
    cout << size_1 << "\n";
    size_2 = sizeof(x * i);
    cout << size_2 << "\n";
    return(0);
}
```

In the first case, the sizeof operator finds the size of x (typically 8) and then multiplies the result by 10 (so that size_1 is assigned 80). In the second case, the sizeof operator finds the size of the expression (x * i), which is the same as the size of x (typically 8). The message should be clear; in order to avoid possible mistakes, use parentheses!

3.4.2   Initialization

It is possible to combine a definition and assignment in a single statement, known as an initialization, as in:

```
int i = 1, j = 2, k = 3;
```

Such initializations are also possible for other fundamental data types. Now that the fundamental data types have been introduced, it is also worth pointing out that identifiers only need to be defined before they are used. Since there is no requirement to have all of the definitions at the start of a program, the following is valid:

```
int i = 1;
++i;
int j = i;
```

Some programmers find that code is more readable if definitions are collected in one place. However, it is a good idea to leave the definition of an identifier until it can be initialized, since this avoids the common error of using uninitialized variables in expressions. In fact, in C++ it would be difficult to insist on collecting all definitions in one place, as is done in some other languages, because the definition of objects is often a run-time decision.

Exercise

Convert the quadratic equation program, given in Section 1.2, so that as far as possible variables are not defined until they can be initialized.

3.4.3 Assignment Operators

It is often necessary to carry out pairs of operations, such as in the following code fragment:

```
int i, j;
double x, y;
i = i + 5;
j = j + i;
x = x * y;
y = y / x;
```

These pairs of operations can all be replaced by the following assignment operators:

```
i += 5;         // Equivalent to i = i + 5
j += i;         // Equivalent to j = j + i
x *= y;         // Equivalent to x = x * y
y /= x;         // Equivalent to y = y / x
```

In fact, there are similar assignment operators corresponding to all arithmetic binary operators for all of the fundamental data types; two more examples are:

```
x -= y;         // Equivalent to x = x - y
i %= j;         // Equivalent to i = i % j
```

Such assignment operators are a convenient shorthand and may help the compiler to produce better code.

It is possible to carry out multiple assignments in one statement, as in:

```
i = j = k = 10;
```

In fact, we are not introducing anything new here. An assignment is an expression (which has a value) and as the assignment operator associates right to left, this example is equivalent to:

```
k = 10;
j = k;
i = j;
```

Multiple assignments can lead to compact code, but it is sometimes better to avoid the technique in order to make the code more readable.

While on the subject of assignments, it worth defining what is meant by *lvalue*, since you may get cryptic compiler messages of the form "Must be lvalue". An lvalue is an expression referring to a named region of storage; i, x and z are all lvalues in the following statements:

```
int i;
float x;
double z;

i = 20;
x = 1.2;
z = 1.222333444;
```

The terminology originally referred to objects that can be the left operand of an assignment operator. For example, an expression of the form:

```
2 = i;        // WRONG!
```

is incorrect because 2 is not an lvalue; we cannot assign the value of i to 2. However, as we will see in Section 3.5, it is possible to have a named region of storage that cannot be assigned to, and these are referred to as *unmodifiable lvalues*.

The analogous term, *rvalue*, is not used very often and originated as referring to an expression on the right-hand side of an assignment statement. An rvalue can be *read* but not assigned to. For instance, in the expression:

```
i = 2;
```

we cannot cannot assign a value to 2 since it is an rvalue.

3.5   const

It is a common requirement in numerical programs to need constants, such as $2.997925 \times 10^8 \, \mathrm{m\,s^{-1}}$ for the speed of light in a vacuum. Instead of having 2.997925e8 scattered throughout a program, we could make the code more readable by using:

```
double speed_of_light = 2.9979e8;
```

However, there is always the possibility that we might inadvertently change the value of speed_of_light, and this would probably have disastrous consequences. A safer technique is to preceed the type double by the specifier const, as in:

```
const double speed_of_light = 2.9979e8;
```

The compiler will then check to make sure that no attempt is made to change the value of `speed_of_light`. Such definitions are usually placed near the start of the relevant file, where they can be readily seen.

3.6 typedef[†]

The `typedef` specifier simply introduces a synonym for a type. For example, if we make the declaration:

```
typedef double DISTANCE, TIME;
```

then instead of:

```
double x, t;
```

we can use:

```
DISTANCE x;
TIME t;
```

to define the variables `x` and `t`. Some programmers consistently use upper case letters for a `typedef`, but this is simply a matter of style. A `typedef` follows the standard rules for constructing any valid identifier and the name space is the same as for other identifiers (apart from labels, which are introduced in Section 4.8). This means, for example, that if we have introduced `DISTANCE` as a synonym for `double`, then:

```
int DISTANCE;    // WRONG!
```

is not allowed.

It is important to realize that a `typedef` does not declare a new type. For example, although x and t are instances of a different `typedef`, the following assignment is valid, since there is no strong type checking:

```
x = t;           // O.K.
```

Common uses of the `typedef` specifier are as follows:

- A `typedef` can isolate compiler-dependent parts of a program. For instance, the type used to represent a variable could be `short`, `int` or `long`, depending on the number of bytes used in their representations. Examples are `size_t` and `ptrdiff_t`, which are each declared as a `typedef` in the library header file, `<cstddef>`. This technique has the advantage that, since the definition only appears in one place, changes are easily made.

- A `typedef` can be helpful for program documentation. The declaration:

  ```
  DISTANCE x;
  ```

  is more meaningful than:

```
double x;
```

Some authors (for example [2]) use this technique extensively. However, a valid alternative is to give meaningful names to variables, as in:

```
double distance, time;
```

- A `typedef` can clarify complicated declarations. An example is given in Section 7.5.

- A `typedef` can be useful as a synonym for a very long identifier. Such long identifiers occur naturally in more advanced C++ techniques, such as templates. (See Chapter 16.)

3.7   Summary

- The integer types are `short`, `int` and `long`. They all have `unsigned` variants.

- The character types are `char`, `signed char` and `unsigned char`.

- The Boolean type is `bool` and can only be assigned the values `true` and `false`.

- The integer, character and Boolean types are known as integral types.

- See Table 3.1 for the basic operators, together with their associativity and precedence.

- The floating point types are `float`, `double` and `long double`.

- An explicit cast (or conversion) can be performed. An example is the cast from `double` to `int` provided by `static_cast<int>(3.142)`. For mixed arithmetic expressions, such conversions are automatically inserted by the compiler.

- The `sizeof` operator gives the size of an object in bytes.

- Use the `const` specifier wherever the contents of an object should not change.

3.8   Exercises

1. Using the `sizeof` operator, write a program that prints the number of bytes used to represent the `char`, `int`, `short int`, and `long int` types on your computer. Work out the maximum and minimum values that can be represented by the integer types. Check your answers by examining the header file, `<climits>`.

2. By examining the header file, `<cfloat>`, find the maximum and minimum values that can be represented by the types `float` and `double` on your system. Check your answers by means of a program. What is the significance of the other compiler-dependent parameters given in `<cfloat>`? (Since the contents of `<cfloat>` are identical to the contents of the C header file `<float.h>`, you may find that on your system `<cfloat>` simply includes `<float.h>`.)

3. Write a program that prompts for the base and height of a triangle and then outputs the area. Write one version that only uses the type `int` and a second that uses the type `double`.

4. Using the techniques covered so far, sum the first four terms in the series:

$$e = \sum_{n=0}^{\infty} \frac{1}{n!}$$

Is the result a reasonable approximation to e?

5. The total relativistic energy, E, of a free particle is given by:

$$E = \frac{Mc^2}{\sqrt{1 - v^2/c^2}}$$

where:

$$M = \text{rest mass of the particle}$$

$$c = \text{speed of light} = 2.99725 \times 10^{10}\,\text{cm s}^{-1}$$

$$v = \text{speed of particle}$$

Write a program to calculate E for sufficient values of v to enable you to plot a graph of E as a function of v, where:

$$0 \le v < c$$

and

$$M = 0.910954 \times 10^{-27}\,\text{g}$$

Chapter 4

Control Structure

In our program to solve a quadratic equation in Section 1.2, the user could enter values of a, b and c, which make:

$$b^2 - 4ac$$

negative. In such cases the program would fail, but there is nothing we can do to prevent this with the techniques we have learned so far. All of our programs have a very simple control structure. In fact control just passes from one statement to the next, with no alternative pathway through a program. Our next task is to introduce iteration and branching techniques, but before we consider these control structures a few preliminaries are needed; in particular, some new operators must be introduced. You can find the precedence and associativity of these operators in Appendix B.

4.1 Relational Operators

There are four binary operators that can be used for comparing the values of arithmetic expressions. They are:

<	less than
<=	less than or equal
>	greater than
>=	greater than or equal

In each case a comparison of the left and right operands is carried out and the result is one of the Boolean values, true or false. An example is given in the following program where b1 is assigned either true or false, according to the results of the four relational operators:

```
#include <iostream>
using namespace std;
int main()
{
    bool b1;
    b1 = 2 > 1;                 // Assigns true
    cout << b1 << "\n";
```

```
    b1 = 2 < 1;                 // Assigns false
    cout << b1 << "\n";
    b1 = 2 >= 2;                // Assigns true
    cout << b1 << "\n";
    b1 = 3.4 <= 2.4;            // Assigns false
    cout << b1 << "\n";
    return(0);
}
```

Exercise

Compile and run the above program. Do you get the expected output?

You may have been surprised to get the output 1 or 0 for the results of the previous exercise. However, this isn't a mistake! The point is that by default the values `true` and `false` are printed as 1 and 0. It is possible to change this default by using the techniques given in Section 18.5.

As pointed out in Appendix B, the relational operators associate left to right. For example, if a, b and c are of type `int`, then:

```
    bool b1 = a < b < c;
```

actually means:

```
    bool b1 = (a < b) < c;
```

This associativity is rarely, if ever, of any use and can lead to some very obscure code. In the above expression, a < b evaluates to `true` or `false`, which is then compared with c. Depending on the result of this comparison, either `true` or `false` is assigned to b1. In general this result is different from that obtained by determining whether or not b lies between a and c, which is what might be expected from the notation of mathematics.

Notice that `<=` and `>=` each constitute a single token, so that the following are invalid:

```
    bool b1 = i > = j;          // White space is not allowed.
    bool b2 = i =< j;           // =< is not a valid operator.
```

The operands of the relational operators can be expressions, as in the following code fragment:

```
    int i = 1, j = 2;
    bool b = (i + 1) >= (j - 72); // 2 is greater than -70, so true
                                  // is assigned to b1.
```

The parentheses used here and throughout most of this chapter are purely for clarity, since the rules for operator precedence make them unnecessary. However, if parentheses make the code more readable, it is worth putting them in.

Exercise

What values would you expect b1 and b2 to take as a result of the following statements:

```
double x = 110.0;
bool b1 = x > (x / 13.0) * 13.0;
bool b2 = x < (x / 13.0) * 13.0;
```

Write a program to test your answer and try other values of x.

4.2   Logical Operators

In addition to the assignment operator, =, the logical operators are:

!	negation
&&	AND
||	OR

These operators all give a result of type bool and a value of true or false.

The logical negation operator, !, is a unary operator turning true into false, and false into true. In the following program, false is assigned to b3 and true is assigned to b4:

```
#include <iostream>
using namespace std;
int main()
{
    bool b1 = true, b2 = false;
    bool b3 = !b1;
    bool b4 = !b2;
    cout << b3 << "\n" << b4 << "\n";
    return(0);
}
```

As with all unary operators, the logical negation operator associates right to left.

The logical AND operator, &&, is a binary operator. If both operands are true, then the result is true, otherwise the result is false. The following code demonstrates using the && operator:

```
#include <iostream>
using namespace std;
int main()
{
    bool b1 = true, b2 = false;
    bool b3 = b1;
    bool b4 = b2;
    bool b5 = b1 && b2;
    bool b6 = b1 && b3;
    bool b7 = b2 && b4;
    cout << b5 << "\n" << b6 << "\n" << b7 << "\n";
    return(0);
}
```

If you run this program, you will find that `false` is assigned to `b5` and `b7`, and `true` to `b6`.

The logical OR operator, `||`, is also a binary operator. If either operand is `true`, then the result is `true`, otherwise the result is `false`. The following code demonstrates the `||` operator:

```
#include <iostream>
using namespace std;
int main()
{
    bool b1 = true, b2 = false;
    bool b3 = b1;
    bool b4 = b2;
    bool b5 = b1 || b2;
    bool b6 = b1 || b3;
    bool b7 = b2 || b4;
    cout << b5 << "\n" << b6 << "\n" << b7 << "\n";
    return(0);
}
```

The output of this program shows that (as expected) `true` is assigned to `b5` and `b6`, and `false` is assigned to `b7`.

For both the `&&` and `||` operators, the evaluation is left to right. Consequently, if either the left operand of `&&` is `false` or the left operand of `||` is `true`, then the right operand is never evaluated.

Exercise

What values are assigned to `b4`, `b5`, `b6`, `b7` in the following statements:

```
bool b1 = false, b2 = true, b3 = false;
bool b4 = b1 || b2 && !b3;
bool b5 = !b1 && b3 || b2;
bool b6 = !(b2 || b3);
bool b7 = !(b2 && !b3);
```

Check your answers by means of a program.

4.3   Equal and Not Equal Operators

The binary operators introduced in this section are:

$$== \quad \text{equal}$$
$$!= \quad \text{not equal}$$

Both operators are used to test operands of arithmetic type, with the result being of type `bool`.

The *equality* operator, `==`, gives a result of `true` if both operands are identical and `false` otherwise. Some examples of using the equality operator are given below.

```
#include <iostream>
using namespace std;
int main()
{
    int i = 0, j = 10;
    double x = 10.0, y = 3.0, z = 10.0 / 3.0;
    bool b1 = i == j;     // Assigns false to b1.
    cout << b1 << "\n";
    bool b2 = y == x;     // Assigns false to b2.
    cout << b2 << "\n";
    bool b3 = z * y == x; // Assigns true or false to b3.
    cout << b3 << "\n";
    bool b4 = y == 3.0;   // Assigns true to b4.
    cout << b4 << "\n";
    return(0);
}
```

The output of this program confirms that b1 and b2 are both assigned false, and that true is assigned to b4. The result for b3 could be either true or false. The reason for this is that rounding errors mean that the result of a floating point calculation is rarely exact. Consequently, using the equality operator in this way is dangerous and a frequent cause of numerical application programs that fail to terminate.

Exercise

Modify the above program so that you can find out by how much z * y differs from x.

The Boolean operations in the following statements are either invalid or don't do what was intended:

```
int i = 1, j = 2;
double z = 3.0;
bool b1 = (i = = j);    // White space is not allowed.
bool b2 == 10;          // b2 = 10 intended.
bool b3 = (z = 5.0);    // Assigns 5.0 to z and true to b3.
                        // z == 5.0 intended.
```

The error in b1 is fairly straightforward, in that white space is not allowed within a token. Also in the statement for b2, the equality operator has been used when the assignment operator was intended. Fortunately, both of these mistakes will be caught by the compiler. However, the error for b3 is more insidious. In this case the assignment operator has been used when the equality operator was intended. The result is that 5.0 is assigned to z and, since this expression is non-zero, true is assigned to b3, whatever the original value of z. In fact, z originally had the value 3.0 so false should have been assigned to b3.

The binary *not equal* operator, !=, again takes arithmetic operands. The result is false if the two operands are identical and true otherwise. Examples of using the != operator are given below.

```cpp
#include <iostream>
using namespace std;
int main()
{
    int i = 1, j = 10;
    double x = 10.0, y = 3.0, z = 10.0 / 3.0;
    bool b1 = i != j;           // Assigns true to b1.
    cout << b1 << "\n";
    bool b2 = y != x;           // Assigns true to b2.
    cout << b2 << "\n";
    bool b3 = z * y != x;       // Assigns true or false to b3.
    cout << x - z * y << "\n";
    cout << b3 << "\n";
    bool b4 = y != 3.0;         // Assigns false to b4.
    cout << b4 << "\n";
    return(0);
}
```

The output of this program shows that b1 and b2 are both assigned true, and that false is assigned to b4. The result for b3 could be either true or false. Once again, the reason for this is floating point rounding errors. Consequently, it is usually important to avoid using the != operator with floating point operands.

The following statements are invalid attempts at using the not equal operator:

```cpp
k = i ! = j;    // White space is not allowed.
k = i =! j;     // =! is not a valid operator. != was intended.
```

Exercise

What is the result of i =! j in the above code fragment?

4.4   Blocks and Scope

A pair of braces, { }, can be used to group statements, definitions and declarations into a *compound statement* or *block*, as it is more commonly known. Any definitions made within a block are only valid within the block and hide definitions made for the same identifiers outside. The following program demonstrates this property:

```cpp
#include <iostream>
using namespace std;
int main()
{
    int x = 10;             // x has type int, value 10.
    // some code
    {
        double x;           // x has type double.
        // more code
        x = 1.772;
```

```
            cout << x << "\n";
        }
        cout << x << "\n";        // x has type int, value 10.
        return(0);
    }
```

Notice that there is no semicolon after the second (terminating) brace of the block.[1]

Some compilers issue helpful warnings when a definition hides a previous definition. The part of a program where a particular identifier is valid (that is where it is visible) is known as the *scope* of that identifier.

Notice how the two braces in the above code have the same indentation and the enclosed statements are all indented by one tab. This layout is not a requirement of C++, but is generally adopted and makes the code more readable.

Blocks can be nested and a block may be followed by one or more further blocks. The following code illustrates how definition hiding works in such circumstances:

```
#include <iostream>
using namespace std;
int main()
{
    int i = 1;
    double x = 1.111;
    cout << i << "   " << x << "\n";
    {
        int x = 2;
        double i = 2.222;
        cout << i << "   " << x << "\n";
    }
    cout << i << "   " << x << "\n";
    {
        char i = 'i';
        char x = 'x';
        cout << i << "   " << x << "\n";
        {
            int x = 3;
            double i = 3.333;
            cout << i << "   " << x << "\n";
        }
        cout << i << "   " << x << "\n";
    }
    return(0);
}
```

Exercise

Explain the output from the above program.

---

[1] A block can be empty. An example of this is when a function needs to be defined but does not perform any operation.

A block can appear anywhere that a single statement is permissible and often occurs in the context of branching and iteration, both of which are considered next.

4.5 Branch Statements

There are three branch statements: the if, if else and switch statements.

4.5.1 if Statement

The if statement takes the form:

```
if (condition)
    statement
```

where the condition is any valid arithmetic expression.[2] The condition is evaluated and, if it is true, the statement is executed; conversely, if the condition evaluates to false then the statement is not executed. An example of using the if statement is given below.

```
if (i == 0)
    x = 100.0;   // 100.0 is assigned to x if i is zero.
```

This code fragment can alternatively be written (perhaps more obscurely) as:

```
if (!i)
    x = 100.0;
```

What happens here is that i is converted to true or false, depending on whether i is non-zero or zero. If i is zero, then the result of the expression !i is true and 100.0 is assigned to x.

Because a block is equivalent to a single statement, it is common for an if statement to involve a block, as shown below.

```
if (i == 2) {
    x = 3.142;      // If i equals 2, all three
    y = 100.0;      // statements are executed.
    z *= x;
}
```

Notice that the closing brace has the same indentation as the if statement. The opening brace is also on the same line as the if statement. Although not a requirement of C++, this style is widely adopted.

It is also possible for any branch statement, such as an if, to involve the bool type directly, as in:

```
if (last_entry == false)
    x = pi;
```

---

[2]The conditions in any of the three branch statements can also involve pointers, which are introduced in Chapter 6.

The following code fragment demonstrates a common error involving the equality operator, which is notoriously difficult to detect:

```
if (i = 4)
    x = 1000.0;
```

The code is syntactically correct, but is equivalent to:

```
i = 4;
x = 1000.0;
```

However, the programmer almost certainly meant to write:

```
if (i == 4)
    x = 1000.0;
```

In general the results of the two code fragments are very different; in the first case, 4 is assigned to i, the expression always evaluates to true and hence 1000.0 is always assigned to x. If the condition for an if statement involves a constant, then it is possible to avoid such errors by writing the constant first, as in:

```
if (4 == i)
    x = 1000.0;
```

Then if we accidentally write:

```
if (4 = i)                     // WRONG
    x = 1000.0;
```

a compiler error occurs. However, care is needed to do this in every case. Some helpful compilers issue a warning if an assignment occurs as the outermost operator in the condition for an if statement.

Another common mistake is to omit the braces for a compound statement following an if. The following statements:

```
if (!i)
    x = 3.142;
    y = 100.0;
    z *= x;
```

look like (the programmer's intention):

```
if (!i) {
    x = 3.142;
    y = 100.0;
    z *= x;
}
```

but execute as:

```
if (!i)
    x = 3.142;
y = 100.0;
z *= x;
```

4.5.2   if else Statement

The if else statement takes the form:

```
if (condition_1)
    statement_1
else if (condition_2)
    statement_2
// more else ifs
else
    statement_n
```

where the conditions are arithmetic expressions. The way this statement works is that if condition_1 is true, then statement_1 is executed and control passes beyond statement_n. Conversely, if condition_1 is false then condition_2 is tested. The sequence continues until one of the statements (which may be statement_n) is executed. In fact, there is no requirement to have the final else statement; there may be no default action to be taken, in which case it is possible for all of the conditions to evaluate to false and for none of the statements to be executed. The following program demonstrates using the if else statement:

```cpp
#include <iostream>
using namespace std;
int main()
{
    double x, y, pi = 3.142;
    int i;
    cout << "Enter an integer: ";
    cin >> i;
    if (i == 0) {
        x = pi;
        y = 2.0 * pi;
    }
    else if (i == 1) {
        x = 2.0 * pi;
        y = 0.0;
    }
    else {
        x = 0.0;
        y = 0.0;
    }
    cout << "x = " << x << "   y = " << y << "\n";
    return(0);
}
```

Exercise

Compile and run the above program. Try entering various integers and verify that you get the output you would expect.

The if else statement has one potential pitfall, known as the *dangling else* trap, which is illustrated in the program below.

```cpp
#include <iostream>
using namespace std;

int main()
{
    int i, j;
    cout << "Enter two integers: ";
    cin >> i >> j;
    if (i == 0)
        if (j == 0)
            cout << "Both i and j are zero\n";
    else {
        cout << "i is non-zero\n";
    }
    cout << "i: " << i << " j: " << j << "\n";
    return(0);
}
```

The intention of the programmer is apparent from the indentation; if i is non-zero then the code in the braces should be executed, whatever the value of j. However, the else is dangling; in other words there is an ambiguity as to whether it is associated with the first or second if. In fact a dangling else is always associated with the nearest preceeding if, so an indentation that more accurately reflects the logic of the code is:

```cpp
#include <iostream>
using namespace std;

int main()
{
    int i, j;
    cout << "Enter two integers: ";
    cin >> i >> j;
    if (i == 0)
        if (j == 0)
            cout << "Both i and j are zero\n";
        else {
            cout << "i is non-zero\n";
        }
    cout << "i: " << i << " j: " << j << "\n";
    return(0);
}
```

This is presumably not what the programmer intended. However, such mistakes can be very difficult to find since the indentation has a very powerful effect on how we read code, even when the indentation is wrong! The dangling else problem can be overcome by using pairs of braces, so the modified program becomes:

```cpp
#include <iostream>
using namespace std;

int main()
{
    int i, j;
    cout << "Enter two integers: ";
    cin >> i >> j;
    if (i == 0) {
        if (j == 0)
            cout << "Both i and j are zero\n";
    }
    else {
        cout << "i is non-zero\n";
    }
    cout << "i: " << i << " j: " << j << "\n";
    return(0);
}
```

As an example of the if else construct, we can now write our quadratic equation program so that complex roots are trapped:

```cpp
#include <iostream>
#include <cmath>          // For sqrt().
using namespace std;
int main()
{
    double a, b, c;
    cout << "Enter the coefficients a, b, c: ";
    cin >> a >> b >> c;
    double temp = b * b - 4.0 * a * c;
    if (temp > 0.0) {
        double root = sqrt(temp);
        double root1 = 0.5 * (root - b) / a;
        double root2 = -0.5 * (root + b) / a;
        cout << "There are two real solutions:  " << root1 <<
            " and " << root2 << "\n";
    }
    else if (temp < 0.0) {
        double root = sqrt(-temp);
        double real_part = -0.5 * b / a;
        double imag_part = 0.5 * root / a;
        cout << "There are two complex solutions: " << real_part <<
            " + i * " << imag_part << " and " << real_part <<
            " - i * " << imag_part << "\n";
    }
    else {
        cout << "Both solutions are: " << -0.5 * b / a << "\n";
```

```
    }
    return(0);
}
```

This is our first complete program with any control structure; the `if else` statement enables us to trap three possible cases resulting from values of a, b, c. Notice how variables, such `root1` and `real_part`, are defined in (and only have scope within) different blocks; this enables us to use appropriate variables while keeping the scope as restricted as possible. Notice also how there is only one `return` statement. It is tempting to put a `return` at the end of each block, but such *alternative returns*, as they are sometimes known, are best avoided if possible. The reason for this is that it is easy to miss a `return` statement in some deeply embedded inner block.

Exercise

Try out our improved quadratic equation program for various values of a, b and c. You should be able to cause the program to fail, in which case, make appropriate further modifications.

4.5.3   switch Statement

Although the `switch` statement takes the general form:

```
switch (expression) {
case constant_1:
    statement_1;
case constant_2:
    statement_2;
// More case, statement pairs.
case constant_n:
    statement_n;
default:
    last_statement;
}
```

our discussion will be clearer if we consider a specific example. Suppose we have a program to solve a differential equation by a variety of iterative methods and we want to be able to choose which one to use; that is we want a menu. A program providing a menu is shown below.[3]

```
#include <iostream>
using namespace std;
int main()
{
    int option;
    cout << "menu:\n\t1 Jacobi\n\t2 Gauss-Seidel\n" <<
        "\t3 Red-black Gauss-Seidel\n" <<
        "Enter a number to choose the required technique.\n";
```

---

[3]Notice the use of \t for a horizontal tab in this program.

```
    cin >> option;

    switch (option) {
    case 1:
        cout << "Starting Jacobi iterations.\n";
        // Jacobi code.
        break;
    case 2:
        cout << "Starting Gauss-Seidel iterations.\n";
        // Gauss-Seidel code.
        break;
    case 3:
        cout << "Starting Red-black Gauss-Seidel iterations.\n";
        // Red-black Gauss-Seidel code.
        break;
    default:
        cout << option << " is not a valid option\n";
        break;
    }
    return(0);
}
```

Here we have used a switch statement to provide a number of different options. Notice that there is no terminating semicolon following the closing brace of the switch statement. In this example, the value of option is tested in turn against the constants appearing after each case keyword. When one of these constants is found to be equal to the value of option, then the subsequent code is executed. The break statement, which we have not met before, causes control to pass to whatever follows the switch statement. The important point to realize is that it is the break statement that alters the flow of control, rather than the case or default statements.[4] If there is no break corresponding to a particular case statement, then the flow of control is unchanged. This drop-through behaviour has both advantages and disadvantages; it sometimes permits elegant code, as in:

```
#include <iostream>
using namespace std;
int main()
{
    char reply;
    cout << "Do you want to continue (Y or N): ";
    cin >> reply;
    switch (reply) {
    case 'Y':
    case 'y':
        cout << "Continuing ...\n";
        // Code for Yes.
```

---

[4]The case and default are actually special types of statement labels that can only appear in a switch statement. Labels are described later in this chapter.

```
        break;
    case 'N':
    case 'n':
        cout << "Exiting ...\n";
        // Code for No.
        break;
    default:
        cout << "'" << reply << "' is not a valid reply.\n";
        // Code to recover from invalid reply.
        break;
    };
    return(0);
}
```

but it is easy to miss out a `break`, with potentially disastrous results. For instance, in the menu example, if the first `break` were omitted, then entering 1 for the menu item would cause both the Jacobi and Gauss–Seidel code to be executed. (Try it!)

If none of the constants match the `switch` expression, then control passes to the statement following the `default` label (if there is one). It is possible for the `break` before the `default` label to be omitted, in which circumstances the final `case` would also lead to the `default`. This is usually the result of a mistake!

The final `break` statement after the `default` label in this example is redundant, but is worth including as it is so easy to subsequently add another `case` without including the necessary `break`. The keywords `case` and `default` can never appear outside of a `switch` statement and there can be at most one `default` label for each `switch` statement. The `case` and `default` labels can appear in any order, but it is better for the `default` label (if any) to come after the `case` statements, since this reflects the flow of control.

Exercise

By modifying the above program, demonstrate that it is indeed possible to put the `default` label anywhere in a `switch` statement, without altering the way in which the program executes.

For large programs, the switch statement can often be avoided by using the object-oriented aspects of C++, resulting in code that is more elegant and maintainable. The key concepts involved are classes, inheritance and polymorphism, which are introduced in later chapters.

4.6 Iteration Statements

In most programs it is necessary to execute one or more statements many times. It is tedious to repeat the statements and, in any case, it is often impossible to predict how many times the execution should be repeated. Such circumstances are handled by the three iteration statements: `while`, `for` and do.

4.6.1   `while` Statement

The `while` statement takes the general form:

```
while (condition)
    statement
```

and has two distinct parts, which we have called `condition` and `statement`.[5] A typical example of a `while` statement is given in the following program:

```cpp
#include <iostream>
using namespace std;
int main()
{
    int n;
    cout << "Enter an integer greater than 1: ";
    cin >> n;
    --n;
    int gamma = n;
    while (n > 2) {
        --n;
        gamma *= n;
    }
    cout << "gamma(" << n << ") is: " << gamma << "\n";
    return(0);
}
```

The `condition` expression (in this case `n > 2`) is evaluated before each execution of what is in this particular case a compound statement. This compound statement is executed if the `condition` expression is `true` and then the `condition` expression is tested again. If the test of the `condition` expression never gives a result of `false`, then the iteration never terminates. This is demonstrated in the following code, but be warned; it will carry on for ever!

```cpp
#include <iostream>
using namespace std;
int main()
{
    int n;
    cout << "Enter an integer greater than 1: ";
    cin >> n;
    --n;
    int gamma = n;
    while (1) {                    // Oops! A mistake here.
        --n;
        gamma *= n;
    }
    cout << "gamma(" << n << ") is: " << gamma << "\n";
```

---

[5]The `condition` in any of the three iteration statements can also involve pointers, which are introduced in Chapter 6.

```
    return(0);
}
```

In this program, the condition expression of the while loop is 1, which always evaluates to true. Consequently, the loop never terminates.

Notice that there is no do associated with the while (unlike some other languages) and that there is no terminating semicolon. Both of these points are illustrated in the following code fragment:

```
while (n > 2) do {      // WRONG: 'do' is not allowed.
    --n;
    gamma *= n;
};               // ; is unnecessary, but does no harm here.
```

Exercise

Use a while loop to sum the first twenty terms in the series:

$$1 - \frac{1}{2} + \frac{1}{3} - \frac{1}{4} \cdots$$

The result should be an approximation to $\ln 2$.

4.6.2   for Statement

The for statement has the general form:

```
for (initialize; condition; change)
    statement
```

and can be seen to consist of four separate parts, which we call initialize, condition, change and statement in order to indicate their respective roles. The initialize statement is executed first and if condition is true, statement is executed. The change expression is then evaluated and if condition is still true, statement is executed again. Control continues to cycle between condition, statement and change, until the condition expression is false. Control then passes beyond the for loop. A typical example is given in the program below.

```
#include <iostream>
using namespace std;
int main()
{
    int n, i;
    cout << "Input a 'small' positive integer: ";
    cin >> n;
    int factorial = 1;
    for (i = 1; i <= n; ++i)
        factorial *= i;
    cout << "factorial(" << n << ") is: " <<
        factorial << "\n";
    return(0);
}
```

Notice that the `initialize` statement is only executed once and, as the name suggests, it performs an initialization. It is possible for the final `statement` part of the `for` statement not to be executed at all, as in:

```
n = 0;
int factorial = 1;
for (i = 1; i <= n; ++i)      // 1 <=0 is false so
        factorial *= i;           // factorial is unchanged.
```

It is also possible for any (or even all) of the expressions (but not the semicolons) to be missing. An example with no `initialize` statement is:

```
int factorial = 1, i = 1;
for (; i <= n; ++i)
        factorial *= i;
```

The following code fragment gives an example of a `for` loop that has no `initialize` statement and no `change` expression:

```
int factorial = 1, i = 1;
for (; i <= n;)
        factorial *= i++;
```

If the second (`condition`) expression is missing then it is taken as evaluating to `true` and the loop continues for ever unless there is some way of breaking out. (See Section 4.7.)

It is often convenient to define the loop variable in the `initialize` statement, as in:

```
#include <iostream>
using namespace std;
int main()
{
    int n;
    cout << "Input a 'small' positive integer: ";
    cin >> n;
    int factorial = 1;
    for (int i = 1; i <= n; ++i)
        factorial *= i;
    cout << "factorial(" << n << ") is: " <<
        factorial << "\n";
    return(0);
}
```

It is important to realize that the scope of a variable defined in the `initialize` statement only lasts until the end of the `for` loop. So the following code fragment is not ANSI C++:

```
for (int i = 1; i <= n; ++i)
        factorial *= i;
cout << "The final value of i is " << i << "\n";
```

This is a significant change from early versions of C++. Code like this fragment may compile on your current compiler (possibly with a warning message) since there is a lot of old C++ code that would fail if this rule were strictly enforced. However, it is better to follow the ANSI standard when writing new code.

Exercise

Convert the above code fragment into a program. If your program fails to compile, modify it appropriately.

It is worth noting that, unlike some other languages, there is nothing special about the variable that controls the iteration in a `for` statement. For instance, the controlling variable (in this case, i) can be assigned to:

```
for (i = 0; i < 10; ++i) {
    x *= 24.0 + pi;
    if (x > 17.0)
        i = 9;
}
```

and can be of floating type:

```
for (double x = 0.0; x != 10.0; ++x)    // Very risky.
    total += x;
```

A test for equality of a floating type is very risky since it is likely that the finite machine precision will mean that the condition never occurs and the iteration continues for ever. For this reason it is very unusual to have a floating point loop counter, although it is possible to use a relational expression, as in:

```
for (double x = 0.0; x < 9.5; ++x)    // Safer.
    total += x;
```

As an example of using `for` loops, the program given below finds some positive integer solutions to the (Diophantine) equation:[6]

$$k^2 = i^2 + j^2$$

```
#include <iostream>
using namespace std;
int main()
{
    int k_max = 40;       // Change this value as required.
    for (int k = 1; k < k_max; ++k) {
        int k_2 = k * k;
        for (int i = 1; i < k; ++i) {
            int test = k_2 - i * i;
            for (int j = 1; j < k; ++j) {
                if (test == j * j)
```

---

[6]Such solutions are known as Pythagorean triples. See [12] for more details, including how to obtain all such triples.

```
            cout << "A solution is: i = " <<
                i << "  j = " << j <<
                "  k = " << k << "\n";
        }
      }
    }
    return(0);
}
```

Exercise

Modify the above program so that it prompts for the value of k_max and
verify some of the solutions that the program provides.

This program is not very efficient since it tests values that cannot possibly
be solutions. Try to improve the efficiency of the program.

4.6.3   do Statement

The do statement has the form:

```
do
    statement
while (condition);
```

where the semicolon is an essential part of the syntax. Here we label the two parts of
the do statement as statement and condition. The first thing that happens is that
statement is executed, after which there are two possibilities. If the condition expression evaluates to false, then control passes to the statement after the do loop. Alternatively, if the condition expression is true, then control passes back to statement
again. The control passes between condition and statement until condition becomes false, at which point control passes beyond the do loop. As an example, we
could change our program in Section 4.5.3 to read:

```
#include <iostream>
using namespace std;
int main()
{
    int option;
    do {
        cout << "menu:\n\t1 Jacobi\n\t2 Gauss-Seidel\n" <<
            "\t3 Red-black Gauss-Seidel\n" <<
            "Enter a number to choose the required " <<
            "technique\n";
        cin >> option;
    } while (option < 1 || option > 3);
    switch (option) {
    case 1:
        cout << "Starting Jacobi iterations.\n";
        // Jacobi code.
```

```
        break;
    case 2:
        cout << "Starting Gauss-Seidel iterations.\n";
        // Gauss-Seidel code.
        break;
    case 3:
        cout << "Starting Red-black Gauss-Seidel iterations.\n";
        // Red-black Gauss-Seidel code.
        break;
    default:
        cout << option << " is not a valid option\n";
        break;
    }
    return(0);
}
```

As before, the menu is sent to the output stream and the user enters an integer. If the integer has the wrong value then the menu is repeated until a correct value is entered.

It is worth including the pair of braces in a do statement, even if a compound statement is not required. If this isn't done then the code may look like a while loop with an empty statement.

With some contortions it is possible to interchangeably use any of the three iteration statements; choosing the most appropriate technique depends on the particular circumstances. In C++ programming (as in C) the for loop seems to be the most common iteration statement. This is probably because the syntax conveniently collects the initialization, loop control and increment all in one place. It is also helpful to read the control expressions before a large block statement is encountered. The while statement is often appropriate when initializations have been performed by the preceding statements. The iteration counter in a while statement is often changed by using an increment or decrement operator, resulting in very compact code, such as:

```
while (i < 10)
    sum += i++;
```

The do iteration statement appears rarely in C++ (or C) programs. However, if there is a requirement for a statement to be executed at least once (as in our menu example) then the do statement may be appropriate, since neither the for nor while loops have this property.

Exercise

The factorial of a positive integer n is defined to be equal to $n(n-1)(n-2)\ldots 1$. Write a program that prompts for a positive integer and then outputs its factorial. You should write three versions of your program using the three types of iteration statement. In each program, all iterations should be done using the same type of iteration statement.

4.7   break and continue Statements

The break statement can only occur inside a switch or iteration statement. We have
already met the break statement in the context where it causes control to exit from
the enclosing switch statement. The break statement can also supply a way of exiting
the three types of iteration statement, such as in the following code fragment:

```
int i;
while (true) {
    cout << "Enter an integer > 0 and < 10 ";
    cin >> i;
    if (i > 0 && i < 10)
        break;
}
```

In this example, the while loop could continue for ever. However, if i is greater than
zero and less than ten, then the break causes control to pass to the first statement
after the end of the while loop. Notice that, since we can only break from a single
enclosing loop, the break statement does not directly provide a way of exiting from
inside deeply nested loops. This is demonstrated by the following program:

```
#include <iostream>
using namespace std;

int main()
{
    int test = 0;
    for (int i = 0; i < 5; ++i) {
        cout << "Testing i = " << i << "\n";
        for (int j = 0; j < 5; ++j) {
            cout << "\tTesting j = " << j << "\n";
            for (int k = 0; k < 5; ++k) {
                test = 10 * k;
                // More code.
                if (test > 20)
                    break;
                cout << "\t\tTesting k = " << k << "\n";
            }   // The break leaves us inside the i and j loops.
        }
    }
    return(0);
}
```

If you run this program, you will find that the inner loop is left when k is equal to
three. However, the remaining i and j loops run their full course (of five iterations
each).

The continue statement can only occur inside an iteration statement and causes
control to pass directly to the next iteration. For instance in the following simple
example:

```
#include <iostream>
using namespace std;
int main()
{
    double x = 0.0, y = 0.0;
    for (int i = 0; i < 10; ++i) {
        ++x;
        if (i == 5)
            continue;
        ++y;
    }
    cout << "x = " << x << "    y = " << y << "\n";
    return(0);
}
```

x and y are both incremented on each pass through the loop, except for when i is equal
to 5. In this case x is incremented, but then control passes directly to the next iteration
(i is equal to 6). The continue statement can play the same role in the while and do
iteration statements.

4.8   goto Statement[††]

C++ even possesses the infamous goto statement. This is an unconditional jump (or
transfer of control) to a labelled statement, such as occurs in the following program:

```
#include <iostream>
using namespace std;
int main()
{
    int i, j, k;
    for (i = 0; i < 100; ++i) {
        for (j = 0; j < 100; ++j) {
            for (k = 0; k < 100; ++k) {
                if (i == 5 && j == 10 && k == 15)
                    goto leap;
            }
        }
    }
    leap:   cout << "Left loops with i = " << i <<
         "   j = " << j << "   k = " << k << "\n";
    return(0);
}
```

Here we have used the identifier, leap, as a label. When i, j and k are equal to the
values specified in the if statement, then control is transferred out of all three loops
to the output stream statement labelled by leap. Any valid identifier is acceptable as
a statement label, but the same label can only be used once in the same function, since
a label has the scope of the function in which it is declared. A label can only be used

by a goto statement and is the *only* identifier whose scope is not local to the block in which it is declared. The following demonstrate valid labels:

```
label9999:  x = y + z;
label1: label2: label3: x = y + z;  // Multiple labels.
a:   a = y - z;        // Labels have their own name space.
```

whereas

```
100error:   x = y + z;  // Must start with letter or _
error 100:  x = y + z;  // White space is not allowed.
```

are invalid.

    Use of the goto is strongly discouraged in modern programming since it makes the flow of control very difficult to follow and is usually a symptom of poor program design. In the previous program, a break statement, together with an extra condition, has the same effect, as shown below.[7]

```
#include <iostream>
using namespace std;
int main()
{
    int i, j, k;
    bool exit_loops = false;
    for (i = 0; i < 100; ++i) {
        for (j = 0; j < 100; ++j) {
            for (k = 0; k < 100; ++k) {
                if (i == 5 && j == 10 && k == 15) {
                    exit_loops = true;
                    break;
                }
            }
            if (exit_loops == true)
                break;
        }
        if (exit_loops == true)
            break;
    }
    cout << "Left loops with i = " << i << "  j = " <<
        j << "  k = " << k << "\n";
    return(0);
}
```

However, there is clearly a trade-off between writing elegant structured code and using the more efficient goto.

---

[7]In this example it would be simpler to change the conditions on the loops. A more realistic application would be calculating something like a three-dimensional integral where the number of iterations would be unknown. Breaking out of the loops might then be governed by the integration appearing to have converged.

4.9 Comma Operator

The *comma* operator can be used to separate a sequence of two or more expressions. The expressions are evaluated from left to right, with the result of each evaluation being discarded before the next expression is evaluated. Consequently, the type and value of a series of expressions, separated by comma operators, is that of the right-most expression, although this is rarely significant.

It is important to realize that many of the commas appearing in C++ programs, such as in definition lists and function arguments, are separators rather than operators. For example, in Chapter 5 we introduce the idea of functions, such as

```
rectangle(width, height)
```

In this context, the comma is used to *separate* the function arguments rather than as an operator and there is no requirement for width to be evaluated before height.

We have already met the comma operator in initialization statements such as

```
int i = 0, j = 0;
```

The comma operator often occurs in for statements, as in the program given below.

```
#include <iostream>
using namespace std;
int main()
{
    int k = 0;
    for (int i = 0, j = 0; i < 10 && j < 10; ++i, ++j)
        k += i + j;
    cout << "k = " << k << "\n";
    return(0);
}
```

In this example, the initialization is equivalent to:

```
i = 0;
j = 0;
```

and the end of each pass through the loop amounts to:

```
++i;
++j;
```

It is clearly possible to use the comma operator to overburden the control part of the for statement with expressions. Therefore, it is a good idea to restrict the comma operator to expressions closely related to loop control.

Spurious comma operators can give rise to errors that are very difficult to find. For instance, the statement:

```
x = y, + 10;
```

is valid C++, but the programmer probably intended:

```
x = y + 10;
```

4.10   Null Statement

The *null* statement consists of nothing but a semicolon. Such statements are useful if the syntax requires a statement, but there is nothing to do. We have already met an example of this in the code fragment:

```
int factorial = 1, i = 1;
for (; i <= n; ++i)
      factorial *= i;
```

In this case the `initialize` part of the `for` loop is a null statement; it does nothing. Another example is where all the calculation occurs in the control part of `for` loop, as in:

```
int i;
for (i = 0; i < 1; cin >> i)
    ;
```

It is worth putting the null statement on a separate line since it makes the intention more obvious.

The null statement is often a mistake. For example the semicolon after the closing brace of the following `for` loop should not be there:

```
for (i = 0; i < 10; ++i) {
    x *= i;
    y += i;
};                      // Semicolon is a mistake, but it does nothing.
```

Sometimes, as here, the null statement doesn't do any harm. However, occasionally the result is a disaster, as in:[8]

```
while (i < 10); // This loop never ends.
    sum += i++;
```

where the semicolon on the first line is actually a null statement. Since nothing is evaluated in this `while` loop, once started it can never finish. Mistakes like this are difficult to detect because the indentation makes the code look superficially correct. A similar mistake can occur with the `if` statement:[9]

```
if (i == 0);                // This statement does nothing.
    x = initial_velocity;   // This statement is always
                            // executed.
```

Such mistakes are unlikely if pairs of braces are used, even when they are not necessary.

---

[8]See Section 4.6.1.
[9]See Section 4.5.1.

4.11 Conditional Expression Operator

The conditional expression operator, `?:`, is the only ternary operator defined in C++ and takes the form:

```
condition ? result_1 : result_2
```

The three operands are conveniently denoted by `condition`, `result_1` and `result_2`. The `condition` is evaluated first and, if `true`, then the whole expression evaluates to `result_1`, otherwise the whole expression evaluates to `result_2`. For instance, in the statement:

```
max = (i > j) ? i : j;
```

if `i` is greater than `j`, then `i` is assigned to `max`, otherwise `j` is assigned. This is equivalent to:

```
if (i > j)
    max = i;
else
    max = j;
```

which some programmers may prefer. However, the conditional expression operator produces compact code and often avoids introducing a temporary variable, as in:

```
cout << "Max. pressure = " << (p1 > p2 ? p1 : p2);
```

Notice that parentheses are necessary in this statement because the precedence of `<<` is higher than the conditional expression operator. (See Appendix B.)

Exercise

Write a program that prompts for pairs of floating point numbers and uses the conditional expression operator to send `true` to the output stream if $x^2 + y^2 \leq 1$ and `false` otherwise.

4.12 Order of Evaluation of Operands

The order of evaluation of operands is undefined for most operators. For example, the value of `k` in the expression:

```
k = ++i + i;
```

is dependent on the particular C++ compiler because either operand of the binary `+` operator may be evaluated first. Since we have now met the only four operators that are exceptions to this rule, it is worth summarizing them here. In each case the operands are evaluated from left to right.[10]

---

[10]Do not confuse the order of evaluation of operands with the precedence of operators. Whereas the former concerns the order in which the operands of *one* operator are evaluated, the latter determines the order in which *several* operators are applied.

- The logical AND operator, `&&`, does not evaluate the right operand if the left one evaluates to `false`.

- The logical OR operator, `||`, does not evaluate the right operand if the left one evaluates to `true`.

- Only one of the second and third operands of the (ternary) conditional expression operator is evaluated, as detailed in the previous section.

- The left operand of the comma operator is evaluated before the right operand.

In the first two cases one of the operands *may* not be evaluated; in the third case one of the operands is *certainly* not evaluated and in the fourth case both operands are *always* evaluated.

4.13   The Preprocessor

In addition to the control statements that are recognized by the C++ compiler, there is a control structure associated with the C++ *preprocessor*. Before a program is compiled, it is passed through a utility known as a preprocessor.[11] This is capable of performing various transformations on a C++ program, but it knows nothing about the syntax of the language and simply makes the requested textual changes. The preprocessor *directives*, or commands, are denoted by a `#` as the first non-blank character in a line and this `#` is followed by a directive. Blanks and horizontal tabs can precede the directive as well as the `#`, but more than one directive on the same line is not allowed. The `#` is a preprocessor operator.[12]

There are many preprocessor directives; the complete list is:

```
#define   #else     #elif      #endif    #error   #if
#ifdef    #ifndef   #include   #line     #undef   #pragma
```

However, not all of these directives will be described in detail; the two most important ones are `#define` and `#include`. The `#if`, `#elif`, `#else` and `#end` are also fairly common.

4.13.1   include Directive

We have already met the `include` directive in the context of standard library facilities. The directive:

```
#include <iostream>
```

tells the preprocessor to replace the directive by the statements contained in the file associated with `iostream`. For instance, the above `include` directive may find the file `/usr/include/g++/iostream`. However, there is nothing in the ANSI C++ Standard

---

[11] A single command usually invokes the preprocessor, compiler and linker, with special compiler options being used in order to omit any of these stages.

[12] There is another preprocessor operator, `##`, which is used for concatenation but is not considered in this book.

to say the file has to exist with the name that appears in the `include` directive; only that the preprocessor must know how to find the file.

It is worth emphasizing that the more up-to-date style is for the `include` directive for library files to have no suffix. This is as specified in the ANSI C++ Standard. However, on older systems you may have to use something like

```
#include <iostream.h>
```

or

```
#include <iostream.hpp>
```

You may even find that `<iostream>` and `<iostream.h>` both work. However, the `.h` file name extension for such library files is a left-over from the C programming language and it is better to use the more up-to-date style.

White space inside the `< >` *is* important, so

```
#include < iostream >
```

will not work. However, white space after the `#include` is ignored, so

```
#include<iostream>
```

is correct. I prefer to leave a space after `include`, but I wouldn't want to advocate any particular style (provided it is used consistently).

There is a second form of the `include` directive that uses the double quotation mark symbol, as in:[13]

```
#include "my_file.h"
```

This is used to tell the preprocessor to include the file `my_file.h`. Such files would *not* be part of the library for a C++ compiler but would typically be files written by application programmers (such as ourselves). Files included in this way should either be in the directory where the program is compiled, or the complete path should be given, as in

```
#include "/home/fred/source/my_file.h"
```

As might be expected, white space within the quotes is important so

```
#include " my_file.h "
```

will not find a file called `my_file.h`.

Files that are copied by the `#include` directive are known as *include files*. Such files are particularly useful for large programs that are split up into files (often known as *source files*) that can be separately compiled. Consistency of any constants and function declarations can be ensured throughout the source files by each source file using the same include file.[14] Include files that are used like this are called *header files*. There is no restriction on user-defined file extensions, but it is advisable to conform to the generally accepted convention of `.h` for header files, as this helps to identify the purpose of such files.

---

[13]Notice that the double quotation mark is one character and must not be confused with two single quotation marks.

[14]Function declarations are introduced in Section 5.1.3.

4.13.2 define Directive[†]

We introduce the `define` directive mainly because it is common in library header files that are shared with the C language, such as `<cfloat>` and `<cmath>`. The `define` directive takes the form:

    #define identifier tokens

and causes all subsequent occurrences of `identifier` to be replaced by `tokens`. This directive can be used to give global constants meaningful names, as in:

    #define SPEED_OF_LIGHT 2.997925e8

It is traditional to use upper case letters for constants defined in this way in order to distinguish them from normal C++ identifiers. Notice that there is neither a semicolon (or statement terminator) nor an assignment operator since the `define` directive simply makes a textual substitution; it is *not* a statement of the C++ language itself. To some extent the `define` directive is a relic of the C programming language. In C++ it is usually much better to use the `const` specifier, together with an initialization, as in:

    const double speed_of_light = 2.997925e8;

With the `const` specifier there is no run-time overhead, but this technique has the advantage of type-checking and scoping.

The define directive can also take the form:

    #define identifier(identifier,..., identifier) token_string

with no space between the first identifier and the opening parenthesis. Using the define directive in this way is a *macro definition*; that is, the directive acts like a function definition, but without the associated call overhead. The preprocessor searches subsequent lines for occurrences of the first identifier and substitutes `token_string`, with the identifiers in parentheses replaced by actual arguments. An example should make this clear. If we have

    #define SQUARE(X) ((X) * (X))

then a subsequent statement of the form:

    y = SQUARE(4.0);

is expanded to

    y = ((4.0) * (4.0));

This is equivalent to:

    y = 4.0 * 4.0;

The use of parentheses may seem a bit excessive, but they are necessary because macros perform purely textual substitutions and, without these parentheses, unexpected expansions may take place. For instance:

    #define SQUARE(X) (X * X)
    y = SQUARE(u + v);

is equivalent to:

```
y = u + v * u + v;
```

which is very different from:

```
y = (u + v) * (u + v);
```

If necessary, any preprocessor directive can be continued to the next line by a \ followed immediately by a carriage return. It is therefore possible to write quite complicated macros that are many lines long. Macros are only mentioned because you may see them in header files derived from C code. You are strongly advised not to use macros. This is because in C++ it is possible to define `inline` functions, which can be inserted directly into the code by the compiler. Such functions have the same advantages as macros, but are also type-checked and cannot give rise to unexpected expansions. Inline functions are introduced in Chapter 5.

4.13.3   Conditional Compilation[†]

It is often convenient to be able to have two versions of the same program; for instance there might be a test version, which gives diagnostic messages, and a production version, which omits such messages. Simultaneously maintaining more than one version of the same program is not easy; it is much better to be able to compile the same program in different ways. Such conditional compilation can be achieved by means of the following sequence of directives:

```
#if condition_1
    // First code segment goes here.
#elif condition_2
    // Second code segment goes here.
#else
    // Third code segment goes here.
#endif
```

The interpretation of these directives is similar to the `if else` statement given in Section 4.5.2. The constant expression, `condition_1`, following the `if` directive is checked to determine whether it evaluates to zero; if it does not then the first code segment is included. If `condition_1` does evaluate to zero, then the constant expression following the `elif` (standing for "else if") directive is evaluated and, if it is non-zero, the second code segment is included. If all of the constant expressions are zero, then none of the code is included. The `endif` directive signifies the end of a sequence of conditional directives, although there may be many sequences in one program. In a given sequence there cannot be more than one `else` directive (there may be none), although there may be any number (including zero) of `elif` directives. There are some restrictions on the constant expressions; they must evaluate to an integral type and cannot contain a cast, `sizeof()` or enumeration constant.

So far we have not really considered programs of sufficient length to justify the use of conditional compilation, but as an example we might modify our quadratic equation solver of Section 1.2 as shown below.

```
// Solves the quadratic equation:
// a * x * x + b * x + c = 0

#include <iostream>
#include <cmath>            // For sqrt() function.
using namespace std;

#define TEST 1

int main()
{
    double root1, root2, a, b, c, root;

    cout << "Enter the coefficients a, b, c: ";
    cin >> a >> b >> c;
#if TEST
    double temp = b * b - 4.0 * a * c;
    cout << "temp = " << temp << "\n";
    root = sqrt(temp);
    cout << "square root of temp = " << sqrt(temp) << "\n";
#else    // TEST
    root = sqrt(b * b - 4.0 * a * c);
#endif   // TEST
    root1 = 0.5 * (root - b) / a;
    root2 = - 0.5 * (root + b) / a;
    cout << "The solutions are " << root1 << " and " << root2 <<
        "\n";

    return(0);
}
```

The test version of this program could be used to discover why incorrect results are obtained for certain values of a, b, c. Because the preprocessor merely makes textual changes, this is a situation where it is essential to use #define. The const specifier, as in :

```
const int TEST = 1;      // Not useful for preprocessing.
```

is not appropriate in this context. Notice that it is useful to include comments to indicate which #if directive goes with each #else or #endif. Moreover, it is worth trying to keep conditional compilation simple since the *dangling else* trap is always waiting for the careless programmer.

In Section 2.2 we mentioned that the /* */ style of comments cannot be used to comment out segments of code that already contain such comments. The #if directive provides a solution to this problem. If we surround the unwanted code with #if 0 and #endif, as in:

```
#if 0
for (int i = 1; i < 10; ++i) {
```

```
        sum_1 += i;          // Evaluates sum_1.
        sum_2 += i * i;      /* Evaluates sum_2 */
}                            // O.K. so far.
#endif  // 0
```

then it is excluded by the preprocessor.

4.14   Enumerations[†]

An *enumeration* is a distinct integral type with a set of named constants. The constants are named by means of an enumeration declaration, which consists of the keyword, enum, followed by a list of the constants, as in:

```
enum day {Sunday, Monday, Tuesday, Wednesday, Thursday, Friday,
    Saturday};
```

In this example, day is the enumeration type and Sunday has the value 0, Monday has the value 1, etc. An individual member of the list (such as Tuesday) is known as an *enumerator*.

Identifiers in an enumeration list can be assigned a particular value, in which case subsequent identifiers increase by one, going from left to right. In the enumeration:

```
enum traffic_light {red = 1, amber, green};
```

amber and green have the values 2 and 3 respectively.

It is also possible for two or more enumerators to have the same numerical value, as in:

```
enum day {Saturday, Sunday = 0, Monday, Tuesday, Wednesday,
    Thursday, Friday};
```

An equivalent but better technique is to use the statement:

```
enum day {Sunday, Monday, Tuesday, Wednesday, Thursday, Friday,
    Saturday = Sunday};
```

since this clearly indicates that Saturday is equivalent to Sunday without distracting us with numerical detail. In both cases, Saturday and Sunday have the value 0 and Monday, Tuesday, Wednesday, ... have the values 1, 2, 3, ...

The motivation for introducing enumerations is that they provide a way of using meaningful names in place of integral constants, as the following program demonstrates:

```
#include <iostream>
using namespace std;
int main()
{
    enum colour {red, amber, green};
    cout << "Enter an integer from " << red << " to " << green <<
        ": ";
    int choice;
    cin >> choice;
```

```
    switch (choice) {
    case red:
        cout << "The signal is red.\n";
        break;
    case amber:
        cout << "The signal is amber.\n";
        break;
    case green:
        cout << "The signal is green.\n";
        break;
    default:
        cout << "Not a valid colour.\n";
        break;
    }
    return(0);
}
```

Although it is necessary to enter an integer in order to choose a colour, once this has been done, the switch can be performed by using meaningful colours in place of numbers.

It is possible to define *enumeration variables*, as in:

```
enum colour {red, amber, green};
colour choice;
```

In this code fragment, choice is an enumeration variable. The only valid operation that is defined on an enumeration variable is assignment. For any other operation (such as equality) the enumeration variables and constants are converted to integers and an integer operation performed. This implies that operations, such as:

```
colour choice;
choice = amber;
if (choice == 1)
    cout << "Colour is amber.\n";
```

are valid. But mixed operations, as in:

```
colour choice;
choice++;         // WRONG!
choice = 2;       // WRONG!
```

are not defined.[15]

There is no need to actually provide a name for the enumeration. For example, the statement:

```
enum {red, amber, green} colour;
```

---

[15]Since an enumeration is a user-defined type, it is possible to define an operator (such as ++) for the type by using the techniques of Chapter 9.

defines colour to be an enumeration variable that can take the values red, amber or green, but does not define an enumeration type. Of course, this technique does not enable us to subsequently define more enumeration variables of the same type, although the original statement can define a list of such variables. For example, the code fragment given below defines colour and another_colour to be enumeration variables.

```
enum {red, amber, green} colour, another_colour;
```

Also, an unnamed enumeration restricts how we can use the enumeration. An example is the following program, where the choice of colour is made within the code instead of by entering a value:

```
#include <iostream>
using namespace std;
int main()
{
    enum {red, amber, green} colour;
    colour = amber;
    switch (colour) {
    case red:
        cout << "The signal is red.\n";
        break;
    case amber:
        cout << "The signal is amber.\n";
        break;
    case green:
        cout << "The signal is green.\n";
        break;
    default:
        cout << "Not a valid colour.\n";
        break;
    }
    return(0);
}
```

Exercise

Define an enumeration for the months of the year. Given a month enumeration variable, what methods could you use to "increment" it to the following month? Try out your ideas in a short program.

4.15 Summary

- The relational operators are: <, <=, >, >=. They are binary operators, returning true or false.

- The negation operator, !, returns true if the operand is false, and false if the operand is true.

- The logical AND operator, `&&`, returns `true` if both operands are `true`, otherwise the operator returns `false`.

- The logical OR operator, `||`, returns `false` if both operands are `false`, otherwise the operator returns `true`.

- The equality operator, `==`, returns `true` if both operands are equal and `false` otherwise.

- The not equal operator, `!=`, returns `false` if both operands are equal and `true` otherwise.

- A compound statement, or block, is denoted by a pair of braces, `{ }`.

- There are three branch statements: `if`, `if else` and `switch`.

- Watch out for the *dangling else* trap.

- The three iteration statements are: `while`, `for` and `do`. Use whichever statement is most natural for a particular problem.

- The `break` statement is used to exit from an iteration.

- The `continue` statement causes control to pass to the next iteration.

- The comma operator separates a sequence of expressions and is mainly used in initialization and `for` statements.

- The conditional expression operator, `?:`, is the only ternary operator and is useful for writing compact code. If the first operand is `true`, the operator evaluates to the second operand, otherwise it evaluates to the third operand.

- Files can be included by using the `#include` preprocessor directive:

  ```
  #include <filename>    // For system filenames.
  #include "filename"    // For user-defined filenames.
  ```

- Avoid using the `#define` preprocessor directive to give constants meaningful names:

  ```
  #define SPEED_OF_LIGHT 2.997925e8
  ```

  Instead use the `const` specifier:

  ```
  const double speed_of_light = 2.997925e8;
  ```

- An enumeration declares a distinct integral type and can be used to give meaningful names to integral constants:

  ```
  enum colour {red, amber, green};
  ```

4.16   Exercises

1. π can be calculated from the series:

$$\sum_{n=1}^{\infty} 1/k^4 = \pi^4/90$$

   Write a program that uses the first five terms of this series to obtain an approximation for π. Compare your result with that given in a standard table of mathematical constants. See how accurate you can make your result by increasing the number of terms in the series.

2. Which of the iterations in the following code segments may never terminate? Justify your conclusions.

   (a)
   ```
   int sum = 1;
   for (unsigned i = 10; i >= 0; --i)
       sum *= 2 * i + 1;
   ```

   (b)
   ```
   double i = 10, sum = 1;
   while (i != 0)
       sum *= 2 * i-- + 1;
   ```

   (c)
   ```
   int i = 0;
   double sum = 1.0;
   while (1) {
       sum *= 2 * i++ + 1;
       if (i = 10)
           break;
   }
   ```

3. Consider the sequence of *integers*: $u_1, u_2, u_3, \ldots$, where $u_1 = 1, u_2 = 1$ and $u_n = u_{n-1} + u_{n-2}$ for $n \geq 3$. These integers constitute what is known as the Fibonacci sequence. Write a program that prompts for a positive integer, n, and lists the first n members of the sequence. Notice how u_n increases very rapidly with n and soon exceeds the largest integer that can be represented as a fundamental type on your computer.

   Verify your results by modifying the program to check that:

   (a) $u_1 + u_2 + \cdots + u_n = u_{n+2} - 1$
   (b) $u_n^2 - u_{n-1}u_{n+1} = (-1)^{n-1}$.

   (If you want to learn what all this has to do with rabbits and number theory, it is well worth consulting [12].)

4. If we define $f(x) \equiv x^n - c$, then solving $f(x) = 0$ is equivalent to finding $c^{1/n}$ for $n > 0$. Given an approximate value of x, the Newton–Raphson method consists of calculating a new approximation, x_{new}, by using:[16]

$$x_{\text{new}} = x - \frac{f(x)}{f'(x)}.$$

---

[16]See [14]

In our case, this reduces to:

$$x_{\text{new}} = \frac{(n-1)x}{n} + \frac{c}{nx^{n-1}}$$

Write a program to find $c^{1/n}$ for positive n and c. The program should prompt for n (as a positive integer), the initial value of x (as a positive double) and c (also a positive double). Incorrect entries should be trapped and a prompt for a new value issued. Use the above formula for x_{new} to iterate towards an approximate value of x and hence $c^{1/n}$. The program should terminate if the number of iterations exceeds a reasonable limit or the difference between two successive iterations is in some sense small. For instance, you might try a limit of twenty iterations and a difference between two successive iterations that is close to the machine precision. Try different values of n and c, and list the approximations to each root.

5. A very simple way of numerically integrating a function in one dimension is the trapezoidal formula, which splits the integration region into strips, as shown in Figure 4.1.[17] The sum of the areas is an approximation to the integral, as given

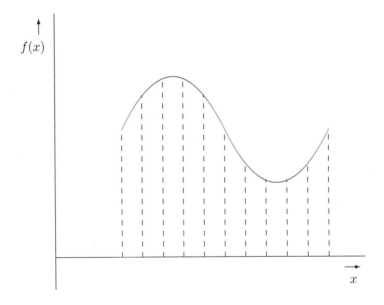

Figure 4.1: Numerical integration using the trapezoidal formula.

by the following formula:

$$\int_{x_0}^{x_m} f(x)\, dx \approx h[f_0/2 + f_1 + \cdots + f_{m-1} + f_m/2]$$

where

$$h = (x_m - x_0)/m$$

---

[17]For a description of better methods, again see [14].

Write a program that uses the trapezoidal formula, with $m = 20$, to evaluate:

$$\int_0^\pi \sin x \, dx$$

and compare your result with the exact answer. Experiment with different values of m and also try to evaluate different integrals.

Chapter 5

Functions

5.1 Introducing Functions

In the previous chapter we introduced the control structures of C++. Although these structures enable us to write very powerful programs, they do not give us a means of controlling this power, nor do they encourage reusable code. The introduction of functions is an essential first step in the writing of modular, maintainable and reusable code. In particular, testing can be carried out on each function in isolation, rather than on the whole program. This is important because the number of different pathways through even a modest program is often very large indeed. Functions provide a way of encapsulating relatively self-contained segments of code. A function typically carries out some well-defined action, such as returning a random number, performing numerical integration or inverting a matrix.

5.1.1 Defining and Calling Functions

In general, a function definition takes the form:

```
return_type function_name(type argument,..., type argument)
{
    // Function body.
}
```

However, it is more instructive to start by considering a simple example. The following code defines a function:

```
int factorial(int n)    // Calculates n*(n-1)*(n-2)*...*1
{
    int result = 1;
    if (n > 0) {
        do {
            result *= n;
            --n;
        } while (n > 1);
    }
```

85

```
    else if (n < 0) {
        cout << "Error in factorial function:\targument = " <<
            n << "\n";
    }
    return result;
}
```

This function calculates the factorial function $(n(n-1)\ldots1$ or $n!$ in the usual mathematical notation) and its structure is shown in Figure 5.1. Notice that there is no

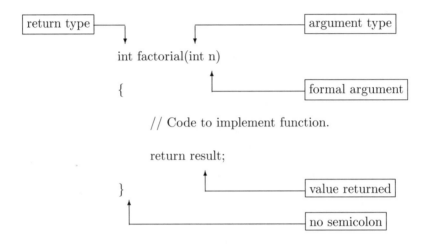

Figure 5.1: Structure of a typical function.

terminating semicolon. The first line, sometimes known as the *function header*, declares that the function returns the type int and defines the name of the function to be factorial, which is the identifier used to invoke this particular function. The identifier, n, is known as a *formal argument* and is again declared to be of type int. A function can have any number of formal arguments, including zero, and these arguments can be a mixture of types. The arguments in a list are separated by commas. In this context the comma is a separator, rather than an operator (in contrast to Section 4.9).

The outermost set of braces contains what is known as the *function body*; it is here that we find all of the code that implements the function. The body of the function contains a return statement, which terminates the function. In this particular case the statement also causes a value to be returned to the calling environment. The value being returned can optionally be enclosed by parentheses, as in:

```
    return(result);
```

Some programmers like to leave a space before the parentheses in a return statement, as in:

```
    return (result);
```

others prefer not to. If the result of evaluating an expression is returned directly, parentheses can help to make the statement more readable. For example, compare:

```
return(a * a + b * b + c * c);
```

with:

```
return a * a + b * b + c * c;
```

A function is *called* (*invoked* or *executed*) by including its name, together with appropriate arguments within parentheses, in a statement.[1] The following code fragment gives an example of calling the `factorial` function:

```
sum = 0.0;
for (i = 0; i < 10; ++i)
    sum += 1.0 / factorial(i);
```

The arguments to the function being called are known as *actual arguments*, distinguishing them from the formal arguments of the function definition. The identifiers used in both cases may or may not be the same; this has no significance since the values of the actual arguments are copied to the formal arguments. The scope of the formal arguments is limited to the function body and changes made to these arguments are not propagated back to the calling program. For instance, in the body of our factorial function, the formal argument, `n`, is decremented, but this has no effect on the actual argument, `i`. This process of copying the value of the actual argument to the formal argument is known as *pass by value*. Some languages use *pass by reference*, in which case any changes to a formal argument that are made by a function are indeed propagated back to the calling environment. As we will see in Chapter 7, C++ functions can also use pass by reference, but pass by value is the usual technique. The reason for this is that pass by value encourages more modular, safer code and helps to prevent unexpected changes in the value of variables.

It is worth pointing out that although function arguments can be expressions, as in:

```
int j = factorial(2 * i + 1);
```

the order of evaluation of expressions in argument lists is not defined. Therefore a function call of the form:

```
double x = too_clever_by_half(++i, (i+5));
```

is likely to give different results with different compilers.

An example of using a function is given in the following program:

```
#include <iostream>
using namespace std;

int factorial(int n)    // Calculates n*(n-1)*(n-2)...1
{
    int result = 1;
```

---

[1]The pair of parentheses in the function call should be regarded as an operator.

```
        if (n > 0) {
            do {
                result *= n;
                --n;
            } while (n > 1);
        }
        else if (n < 0) {
            cout << "Error in factorial function:\targument = " <<
                n << "\n";
        }
        return result;
    }

    int main()
    {
        double value = 0.0;
        for (int loop = 0; loop <= 10; ++ loop)
            value += 1.0 / factorial(loop);
        cout << "The value of e is: " << value << "\n";
        return(0);
    }
```

The program uses the `factorial` function to calculate an approximation to the mathematical constant, *e*.

5.1.2   Return Type

The `return` statement causes a function to terminate, but in many cases the statement also causes a value to be passed back to the calling environment. The type returned by a function must agree with that specified in the function header. Any attempt to return a type differing from that declared should be trapped by the compiler. However, implicit type conversions are allowed so the `return` in the following code fragment is valid:

```
    double sum(int n)
    {
        int result = 0;
        for (int i = 1; i <= n; ++i)
                result += i * i;
        return result;
    }
```

In the above `return` statement, `result` is of type `int` and is converted to a `double`. But the following is not valid since the first `return` does not return a value:

```
    double sum(int n)
    {
        int result = 0;
        if (n < 0)
```

```
        return;                       // WRONG!
    for (int i = 1; i <= n; ++i)
            result += i * i;
    return result;
}
```

In this example the compiler prevents an arbitrary value being returned as an apparently valid result. We should, of course, replace the incorrect return statement by some action that recognizes that if **n** is less than zero then an error has occurred. Something should also be done to limit the consequences of the error; one possible course of action is suggested later in this chapter (in Section 5.8).

It is quite common for there to be no requirement for a function to return a value, in which case a special type, called **void**, should be used to specify the return type in the function header. For example, the following function simply sends a message to **cout** and does not return a value:

```
void print_welcome()
{
    cout << "Welcome to the interactive C++ tutorial.\n";
    return;     // This statement can be omitted.
}
```

If no value is returned by a function (as above) then the **return** statement at the end of the function body can be omitted.

ANSI C++ (unlike earlier versions) does not allow us to omit the return type in the function header, so the following function is incorrect and gives a compiler error:

```
print_welcome()             // WRONG: must specify return type.
{
    cout << "Welcome to the interactive C++ tutorial.\n";
}
```

If a function has no arguments, then **void** may be used, as in:

```
int rand(void)
{
    // Return a pseudo-random integer.
}
```

The use of **void** is preferable to empty parentheses, since it emphasizes that the programmer intends the function to take no arguments. Apart from this equivalence to an empty argument list, **void** cannot be used as an argument type, nor can an identifier of type **void** be declared:

```
int factorial(int n, void)  // WRONG: void is not valid here.
{
    void empty;                 // WRONG: cannot declare a variable
                                // to have type void.
    // Function body
}
```

It is also invalid to have `void` as the operand of the `sizeof` operator.

Notice that even if a function takes no arguments, the calling program must use parentheses, as in:

```
print_welcome();
```

rather than:

```
print_welcome;               // This does nothing useful.
```

Although the statement does not produce a compiler error, using the function name without parentheses does nothing useful in this context.

5.1.3   Function Declarations

A function can be declared without specifying how it is implemented. A function declaration for the `factorial` function is given below.

```
int factorial(int n);
```

Such a statement is known as a *function declaration*. Notice that, unlike a function definition, a function declaration must end with a semicolon and has no function body. A function with an empty body is *not* a declaration, but instead defines a function which does nothing, as in:

```
int lazy_function(int n) { }
```

Any argument names given in the declaration are ignored by the compiler, but including these names is a useful documentation technique. For complicated programs, possibly involving use of the same functions in different source files, function declarations are a considerable help in program maintenance and modularity. This is because the code implementing a function may be compiled independently to code using it. Such declarations are often collected together in a header file and since a function is only known to the compiler after it has been declared, it makes sense to include the header file at the start of all files that either use functions or implement them. As we will see when we come to the specifically object-oriented aspects of C++, declarations have a fundamental role to play in defining classes and therefore objects. In fact, C++ requires either that a function is declared or else its definition is seen before the function is invoked.

5.1.4   Functions Cannot Declare Functions

In some languages, such as Pascal, it is possible to declare functions within other functions. It is important to realize that this feature is not available in C++. So a program to print out the first ten factorials is:

```
#include <iostream>
using namespace std;

int factorial(int n)
{
```

```
            int result = 1;
            if (n > 0) {
                do {
                    result *= n;
                    --n;
                } while (n > 1);
            }
            else if (n < 0) {
                cout << "Error in factorial function\targument = " <<
                    n << "\n";
            }
            return result;
        }

        int main()
        {
            for (int i = 0; i < 10; ++i)
                cout << i << "!  =  " << factorial(i) << "\n";
            return 0;
        }
```

rather than:

```
    #include <iostream>
    using namespace std;

    int main()
    {
        int factorial(int n)    // WRONG: cannot declare function here.
        {
            // Body of function as in previous example.
        }
        for (int i = 0; i < 10; ++i)
            cout << i << "!  =  " << factorial(i) << "\n";
        return 0;
    }
```

Since every program must have a function called `main()` and a function cannot be defined inside another function, the general structure of a program is a list of function definitions. We might expect the relationship between the functions to look like a tree with the function `main()` as the root. Unfortunately programs can be much more complicated than this, being general graphs, complete with cycles. For some problems, functions are a very effective way of controlling complexity. However, for really complicated problems it becomes difficult to control the relationships between functions and more powerful techniques are needed. It is by encapsulating related functions and the data on which they operate into a single class, that C++ can bring some order to these more complicated problems. You are probably wondering why we don't go directly to these more powerful methods. There are really two reasons: firstly, many small problems can be solved most appropriately without introducing classes;

secondly, classes at their most fundamental level are partly built from functions and so we do need a good understanding of how to use functions before we can progress to object-oriented techniques.

5.1.5   Unused Arguments[†]

There is no requirement for all of the formal arguments in a function declaration to be used, in which case an identifier does not have to be specified. An example of a function definition with an unused argument is given below.

```
void print_error(int)
{
    cout << "An error has occurred.\n";
}
```

However, the function cannot be called without an argument, even though the argument is not used. So in the following code fragment, the first statement is correct, even though the constant argument has no significance:

```
print_error(1); // O.K.
print_error();  // WRONG: incorrect number of arguments.
```

Unused arguments may serve to reserve a place in the argument list for future use. For instance, in the previous example we might want to have a list of the various possible errors, as in:

```
void print_error(int error_number)
{
    cout << "An error has occurred:\n";
    switch (error_number) {
        case 1:
            cout << "\tThe matrix is too large for " <<
                "the available memory.\n";
            break;
        case 2:
            cout << "\tThe matrix is singular.\n";
            break;
        case 3:
            cout <<"\tThe iteration is not converging.\n";
            break;
        default:
            cout << "\tUnknown error.\n";
            break;
    }
    // Perhaps some code to handle the error.
}
```

An argument type without an identifier can also be useful when an argument has been made redundant by a changed function implementation, but the calling functions have not yet been brought up to date.

5.1.6 Default Arguments

A function declaration can specify values that are to be used as the defaults for one or more arguments. For instance, if we define a function to return the area of a rectangle, with defaults for the height and width, as in:

```
double rectangle(double height = 1.0, double width = 2.0)
{
    return(height * width);
}
```

then we could make the function calls given in the following code fragment:

```
area1 = rectangle();        // 1.0 * 2.0 = 2.0 assigned.
area2 = rectangle(3.0);     // 3.0 * 2.0 = 6.0 assigned.
area3 = rectangle(2.0, 4.0); // 2.0 * 4.0 = 8.0 assigned.
```

Notice that it is the trailing arguments that are assumed to be missing, which means that default arguments must be supplied right to left in the declaration. For instance, we can make the declaration:

```
double rectangle(double height, double width = 2.0);
```

but not:

```
double rectangle(double height = 2.0, double width);
```

According to the ANSI C++ Standard, a default argument cannot be redefined by a subsequent declaration, not even to the same value. A common error is to give the same default arguments in a function definition, as in the corresponding declaration:

```
double rectangle(double height = 1.0, double width = 2.0);

// Perhaps some other code.

// The following function header is WRONG:
double rectangle(double height = 1.0, double width = 2.0)
{
    return(height * width);
}
```

This condition is not enforced by all compilers. Nevertheless, since default argument values are a feature of the function interface, rather than the implementation, the defaults should be in the header (.h) file rather than the source (.cxx) file.

However, a subsequent declaration can introduce one or more additional default arguments, as demonstrated in the following program:

```
#include <iostream>
using namespace std;
const double PI = 3.142;
double disc_area(double radius);
```

```
// Some other code.

double disc_area(double radius = 10.0)
{
    return(PI * radius * radius);
}

int main()
{
    cout << disc_area() << "\n";
    return(0);
}
```

It is worth emphasizing that default arguments *must* be provided for any arguments that are omitted in the function call, as demonstrated by the following code fragment:

```
double rectangle(double height, double width = 2.0)
{
    return height * width;
}

int main()
{
    double rect1 = rectangle(5.0);   // O.K.
                                     // Default width = 2.0
    double rect2 = rectangle();       // WRONG: insufficient
                                     // function arguments.
    // More code.
}
```

Since default arguments do not need to be named in a function declaration, we could use the following for the rectangle function:

```
double rectangle(double, double = 2.0);
```

However, omitting argument names it not a good idea since they are a valuable method of documenting code. Also, notice the distinction between declarations and definitions; argument names cannot be omitted from function definitions, even if they do have default values. This is illustrated by the following function where width is not defined:

```
double rectangle(double height, double = 10.0)  // WRONG!
{
    return height * width;           // width is not defined.
}
```

Exercise

Implement and test a function that has the base and height of a triangle as arguments and returns the area. If the function is called with a single argument, then the value of the base should default to 1.0. In the case of no arguments, the function should return a result of 1.0.

5.1.7 Ignoring the Return Value

The calling function can ignore the value returned by a function. For instance, consider a function `mkdir(char*)` that creates a file system directory. In general it is important to know whether a directory is successfully created or not, so such a function could return `true` to indicate success and `false` to indicate failure. A program might contain the following:

```
bool status = mkdir("test"); // Creates directory "test".
if (status)
    cout << "Directory 'test' created.\n";
else
    cout << "Failed to create directory 'test'.\n";
```

However, we may be willing to ignore the risk that the directory creation could fail and to simply use:

```
mkdir ("test");      // Return status is ignored.
```

In general it is safer to program defensively; assume that anything which may fail is worth testing, at least in the development phase.

5.2   Recursion

A function can call another function, as in:

```
double binomial_coef(int n, int k)
// calculates binomial coefficient:   n!/(k! (n-k)!)
{
    return(factorial(n) / static_cast<double>(factorial(k) *
        factorial(n - k)));
}
```

(The cast to `double` in this function is necessary to force floating point rather than integer division; otherwise integer truncation would occur.)

A function can also call itself and this is known as *recursion*. For example, we could rewrite our factorial function as shown in the program below.

```
#include <iostream>
using namespace std;
int factorial(int n)
{
    int result;
    if (n == 1 || n == 0)
        result = 1;
    else if (n < 0)
        cout << "Error in factorial function: argument = " <<
            n << "\n";
    else
        result = n * factorial(n - 1);
```

```
        return result;
    }

    int main()
    {
        int j;
        cout << "Enter an integer: ";
        cin >> j;
        for (int i = 0; i < j; ++i)
            cout << i << "!  =  " << factorial(i) << "\n";
        return(0);
    }
```

A recursive function invariably follows the same pattern; it has a base case and always calls itself. In this example the base case occurs when n is one or zero; the function then returns without calling itself. When the function does call itself, then the value of the argument is different and must in some sense be changing in the direction of the base case. Care should be taken to trap any possible call errors (such as n less than zero in this example) otherwise the function may call itself for ever. It is also possible to have a set of mutually recursive functions. We don't consider mutually recursive functions in this book, but the general ideas are the same as given here.

Recursion should be used with care; it is sometimes a very effective way of solving problems that are otherwise difficult. However, every use of a function carries a call overhead and if the recursion depth is large and little calculation is done on each call, then recursion can be very inefficient. Another disadvantage is that each function call makes temporary use of a fixed quantity of memory, known as the stack.[2] If the depth of a recursive function call is large, it is possible to use up all of this memory, in which case the program will fail.

Exercise

Implement a function, with the declaration:

```
double u(int n);
```

to recursively generate the Fibonacci sequence defined in Exercise 3 of Chapter 4. Check that the results given by the two exercises are consistent.

5.3   Inline Functions

It is often desirable to write short functions in order to improve the readability of a program, but the call overhead may not justify the use of a function definition. In such cases it is possible to define *inline* functions by means of the `inline` keyword:

```
    inline int min(int i, int j)
    {
        return((i < j) ? i : j);     // Returns the minimum of i and j.
    }
```

---

[2]We will describe stacks in more detail in Section 10.3.1.

The `inline` keyword is a suggestion to the compiler that the body of the function should be substituted directly in the code, instead of the program making a run-time function call. The suggestion may be ignored and probably will be if the function is complicated. A recursive function could be declared `inline`, but even a simple function would not be completely expanded, since the recursion depth is not known at compile-time.

Inline functions usually (but not always) increase the size of the generated code and decrease the execution time. For most functions, the call overhead is insignificant compared with the time taken to execute the body of the function and these functions should not be declared inline. In general it is only very short functions, consisting of one or two statements, that are worth declaring inline. Such functions are common when implementing the data-hiding techniques associated with classes, as described in Chapter 8.

Functions declared `inline` should be *defined* in a header file, rather than a `.cxx` file, since the definition (in contrast to the declaration) must be seen before every use of the function. This is particularly important when a program is split into separately compiled source files, as each file using an `inline` function needs the definition and files `#include` the `.h` (not the `.cxx`) file.

Exercise

Try changing some of your previous programs to declare `inline` functions (for example, any programs using the `factorial()` function).

5.4   More on Scope Rules

Now that we have introduced functions, it is worth considering the concept of scope in more detail. There are three kinds of scope: local, class and file. Classes will not be introduced until Chapter 8, but for the moment we can state that any member of a class has class scope.

A variable defined within a function has *local scope* and each function has a distinct scope. Such variables hide definitions made outside of the function. (In fact, as we noted in Section 4.4, the same is true of any block.) For instance, in the program given below we define a function to calculate the sum of the squares of the first n integers and this function is called from `main()`. The two declarations of `i` (and the two declarations of `result`) in the functions are hidden from each other. This is true whatever the relative positions of the functions within a single file.

```
#include <iostream>
using namespace std;

int sum(int n)
{
    int result = 0;
    for (int i = 1; i <= n; ++i)
        result += i * i;
    return result;
```

```
    }

    int main()
    {
        int result;
        for (int i = 1; i <= 10; ++i) {
            result = sum(i);
            cout << "sum of the first " << i << " squares is " <<
                result << "\n";
        }
        return(0);
    }
```

Reuse of the same identifier in different scopes in the same file should not be over-done as it can lead to programs that are very difficult to read. The reuse of nondescript identifiers such as i (which may typically be used as a generic loop counter) is quite acceptable. But consider a file using square matrices and suppose the file contains identifiers called matrix_size. In one function it could mean the number of rows (or columns) of a matrix; in another, the total number of elements in a matrix and in a third, the total number of bytes needed to store a matrix. The possibility of errors would be greatly reduced if we used three different and more descriptive identifiers.

Any object not having local or class scope has *file* (or *global*) *scope* and is visible throughout the file. Although this visibility is useful, it is also dangerous, which means that global variables should be kept to a minimum. Since any function has access to a variable with file scope, it is well worth using the const specifier if at all possible. A simple example of file scope is given in the following program:

```
    #include <iostream>
    using namespace std;

    const int i_max = 10;        // i_max has file scope.

    int main()
    {
        int sum = 0;
        for (int i = 1; i <= i_max; ++i) {
            sum += i;
            cout << "sum of the first " << i << " numbers is " <<
                sum << "\n";
        }
        return(0);
    }
```

If a local variable and a global variable have the same name, then the local variable hides the global variable. However, the global variable can be accessed by prefixing the name with what is known as the *scope resolution operator*. This is a single token, consisting of two colons, as in ::x. We will make more use of the scope resolution operator later, but the following program demonstrates its use in this situation. However, this shouldn't be taken as encouragement for the use of global variables.

```
#include <iostream>
using namespace std;

double x = 2.7183;

int main()
{
    cout << "x = " << x << "\n";
    double x = 0.6931;
    cout << "x = " << x << "\n";
    cout << "::x = " << ::x << "\n";
    return(0);
}
```

The first statement in function **main()** sends the value of the global variable to the output stream. The second statement hides the global variable, and so the third statement in **main()** sends 0.6931 to the output stream. The fourth statement demonstrates that use of the scope resolution operator accesses the global variable.

5.5   Storage Class static[†]

In the following simple function:

```
int sum(int n)
{
    int result = 0;
    for (int i = 1; i <= n; ++i)
        result += i * i;
    return result;
}
```

the variables i, result and n are all *automatic*. That is each time control passes to the body of this function, memory is allocated for these variables and then deallocated when control leaves the function. There is no guarantee that the same memory will be allocated each time control passes to the function body and that the memory won't be used to store something else when control passes elsewhere. All local initializations are performed each time control passes to a block; in this example result is set to zero each time the function is invoked.

If no explicit initialization is performed, then the value of an automatic variable is initially arbitrary:

```
int sum(int n)
{
    int result;
    for (int i = 1; i <= n; ++i)
        result += i * i;
    return result;          // WRONG: result is arbitrary.
}
```

Identifiers that are local to a block (which includes a function body) have automatic storage by default.

Sometimes it is convenient to have a variable that retains its value when control leaves the block where the variable is defined; this can be achieved by declaring the storage class to be `static`. An example of declaring static storage is given in the following program:

```cpp
#include <iostream>
using namespace std;

int sum(int n)
{
    static int grand_total;
    int result = 0;
    for (int i = 1; i <= n; ++i)
        result += i * i;
    grand_total += result;
    cout << "Total so far = " << grand_total << "\n";
    return result;
}

int main()
{
    cout << "First sum = " << sum(2) << "\n\n";
    cout << "Second sum = " << sum(2) << "\n\n";
    cout << "Third sum = " << sum(3) << "\n";
    return(0);
}
```

In this example the `static` variable, `grand_total`, accumulates the values returned by `sum()`.[3] This is because a `static` variable retains its value even when it goes out of scope. Notice that we didn't need to initialize `grand_total` since, by default, a static object is initialized to zero. This initialization is only performed once. Any explicit initialization is also only performed once, so in the following example `grand_total` correctly accumulates the values returned by sum:

```cpp
int sum(int n)
{
    static int grand_total = 0;      // Initialized once.
    int result = 0;
    for (int i = 1; i <= n; ++i)
        result += i * i;
    grand_total += result;
    cout << "Total so far = " << grand_total << "\t";
    return result;
}
```

---

[3] Within the text we often refer to a function by its name followed by a pair of empty parentheses. The parentheses are included to emphasize that we are dealing with a function and the fact that they are empty is not meant to imply that the function doesn't take any arguments.

An initializer for a `static` variable does *not* need to be a constant expression and can involve any previously declared variables and functions:

```
static int grand_total = total_for_last_week *
    weighting_factor(current_week);
```

It is a good idea to keep such initializers simple since the order in which initializations are performed may not be obvious; a local `static` variable is initialized when the function is first invoked and a `static` variable with file scope is initialized before any functions or objects are used.

Global variables (that is variables having file scope) have `static` storage by default and are therefore initialized to zero, unless an explicit initializer is given:

```
#include <iostream>
using namespace std;
int grand_total; // Static variable, initialized to zero.

int sum(int n)
{
    int result = 0;
    for (int i = 1; i <= n; ++i)
        result += i * i;
    grand_total += result;
    return result;
}

int main()
{
    for (int i = 1; i <= 10; ++i)
        cout << "sum of the first " << i << " squares is " <<
            sum(i) << "\n";
    cout << "The sum of the sum of the first 10 squares is " <<
        grand_total << "\n";
    return(0);
}
```

Exercise

Why is the use of `grand_total` in the above program considered to be bad style? How could you improve the program?

5.6 Overloading Function Names

In C++ it is possible to use the same function name for functions that actually have different function bodies. This is known as *function overloading*. The functions must be distinguished by having different numbers of arguments or different argument types, such as:

```
    double norm(double a, double b, double c)
    {
        return sqrt(a * a + b * b + c * c);
    }              // Returns the norm of a 3-dimensional vector.

    double norm(double a, double b)
    {
        return sqrt(a * a + b * b);
    }              // Returns the norm of a 2-dimensional vector.

    float norm(float a, float b)
    {
        return(a * a + b * b);
    }              // Correct but confusing. See below.
```

but not:

```
    float norm(double a, double b, double c)     // WRONG!
    {
        return(a * a + b * b + c * c);
    }
```

The fourth function does not correctly overload the first, since functions must be distinguished by their argument types and not by their **return** types. The third function is correct, but confusing; the square root has been omitted and therefore what the function actually does is different from the first two examples. It would be better to give the third function a different name instead of using function overloading.

Remember that floating constants are of type **double** by default, so the first and second definitions of the **norm** function would give the following (distinct) function calls:

```
    double norm2 = norm(3.0, 4.0);          // norm2 is 5.0
    double norm3 = norm(3.0f, 4.0f);        // norm3 is 25.0
```

It is clearly possible to write very confusing code by overloading the same function name with functions that perform completely different actions. This should certainly be avoided. However, there are circumstances in which overloaded functions are very useful. For instance we might want to have both **float** and **double** versions of the mathematical functions, such as sine, cosine, square root etc. The **float** version would be used where speed is more important than accuracy. We could then have function declarations of the form:

```
    float log(float x);
    double log(double x);
    float sin(float x);
    double sin(double x);
```

Typically these functions are approximated by power series expansions and so the **float** version would use fewer terms in the series. In fact, the C++ Standard Library implements overloaded versions of the common mathematical functions. For each function there is a **float**, **double** and **long double** version, declared in **<cmath>**. Once

again, don't forget that constants are double by default so, for instance, to invoke the
`float` version of `log` we need:

```
float x = log(4.0f);
```

rather than:

```
float x = log(4.0);
```

Exercise

Implement overloaded functions to calculate an approximation to:

$$e^x = \sum_{n=0}^{\infty} \frac{x}{n!}$$

One version should have an argument and return type of `double` and a
second version should use the `float` type. The number of terms used in
each function should be appropriate for the type. Test the functions either
against the library function, `exp()`, which is declared in `<cmath>`, or by
calculating $e^x e^{-x} - 1$. Check that the test program really does invoke the
two different functions.

5.6.1 Ambiguity Resolution[††]

The rules used by a C++ compiler to resolve which overloaded function is invoked
by a particular function call are quite complicated. The resolution is done entirely
by comparing the type of each argument in the function call with the corresponding
argument in the function declarations that have the same function name. Functions
with default arguments can be considered as a set of overloaded functions, so that:

```
double rectangle(double height = 1.0, double width = 10.0);
```

is equivalent to:

```
double rectangle(double, double);
double rectangle(double);
double rectangle(void);
```

The compiler constructs the *intersection* of the sets of functions that "best" match
each argument (that is the match is done on an argument-by-argument basis). In
order that the function overloading is resolved, this intersection must have one (and
only one) member. It is the interpretation of the adjective, "best", which gives rise
to some complication. A good description of the rules is given in [10], but briefly, the
fewer promotions and conversions that are needed to achieve a match, the better. As
a very simple example, suppose we have the following function declarations:

```
void f(float x);
void f(double x);
```

Then the statement:

```
f(2.0f);
```

invokes the first, rather than the second function; there is an exact match for the first function whereas the second requires a promotion from `float` to `double`. However, the statement:

```
f(2);
```

is ambiguous since an `int` could be promoted to either `float` or `double`.

5.7   Function `main()`

We first encountered the function `main()` in Chapter 1, but now that we have introduced the general features of functions it is worth making some further comments. As far as the programmer is concerned, this function is where the program actually starts. There are some restrictions on what we can do with `main()`; in particular `main()` cannot be declared inline, cannot be overloaded and cannot call itself. Every C++ program must have one and only one function called `main()`. The body of `main()` is entirely defined by the programmer, but an ANSI compliant C++ compiler must allow a definition of the form:

```
int main()
{
    // code
}
```

where the return value is used to indicate the termination status of the program.

There is a second form of `main()`, involving function arguments, and this is considered in Chapter 7.

5.8   Standard Library

Many numerical application programs make frequent use of mathematical functions, such as sine, cosine, square root etc. These functions are not part of the language as such, but are provided by any ANSI C++ implementation as part of a standard library. The available functions extend far beyond the obvious mathematical ones and it is not intended to describe the functions in any detail; entire books have been written on the ANSI C++ Standard Library. Rather, a brief description of what is available is given, in order to encourage you to look through the function manual for your own system and to indicate the extent of the Library. Some of the functions will be introduced during the course of this book.

It must be emphasized that in this subsection we only describe those parts of the Library that have been acquired from the C library. The headers for these parts of the Library are distinguished by having the letter "c" as the first part of their name. For example, the standard mathematical functions are declared in `<cmath>`. Some parts of the C++ Standard Library will only be relevant when we have learned more advanced techniques (such as templates) so we will revisit the Standard Library in Chapter 17.

All library functions require the inclusion of a header file, which typically contains function declarations and constant definitions. The header is normally included near the top of the file in which the function is used, as demonstrated by the following program:

```
#include <cmath>          // For sqrt()
#include <cstdlib>        // For EXIT_SUCCESS
#include <iostream>       // For cout
using namespace std;

int main()
{
    for (int i = 1; i <= 20; ++i) {
        cout << "square root of " << i << " is " <<
            sqrt(double(i)) << "\n";
    }
    return(EXIT_SUCCESS);
}
```

In this example `<cmath>` is required for the `sqrt()` function and `<cstdlib>` is required for the definition of the `EXIT_SUCCESS` constant. This constant is used to tell the operating system that the program has terminated normally. There is another constant, `EXIT_FAILURE`, that can be used to indicate that an error has occurred. Rather than returning 0 or 1, it is more meaningful to use named global constants (such as `EXIT_SUCCESS` and `EXIT_FAILURE`). Using names rather than numbers also helps to avoid the mistake of using the wrong constant.

As a further example, `<cstdlib>` also contains a declaration for the `exit()` function. Invoking this function terminates a program and returns control to the operating system, after first doing any necessary tidying up (flushing output buffers etc.). The `exit()` function takes an `int` argument that can be used to indicate how a program has terminated. By convention, either `EXIT_SUCCESS` or `EXIT_FAILURE` (as appropriate) is used for this argument.

We can use `exit()` to improve our `factorial()` function in Section 5.1.1, as shown below.

```
int factorial(int n)          // calculates n*(n-1)*...*1
{
    int result=1;
    if (n > 0) {
        do {
            result *= n;
            --n;
        } while (n > 1);
    }
    else if (n < 0) {
        cout << "Error in factorial function:  " <<
            "argument = " << n << "\n";
        exit(EXIT_FAILURE);
    }
    return result;
}
```

Now if we call the `factorial` function with a negative argument, the program terminates cleanly, rather than continuing an invalid calculation.

The `exit()` function can also be used to improve our function to sum squares in
Section 5.4:

```
double sum(int n)
{
    int result = 0;
    if (n < 0) {
        cout << "Error: sum(n) called with n = " << n << "\n";
        exit(EXIT_FAILURE);
    }
    for (int i = 1; i <= n; ++i)
        result += i * i;
    return result;
}
```

This modified function terminates the program if we call `sum()` with a negative argument since there would not be much purpose in continuing with an invalid return
value.

The same header file is often required for different but related functions in the
Standard Library; for instance `<cmath>` is needed for the trigonometric, hyperbolic,
square root, logarithmic and many other mathematical functions. For this reason it
is convenient to group the functions by their header files. The more useful groups of
standard functions are briefly described below.

cctype These functions are used to determine the type of a single character. For
example, there are functions to determine whether a character is upper or lower
case, a digit, a letter etc. and to convert between upper and lower case.

cfloat This file contains definitions of constants that give the compiler-dependent limits of the floating point types.

climits This file contains definitions of constants that give the compiler-dependent
limits of the integer types.

cmath Here are the common mathematical functions that are provided by the Standard Library, including trigonometric, hyperbolic, logarithmic and exponential
functions. There are also useful mathematical constants, such as π, e and $\log_e 2$
(represented by `M_PI`, `M_E` and `M_LN2`).

cstddef The main use of this header file is for the definition of `size_t`, which is the
type returned by the `sizeof` operator.

cstdlib A few types, such as `size_t` (which is also defined in `<cstddef>`) are defined
in this file, together with declarations for many commonly used functions, such
as `exit()`, `rand()` (an integer random number generator), `system()` (to call a
system function) etc.

cstring The declarations for many useful functions that manipulate strings and memory are contained here. These functions can copy, concatenate, compare and

determine the length of strings. They can also copy, compare and search areas of memory. The strings in this part of the Library are taken over from the C language strings, and are not to be confused with the C++ string template class, described in Chapter 17.

ctime The functions associated with this header are all connected with time; for example it is possible to get the current time.

All the above groups of functions are directly taken over from C in order to provide compatibility. The more advanced aspects of C++ mean it is an excellent language for writing library functions. Apart from the ANSI Standard Library functions described in Chapter 17, there are many commercially available libraries.[4]

Exercise

What mathematical functions are declared in `<cmath>`? Write a program that evaluates:
$$\cosh^2 x - \sinh^2 x - 1$$
for 1000 different values of x. Explain any unexpected results.

5.9 Using Functions

Now that we have introduced functions in some detail, we can use them to write more interesting programs.

5.9.1 A Benchmark

There have been a large number of attempts to find programs that can be used to compare the performance of different computers. Such programs are commonly called *benchmarks*. One of these programs, called the Savage benchmark (after its author), tests the speed and accuracy of the common mathematical functions.[5] A version of the test is given below and you should try it on whatever systems you have access to.

```
// The Savage benchmark: tests the speed and accuracy
// of some common mathematical functions.
#include <iostream>    // For cout
#include <cmath>       // For tan(), atan(), exp(), log(), sqrt()
#include <ctime>       // For clock(), CLOCKS_PER_SEC
#include <cstdlib>     // For EXIT_SUCCESS
using namespace std;

int main()
{
    int loops;
    for (;;) {
```

---

[4]A description of useful packages for scientists and engineers is contained in [9].

[5]See: B. Savage, Dr. Dobb's Journal, 120, September 1983.

```
            cout << "Input a positive integer: ";
            cin >> loops;
            if (loops > 0)
                break;
        }
        double test = 1.0;
        clock_t start_time = clock();
        for (int i = 1; i < loops; ++i)
            test = tan(atan(exp(log(sqrt(test * test))))) + 1.0;
        clock_t stop_time = clock();
        cout << "test = " << test << "\n";
        cout << loops << " - test = " << loops - test << "\n";
        cout << "Time taken = " <<
            static_cast<double>(stop_time - start_time) /
            CLOCKS_PER_SEC << " secs.\n";
        return(EXIT_SUCCESS);
    }
```

The program starts with a simple comment about what it actually does. As far as possible it is better if code can be made self-documenting by using well-chosen identifiers since too many comments are distracting. Of course, a balance should be kept between too few and too many comments, since either extreme can lead to unreadable code. Typical circumstances in which comments are useful are:

- giving references to books, algorithms and maintenance manuals:

  ```
  // See Press et al., Numerical Recipes in C, page 255.
  ```

- explaining the purpose of a function where the function name is insufficient:

  ```
  // Finds roots of a polynomial using Newton-Raphson.
  ```

- drawing attention to any compiler-dependent or particularly tricky segment of code:

  ```
  // Assumes the int type is at least 32 bits.
  ```

The required header files follow the comment. These are all files supplied by the system so the names appear inside < > pairs. We have noted which of the constants and functions, declared in these files, are used in this program. It is not usual to include such comments, but you may find them helpful until you are familiar with the contents of the Standard Library. Also, although the only requirement is that the include directive must appear before the associated functions or constants are first used, it is normal practice to collect all such directives at the start of the file.

In this program there are no user-declared functions; if there were, then the appropriate function definitions would follow the include directives.[6] Inside the body

---

[6]If all functions are declared first, then no restriction is placed on the order in which functions are defined.

of `main()` we define variables as they are first used. The type `clock_t` is defined in `<ctime>` and its use avoids needing to know the type returned by the `clock()` function for a particular compiler. Likewise, use of the `CLOCKS_PER_SEC` constant, to convert the result of `clock()` to seconds, hides compiler-dependent details. It is important to realize that on a time-sharing operating system, such as UNIX or Linux, the elapsed time may not be the same as the time measured by the number of CPU ticks. Consequently, for benchmarking purposes it is important to use the `clock()` function rather than `time()`, since the latter returns the calendar time.

The program tests for speed and accuracy by making repeated calls to the tangent, inverse tangent, exponential, natural logarithm and square root functions. In fact, it is easy to show that the final result would be 2500 for infinite precision arithmetic. Notice how the function calls can be nested quite deeply.

Our final statement returns a *normal termination* flag. This may not be necessary on some systems, but others may actually care whether a program fails or not! Again, the `EXIT_SUCCESS` constant avoids having to know the numerical value of the flag.

5.9.2 Root Finding by Bisection

We now have sufficient experience to make it worthwhile examining a more complicated example of using functions. A frequent requirement in numerical analysis is to find the roots of a function of a single real variable; specifically to solve:

$$f(x) = 0$$

for x in the interval:

$$x_1 \leq x \leq x_2$$

If $f(x_1)$ and $f(x_2)$ have opposite signs, then at least one zero or singularity must lie between these limits, as shown in Figure 5.2. The method of bisection consists of finding the value of $f(x)$ at the midpoint of the interval. The segment of the interval for which $f(x)$ changes sign again brackets a zero (or singularity) and this interval can again be bisected. The procedure continues until the zero is known to be within a sufficiently small interval. What constitutes a sufficiently small interval can be quite subtle, but in the example given below this stopping criterion is kept deliberately simple. If you do need a more detailed discussion of when to stop the procedure, then see [14].

It is necessary to distinguish a zero from a singularity. This can be done by evaluating the function at the midpoint of the final interval. The method of bisection is not the fastest technique for finding a root, but it does have the advantage of simplicity.

Recursion is clearly a natural way to implement the bisection algorithm and this is done in the program listed below.

```
#include <iostream>
#include <cmath>        // For exp(), pow(), cos()
#include <cstdlib>      // For exit()
using namespace std;

// Function declarations:
double f(double x);
double root(double x1, double x2, double f1, double f2);
```

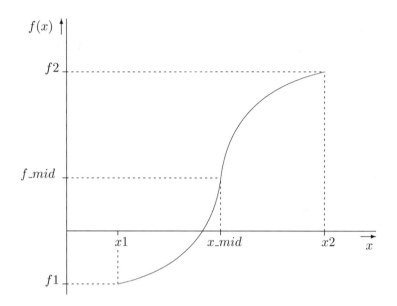

Figure 5.2: Root finding using bisection.

```
double find_root(double x1, double x2);

double f(double x)   // A solution of f(x) = 0 is required.
{
    return(exp(x) + pow(2.0, -x) + 2.0 * cos(x) - 6.0);
}

double root(double x1, double x2, double f1, double f2)
// Finds a root of f(x) = 0 using bisection.
// Assumes that x2 >= x1 and f1 * f2 < 0.0
{
    const int max_depth = 50;
    const double x_limit = 1e-5;
    static int depth;
    double estimated_root;
    double x_mid = 0.5 * (x1 + x2);
    if (x2 - x1 <= x_limit) {
            cout << "Root found at recursion depth = " << depth <<
                "\n";
            estimated_root = x_mid;
    }
    else if (++depth > max_depth) {
        cout << "WARNING: maximum limit of " << max_depth <<
            " bisections reached\n";
```

```
                estimated_root = x_mid;
        }
        else {
                double f_mid = f(x_mid);
                if (f_mid == 0.0) {
                        // Zero at x_mid.
                        estimated_root = x_mid;
                }
                else if (f(x1) * f_mid < 0.0) {
                        // Zero in first segment.
                        estimated_root = root(x1, x_mid, f1, f_mid);
                }
                else {
                        // Zero in second segment.
                        estimated_root = root(x_mid, x2, f_mid, f2);
                }
        }
        return estimated_root;
}

double find_root(double x1, double x2)
{
        double f1 = f(x1);
        double f2 = f(x2);
        if (f1 * f2 > 0.0) {
                cout << "Error in find_root(): " <<
                        "endpoints have same sign\n";
                exit(EXIT_FAILURE);
        }
        if (x2 - x1 > 0.0)
                return root(x1, x2, f1, f2);
        else
                return root(x2, x1, f2, f1);
}

int main()
{
        double x = find_root(1.0, 2.0);
        cout << "Root is " << x << "\nf(x) at root = " << f(x) << "\n";
        return(EXIT_SUCCESS);
}
```

The overhead incurred by using recursion for this algorithm is probably small since for most functions, f(x), the program will be dominated by the evaluations of f(x) itself.

In this example, we use bisection to search for a zero of $f(x)$, where

$$f(x) = e^x + 2^{-x} + 2\cos x - 6$$

in the range

$$1 \leq x \leq 2$$

The function, $f(x)$, is implemented by the first function in our program. Three library functions are used, which are all declared in the header file <cmath>. The exp() and cos() functions are self-explanatory; pow(x,y) returns x^y. (Unlike some other languages, C++ has no built-in operator that raises a number to a power.)

The analysis of the remainder of this program starts at the function main(). As can be seen, main() is kept very simple and calls the function find_root(), which returns the calculated value of the root. This value is assigned to x.

The function, find_root(), takes as its arguments the interval limits and the maximum recursion depth. It would be useful if we could specify f(x) as a fourth argument to find_root(), in order to facilitate root finding for other functions. However, we must wait until Chapter 7 to learn how to do this.

Examining the body of find_root(), we see that the current depth is set to zero and the maximum recursion depth is set to the user-defined value. If find_root were called with only two arguments, then the maximum recursion depth would be set to a default value (in this case 100). The function, find_root() also checks that f(x) has opposite signs at the two endpoints.

The function root() has the declaration:

```
double root(double x1, double x2, double f1, double f2);
```

where x1, x2 are the endpoints of the interval and f1, f2 are the corresponding values of f(x). The implementation of root() assumes that x2 is not less than x1 and that f1 and f2 have opposite signs. Both of these assumptions are enforced by find_root(). The function, root(), first finds x_mid, the midpoint of the interval. If (x2 - x1) is less than the required accuracy, or the maximum recursion depth has been reached, then the value x_mid is returned. Next, f_mid, the value of f(x) at the midpoint, is evaluated. If f_mid is zero, a root has been found and x_mid is returned. Finally, if none of the previous cases has occurred, then the value of root() is returned for the segment of the interval that shows a sign change for f(x). It is this call of the root() function that constitutes recursion.

It is worth pointing out that this program can only find the root of a single function since there is no way of resetting the value of depth. This restriction is removed in Section 7.5.

5.10   Summary

- A function definition implements the function:

```
double f(int n)
{
    // Code implementing function.
}
```

- Functions are executed (called or invoked) by statements such as:

```
x = f(m);
```

or

```
f(m);
```

- If a function is declared to return a value it must do so:

```
double f(int n)
{
    // Code.
    return;              // WRONG!
}
```

- The order of evaluation of expressions in argument lists is not defined:

```
x = f(++i, (i + 5));    // Compiler-dependent.
```

- A function declaration defines the function interface:

```
double f(int n);
```

- A function cannot be declared inside another function:

```
int main()
{
    int f(int n);        // WRONG!
    // Code.
}
```

- Functions can have default arguments. These are supplied right to left, as in:

```
double rectangle(double height, double width = 10.0);
```

which can be called by either:

```
z = rectangle(x);
```

or

```
z = rectangle(x, y);
```

- Simple functions may be worth declaring inline:

```
inline int theta(int i)
{
    return((i > 0) ? 1 : 0);
}
```

- Function names can be overloaded:

```
float sqrt(float x);
double sqrt(double x);
```

- Consult the Standard Library; it may have the function you are about to write!

```
x = cosh(y);            // Returns hyperbolic cosine of y.
```

5.11   Exercises

1. Using a suitable function from the Standard Library supplied with your system, write a program that measures how long it takes to evaluate the factorial function. Use various function arguments and compare the times for both the recursive and iterative versions of the function. It is worth using at least four-byte integers for this example, since the factorial function grows rapidly with the magnitude of its argument. (On a very fast computer you may find it difficult to get a non-zero elapsed time with a sensible value for the factorial function argument. In this case you should time the same calculation carried out many times.)

2. Repeat the previous exercise for both recursive and iterative functions that sum the first n positive integers.

3. Write a function that converts degrees Celsius to degrees Fahrenheit and a second function that converts degrees Fahrenheit to Celsius. Both functions should take a `float` argument and return a `float`. They should also trap impossible values of the arguments.

   Check your functions by listing some temperatures using both scales. Also try one function with the other as its argument; do you get the original temperature?

4. Use the power series expansion:

$$\sin x = x - \frac{x^3}{3!} + \frac{x^5}{5!} - \frac{x^7}{7!} + \frac{x^9}{9!} - \cdots$$

   for $|x| < \infty$ to implement a sine function with the declaration:

   ```
   float sin(float x);
   ```

   (You should truncate the series and only consider arguments satisfying $|x| \le \pi/2$.)

   Compare the accuracy and time taken to evaluate `sin(x)` with the Standard Library function in `<cmath>` for $|x| \le \pi/2$. Make sure that you really do invoke the two overloaded `sin()` functions. (Beware of implicit conversions!)

   Try to make your version of `sin(x)` as accurate and efficient as possible. For instance:

   (a) Don't evaluate unnecessary terms in the power series.
   (b) Instead of evaluating the factorials for every function call, use pre-calculated values for the inverses. In fact, for a truncated power series, more accurate values of the coefficients are given in standard mathematical tables such as [11].
   (c) Reduce the number of multiplications by using nested parentheses (Horner's method) as in:

$$\sin x \approx ((((a_8 x^2 + a_6)x^2 + a_4)x^2 + a_2)x^2 + 1)x$$

5. Rewrite our root finding by bisection program (Section 5.9.2) so that an iterative, rather than recursive technique, is used. Compare the time taken for both methods.

6. The Monte Carlo technique is often used to evaluate multi-dimensional integrals. As an illustrative example, we can generate random points in the region of the x-y plane given by $0 \le x \le 1$ and $0 \le y \le 1$. If we count those points satisfying $x^2 + y^2 \le 1$ as "hits", then the hits divided by the total number of points should be an approximation to the area of a quadrant of a disc with unit radius. Since the total area of such a disc is π, this gives us a (rather inefficient) way of calculating π.

   The random number generator, `rand()`, declared in `<cstdlib>`, returns random integers in the range 0 to `RAND_MAX`, which is also given in `<cstdlib>`. Use this library function to write a Monte Carlo program that calculates π and lists deviations from the tabulated value as the number of points increases. (For example, give the deviation for every 1000 points.) Try to explain any pattern you find in the deviations. (The constant, `M_PI`, is conveniently defined in `<cmath>`.)

7. Write a function that tests whether a positive integer is prime.[7] The function should have the declaration:

   ```
   bool test_prime(unsigned n)
   ```

   where n is the integer to be tested and the function returns `true` if n is prime and `false` if n is not prime.

   Write a second function, with the declaration:

   ```
   void list_primes(unsigned m)
   ```

   that uses `test_prime()` to list all prime numbers up to m. Write a program, using the `test_prime()` and `list_prime()` functions, that prompts for a positive integer and then lists all primes less than or equal to this integer. Check your results against a table of prime numbers.

8. Calculation shows that $f(i)$, given by:

   $$f(i) = i^2 + i + 41$$

   generates prime numbers for $i = 0, 1, 2, 3, 4, 5, 6$. Write a simple function that implements $f(i)$ and use `test_prime()` from the previous exercise to check whether or not primes are generated for larger values of i.

9. The $J_0(x)$ Bessel function is given to at least seven decimal places by:[8]

   $$\begin{aligned} J_0(x) \;=\; & 1 - a_1(x/3)^2 + a_2(x/3)^4 + a_3(x/3)^6 + a_4(x/3)^8 \\ & + a_5(x/3)^{10} + a_6(x/3)^{12} \end{aligned}$$

   where:

   $$\begin{array}{ll} a_1 = 2.2499997 & a_4 = 0.0444479 \\ a_2 = 1.2656208 & a_5 = 0.0039444 \\ a_3 = 0.3163866 & a_6 = 0.0002100 \end{array}$$

---

[7]Straightforward division by successive integers is sufficient for this exercise.
[8]See [11].

and:
$$|x| \leq 3$$

Write a function that returns $J_0(x)$ and show that $J_0(x)$ changes sign for $0 < x < 3$.

Modify the bisection program given in Section 5.9.2 to find a solution of:
$$J_0(x) = 0$$

within the stated range of x. You should check your result by directly evaluating the function.

10. Legendre polynomials, $P_n(x)$, can be defined by:
$$P_n(x) = \sum_{k=0}^{[n/2]} \frac{(-1)^k(2n-2k)!x^{n-2k}}{2^n k!(n-2k)!(n-k)!}$$

where n is an integer, with:
$$n \geq 0$$
$$-1 \leq x \leq 1$$

and:
$$[n/2] = \begin{cases} n/2 & \text{if } n \text{ is even} \\ (n-1)/2 & \text{if } n \text{ is odd} \end{cases}$$

Write a function, taking n and x as arguments, that returns the value of $P_n(x)$.

Legendre polynomials satisfy many identities, including the so-called *pure recursion relation*:
$$(2n+1)xP_n(x) = (n+1)P_{n+1}(x) + nP_{n-1}(x)$$

for $n = 1, 2, 3, \ldots$

Use this relation to write another function that tests your Legendre polynomial function. (Notice that if we were given $P_0(x)$ and $P_1(x)$, then this relation could be used to recursively generate $P_n(x)$.) You may come across two problems in this example; the factorial function given earlier in this chapter suffers from integer overflow for values of n which are not very large and, secondly, you may encounter rather large rounding errors for x close to zero. You could solve the first problem by converting the factorial function to return a **double** and the second by ignoring it, once you have convinced yourself that the cause really is rounding error!

11. The logistic map, defined by
$$x_{n+1} = cx_n(1 - x_n)$$

where
$$0 \leq x_n \leq 1$$

and
$$0 < c \leq 4$$
can be used to model a simple biological population at successive time intervals. x_n is the population, normalized so that the maximum sustainable population is one. The motivation for the equation is that as x_n is increased from zero, the next generation will be larger since there are more parents. However, as the normalized population approaches one, survival is less likely and the size of the next generation actually decreases. The variable, c, is the control parameter for both of these effects.

Starting from some initial population, the logistic map can be used to generate successive populations. Write a function that implements the logistic map. The function should have c and x as its arguments and return the value of x after a large number of generations, such as 150.

Write a function that has c as its argument and uses the first function to list the final values of x for a selection of initial values of x. Using these two functions, write a program that continues to prompt for a value of c and then lists a number of final populations.

For c between 1 and 3 you should find that the population tends to a single value. In fact, if you plot these *fixed points* against the values of c you should get a smooth curve. (It would be convenient if a C++ program could plot the curve, but this is beyond the techniques that we have covered so far.)

Study of the logistic map has given rise to an enormous number of research papers. If you investigate successive iterations for the two regions

$$3 < c < 3.449499$$

and

$$3.449499 < c < 3.544090$$

you should get a glimpse of the complex structure that arises from this simple map.[9]

---

[9]There is a large literature on the logistic map. A good start is [13].

Chapter 6

Pointers and Arrays

It is difficult to imagine performing many scientific calculations without the ability to manipulate arrays of one or more dimensions. Arrays are needed for vectors, matrices and the convenient storage of related data. In this chapter we introduce a simple picture of the way memory is organized, together with powerful methods for using contiguous areas of memory.

6.1 Memory, Addressing and Pointers

The memory of a computer can be regarded as a collection of labelled storage locations, as shown in Figure 6.1. In our discussions we assume that we can access these memory locations in any order, and we visualize memory as a linear sequence of storage locations, one byte in size, which are labelled 1, 2, 3 ... etc. A particular label is known as the address of the corresponding memory element. The two operations of interest are to read what is stored at a particular address and to write data to the memory labelled by an address.

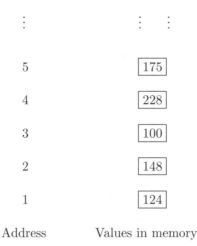

Address	Values in memory
5	175
4	228
3	100
2	148
1	124

Figure 6.1: Labelled storage locations.

6.1.1   Address-of Operator

In C++ the address of any variable can always be found by preceding the variable with
the *address-of* operator, denoted by &. The address-of operator has the same precedence and right to left associativity as other unary operators (apart from the postfix
increment and decrement operators). Several new operators are defined during the
course of this chapter, and their precedence and associativity are given in Appendix B.
In the present context, the & operator simply means "return the address of the variable to the right". The following program illustrates the difference between the values
stored by variables of various types and their corresponding addresses:

```
#include <iostream>
#include <cstdlib>        // For EXIT_SUCCESS
using namespace std;
int main()
{
    int i = 1;
    double x = 3.0;
    float z = 4.0;
    cout << "Address of i is " << &i <<
        "  Value of i is " << i << "\n";
    cout << "Address of x is " << &x <<
        "  Value of x is " << x << "\n";
    cout << "Address of z is " << &z <<
        "  Value of z is " << z << "\n";
    return(EXIT_SUCCESS);
}
```

When you run this program you will probably find that the addresses are given in
hexadecimal format, but this depends on the compiler. You may also notice that the
addresses differ by the size of the various types. However, whether or not this occurs
depends on how the compiler allocates storage.

There are a few things that we cannot do with the address-of operator. It is illegal
to take the address of a constant, as in:

```
&10;              // WRONG: cannot take address of a constant.
&3.142;           // WRONG: cannot take address of a constant.
```

It is also illegal to take the address of an expression, as in:

```
float x = 302.8;
&(x + 73.6);      // WRONG: cannot take the address of an expression.
```

6.1.2   Dereferencing or Indirection Operator

Given an address, there is an operator, known as the *dereferencing operator*, denoted
by *, which returns the value stored at that particular address. The phrase, *indirection
operator*, is often used in place of dereferencing operator. In some sense the address-of
and indirection operators are inverses of each other; if j is a variable of type int, then

&j is the address in memory where that variable is stored. The value stored at the memory location, &j, is given by, *&j; in other words, j:

```
int i, j = 1;
i = *&j;    // This is an over-complicated way of writing:
i = j;
```

Notice that the address-of and indirection operators are only inverses in a limited sense; for example, &*j would be meaningless in this context since j is not an address and so cannot be dereferenced.

The indirection operator, *, should not be confused with the multiplication operator, *. Whereas indirection is a unary operator, multiplication is a binary operator; their meanings are completely unrelated. Both the compiler and human reader can unambiguously distinguish these operators by their context.

6.1.3 Pointers

So far we have come across many different types of variable. In general, the value of a particular variable in one of our programs is held at a location in memory with some specific address. It is the *starting address* that is significant; the actual number of bytes used varies with the variable type and the C++ compiler. For instance a char may be stored in one byte and a double in eight bytes. It is useful to have variables that can hold the address of a storage location. For instance we may have many large contiguous areas of memory, each one storing a large amount of data concerning observations on galaxies. If we know the addresses of these areas of memory, we can efficiently sort the data by manipulating addresses rather than copying the data itself. Variables for storing memory addresses are known as pointers. There is a special pointer type corresponding to each variable type. The different pointer types are needed because it is usually necessary to know how much memory is being pointed to for each type of data and how the bits stored in memory are to be interpreted.

Examples of pointer declarations are given below.

```
int *p1;        // p1 is a pointer to the type int.
double *p2;     // p2 is a pointer to the type double.
float *p3;      // p3 is a pointer to the type float.
```

The notation for pointer declarations may seem peculiar, but it can be understood by recalling that * is the dereferencing operator. This operator returns the value stored at the address held by its (right) operand. Since the dereferencing operator acting on p1 gives an int type, p1 must have the correct type for storing the address of an int; that is p1 is a pointer to an int. Remember that white space is ignored so the above could be rewritten as:

```
int*p1;
```

or

```
int * p1;
```

or

```
int* p1;
```

but all of these variations are unusual. Similar remarks apply to the declarations for p2 and p3.

In order to declare a number of integer pointers we might be tempted to use:

```
int *p1, p2, p3;            // WRONG: p2, p3 are not pointers.
```

However, the indirection operator, like all unary operators binds to the right, not the left. In the above case the indirection operator binds only to p1, leaving p2 and p3 defined as having type int. To correctly implement our intention we need to use the following declaration:

```
int *p1, *p2, *p3;
```

If we do indeed wish to use a single statement to simultaneously declare both identifiers of a type and pointers to that type, then we can do so and the order of the declarations is not significant. For example, in the following declaration x, y are both of type int, and pt_x and pt_y are both pointers to the type int:

```
int x, y, *pt_x, *pt_y;     // pt_x and pt_y are pointers.
```

However, it is best to avoid mixing declarations like this, since it is so easy to make mistakes.

Dereferenced pointers can be used anywhere that an identifier of the corresponding type would be valid. The following program demonstrates how dereferenced pointers can be used to make an assignment to an object and to perform operations (in this case addition) on objects:

```
#include <iostream>
#include <cstdlib>               // For EXIT_SUCCESS
using namespace std;
int main()
{
    int i, j, k;
    int *pt_i, *pt_j;
    pt_i = &i;                   // Assigns address of i to pt_i.
    pt_j = &j;                   // Assigns address of j to pt_j.
    i = 1;
    j = 2;
    k = *pt_i + *pt_j;           // 3 is assigned to k.
    *pt_i = 10;                  // 10 is assigned to i.
    cout << "k = " << k << "\ni = " << i << "\n";
    return(EXIT_SUCCESS);
}
```

Notice that a dereferenced pointer can appear on the left of the assignment operator; that is assignments can be made to the object that the pointer points to. As stated in Appendix B, the indirection operator has a higher precedence than either the assignment or addition operator. Therefore the assignment to k in the above program is equivalent to:

```
k = (*pt_i) + (*pt_j);
```

In this example pt_i and pt_j are assigned the addresses of the memory where i and j are stored. The values stored in these memory locations can then be accessed by using i and j or *pt_i and *pt_j, as shown in Figure 6.2. (The addresses are decimal and do not refer to any particular computer.) Again, recall that white space is ignored so that:

```
pt_i = &i;
```

could equivalently be written as:

```
pt_i = & i;
```

although this style is quite unusual.

Exercise

Rewrite the quadratic equation program, given in Section 1.2, so that as many operations as possible are performed by using dereferenced pointers. Since there is considerable scope for mistakes, you may find it worthwhile to work in stages, modifying a small part of the original program at a time. (Note that this is purely an exercise and is certainly *not* advocated as a style of programming.)

A pointer can point to a const type, but the pointer definition must also include the const specifier:

```
const int w = 100;
const int *pt_c;
int *pt_i;
pt_c = &w;        // O.K.
++pt_c;           // O.K.
pt_i = &w;        // WRONG: a non-const pointer cannot point to const.
```

The fact that it is illegal to assign the address of a const type to an unqualified pointer prevents us from using a dereferenced pointer to attempt to modify a const object.

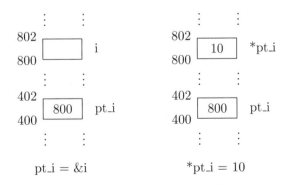

Figure 6.2: The address-of and dereferencing operators.

Notice that, although we cannot take the address of a constant (such as 3.142), we can take the address of a const type. Also, it is essential to distinguish between a pointer to a const type (described above) and a const pointer, an example of which is given in the following code fragment:

```
double x = 127;
double *const pt = &x;
++*pt;        // O.K:   x is not constant.
++pt;         // WRONG: pt is constant.
```

Although a const pointer is initialized to an address that cannot be changed, the data stored in the memory corresponding to this address *can* be changed. Notice that a const pointer *must* be initialized, since it cannot be assigned to.

A pointer is simply a variable that is used to hold the address of another variable, so it is also possible to store the address of a pointer. A variable used to store the address of a pointer is known as a pointer to a pointer. The following program illustrates using a pointer to a pointer, although it should be emphasized that the program is given for demonstration purposes only:

```
#include <iostream>
#include <cstdlib>        // For EXIT_SUCCESS
using namespace std;
int main()
{
    double x;            // Memory defined to store a double.
    double *pt;          // Memory defined to store the
                         // address of a double.
    double **pt_pt;      // Memory defined to store the address
                         // of a pointer to a double.
    pt = &x;
    pt_pt = &pt;
    x = 11.11;
    // Try dereferencing pt_pt to access x:
    cout << "x = " << **pt_pt << "\n";
    return(EXIT_SUCCESS);
}
```

Although the value stored by pt can be accessed by dereferencing pt_pt, to access x we need to dereference pt_pt twice. The relation between the three storage locations is shown in Figure 6.3. In numerical applications, pointers to pointers usually occur when we have multi-dimensional arrays. The chain of pointing can continue indefinitely (pointer to pointer to pointer to ...). This is one effective way of implementing linked lists, although it is rare to see explicit multiple dereferencing of the form ***pt.

Exercise

Try out the above code on your system. Modify it to multiply two numbers using pointers to pointers and check that you get the correct result.

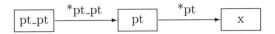

Figure 6.3: Pointer to a pointer.

6.1.4 Pointers Do Not Declare Memory

We make considerable use of pointers throughout this book and you will come to appreciate that pointers are an extremely powerful tool. However, like most powerful tools, pointers can also be abused. One problem is that you may make a programming error so that an incorrect address is dereferenced. For instance, suppose we made an error in the example on page 122, resulting in the following program:

```cpp
#include <iostream>
#include <cstdlib>        // For EXIT_SUCCESS
using namespace std;
int main()
{
    int i, j, k;
    int *pt_i, *pt_j;
    // Forget to assign &i to pt_i here.
    pt_j = &j;
    i = 1;
    j = 2;
    k = *pt_i + *pt_j;        // WRONG: pt_i is arbitrary.
    *pt_i = 10;               // WRONG: pt_i is arbitrary.
    cout << "k should be 3, it is: " << k << "\n";
    cout << "*pt_i should be 10, it is: " << *pt_i << "\n";
    return(EXIT_SUCCESS);
}
```

Here, we have not assigned the address of i to `pt_i` and hence `pt_i` stores some arbitrary address, not necessarily zero. When we use `*pt_i` in the statement:

```cpp
k = *pt_i + *pt_j;
```

an integer is read from whatever area of memory `pt_i` happens to point to. The memory may not be storing an `int`; it could be a `double` or, on some systems, even part of an executable program or part of the operating system. Nevertheless the bit pattern will be interpreted and returned by `*pt_i` as an integer. A mistake of this kind may well be fatal to your program, but is no worse than many typical programming errors in C++ or other languages.

The real disaster is the final statement:

```cpp
*pt_i = 10;
```

Since we don't know where `pt_i` points, some arbitrary area of memory gets assigned the value 10. This may mean that your program fails at some apparently unrelated statement and the cause of the error may be very difficult to find. You may even attempt to write to an area of memory which doesn't exist. On some computers you may actually write over the operating system, causing the entire system to "crash".[1]

A related and quite common error is to forget to allocate any memory for the pointer to point to. For instance, consider the two statements:

```
int *pt_k;
*pt_k = 10;              // WRONG: pt_k is arbitrary.
```

In this example `pt_k` is again the address of some arbitrary area of memory, to which we attempt to assign the value 10. Here the mistake is readily apparent, but analogous mistakes are often made in complicated programs; typically the indirection takes place in a function that may be several nested function calls removed from where the memory should have been defined.

Exercise

Try running the above code, which manipulates `i`, `j` and `k`, on your system. You should output the values of `k` and `*pt_i` so that the results can be checked. If, by good fortune, no disaster occurs, try incrementing the value of `pt_i` before it is used; you should be able to produce something like a "memory error" message. Now modify the code so that it runs and gives the correct results.

6.1.5   Null Pointer

There is a special constant pointer, known as the *null* pointer, that is guaranteed not to be a valid memory address. If we assign zero to a pointer, then the zero is converted to the null pointer. This pointer may not have the same bit pattern as 0, but this isn't of any significance.[2] Since the null pointer cannot be a valid address, we have a very useful way of signalling certain error conditions involving memory locations.

Some programmers make a habit of initializing all pointers to zero (if the pointers can't be initialized by anything else) and assigning zero to any pointer when its current value is no longer of any use. For example, in the following program we have forgotten to assign the address of `i` to `pt_i`. However, since we have initialized all pointers to zero, the attempt to calculate the value of `k` by dereferencing `pt_i` results in a run-time error. So instead of the program giving an invalid result for `k`, the program terminates with an appropriate error message, such as "Segmentation fault".

```
#include <iostream>
#include <cstdlib>        // For EXIT_SUCCESS
using namespace std;
int main()
```

---

[1]On most modern computers, using a modern operating system, attempts by a program to access memory not allocated to it are likely to be trapped by the operating system. The program will terminate with an error message (such as "segmentation error"). Nevertheless, it cannot be overemphasized that when using pointers you *must* make certain you get them right!

[2]Any integer, floating point or pointer type can be assigned 0. An appropriate zero will be assigned, which typically, but not necessarily, has all bits set zero.

```
{
    int i, j, k;
    int *pt_i = 0, *pt_j = 0;
    // Forget to assign &i to pt_i here.
    pt_j = &j;
    i = 1;
    j = 2;
    k = *pt_i + *pt_j;
    cout << "k should be 3, it is: " << k << "\n";
    return(EXIT_SUCCESS);
}
```

In C and pre-ANSI versions of C++ it was popular to define the null pointer by using NULL, given in `<stddef.h>`. This is best avoided when writing new code or else you may run into problems with the better type checking of modern C++. Simply use the technique described here and set a pointer to zero to get the null pointer.

6.2 One-dimensional Arrays

One-dimensional arrays use a single integral parameter to access contiguous areas of memory. A square bracket pair is used to declare a one-dimensional array. Some examples of declarations of one-dimensional arrays are given below.

```
int i[10];              // Defines an array of 10 ints.
double x[100];          // Defines an array of 100 doubles.
char c[80];             // Defines an array of 80 chars.
```

In this code fragment, ten contiguous memory locations, each large enough to hold one int, are allocated for the array named i. The individual values that are stored in the memory are known as *elements*. The array named i has 10 elements, x has 100 elements and c has 80 elements. The number of elements in an array is known as the *size* of the array. The array size can be specified by a constant expression; that is an expression that evaluates to an integral constant at compile-time, as in:

```
const int MAX_DATA = 100;
double data[3 * MAX_DATA];
```

but not:

```
int MAX_DATA = 100;
double data[3 * MAX_DATA];   // WRONG!
```

The array elements can be accessed (that is assigned to and read from) by specifying an *index* (also known as a *subscript*). The same square bracket notation (the *subscripting operator*) is used, but the context is slightly different. The following program demonstrates assigning a value to an element and accessing the value assigned:

```
#include <iostream>
#include <cstdlib>        // For EXIT_SUCCESS
using namespace std;
```

```
int main()
{
    int i[10];            // Defines an array of 10 ints.
    int j;
    i[6] = 24;            // Assigns 24 to the element with index 6.
    j = i[6];             // Assigns i[6] (that is 24) to j.
    cout << "j = " << j << "\n";
    return(EXIT_SUCCESS);
}
```

The array index must be of integral type (apart from a rather bizarre exception demonstrated in Section 6.2.1). For example, the following code fragment is not correct because x in y[x] has type double:

```
double x, y[20];
```

```
x = 10;
y[x] = 42.0;             // WRONG: x is not of integral type.
```

In contrast to some languages, such as FORTRAN, the indexing starts from zero, so that the array defined by i[10] has elements i[0], i[1], i[2], ..., i[9]. A very common mistake is assigning to an element with an index equal to the size of the array. For instance, in the following code fragment, we declare an array of size ten and then attempt to assign 24 to an area of memory just beyond that used for storing the array. In fact the memory accessed by i[10] may very well be assigned to store another variable (possibly of a different type). Since there is no array bounds checking in C++, the results of the assignment are unpredictable.

```
int i[10];
i[10] = 24;             // WRONG: i[10] is not in the array.
```

This type of error often occurs in a for statement that attempts to use indices going from one up to the size of the array, as demonstrated in the following code fragment:

```
double x[5];
for (int i = 0; i <= 5; ++i)    // WRONG: x[5] is not in the array.
    x[i] = double(i);
```

The result of such mistakes is exactly the same as assigning to a dereferenced pointer that points to an arbitrary address. The results are likely to be unpredictable and erratic, with the cause difficult to trace.

One-dimensional arrays have many obvious uses in engineering and science. For instance:

```
double pressure[1000];
```

could be used to store pressure measurements at 1000 different time intervals and the array defined by:

```
double velocity[3];
```

could be used to hold the components of a 3-dimensional velocity vector. However, you should realize that the C++ definition of an array is simply a sequence of contiguous memory locations; there is no concept of adding or multiplying arrays, unlike the vectors of mathematics. As we will see later, this is not much of a limitation since C++ allows us to define our own types. This means that it is possible to impose more structure on the idea of an array and to create types closer to what we want in our own applications. We return to this theme in Chapter 8, but for the present, if an array represents a vector, then vector operations must be performed element by element. An example demonstrating vector addition, followed by calculating a scalar product is given in the following program:

```cpp
#include <iostream>
#include <cstdlib>        // For EXIT_SUCCESS
#include <cmath>          // For sqrt()
using namespace std;

int main()
{
    double velocity_1[3], velocity_2[3], total_velocity[3];
    // Assign values:
    for (int i = 0; i < 3; ++i) {
        velocity_1[i] = i + 1.0;
        velocity_2[i] = 2.0 * i + 2.0;
    }
    // Vector addition:
    for (int i = 0; i < 3; ++i)
        total_velocity[i] = velocity_1[i] + velocity_2[i];
    // Calculate speed:
    double temp = 0.0;
    for (int i = 0; i < 3; ++i)
        temp += total_velocity[i] * total_velocity[i];
    double velocity_norm = sqrt(temp);
    cout << "The total speed is: " << velocity_norm << "\n";
    return(EXIT_SUCCESS);
}
```

Exercise

Run the above code on your system. Check the value of `velocity_norm` for various assignments to the arrays, `velocity_1[]` and `velocity_2[]`.

(Note that in order to emphasize that we are dealing with an array, we place a pair of empty square brackets after an array name appearing in the text, as in `velocity_1[]`.)

6.2.1 Pointers and One-dimensional Arrays

The concept of an array in C++ is very primitive and directly related to pointers. If we define an array by:

```
int a[20];
```

then a[i] means:

```
*(&a[0] + i)
```

and nothing more. Let's explain this in more detail. The expression &a[0] is the base address of the array, so (&a[0] + i) is the address of the element i locations up from the base address. However, the element i locations up from the base element is simply a[i]. Therefore, (&a[0] + i) is the address of a[i]. Applying the dereferencing operator, we get that *(&a[0] + i) is equivalent to a[i].

In C++ it is possible to perform arithmetic on pointers. The arithmetic that can be done is often very useful even though it is restricted to the following operations:

1. **Addition of a pointer and an integral type or subtraction of a pointer and an integral type**

   This is demonstrated in the following code fragment:

   ```
   double *pt;
   double a[10];
   pt = &a[0];      // pt points to element 0.
   ++pt;            // pt points to element 1.
   pt += 4;         // pt points to element 5.
   pt = pt - 2;     // pt points to element 3.
   ```

   Exercise
   Use the standard subscripting notation to assign the values 1, 2, 3 ... to the elements of a[] in the above code. By adding integers to the pointer, pt, list these elements. Also list the elements in reverse order by subtracting integers from pt.

Notice that the above expressions don't involve the number of bytes used to represent a double; pt is declared to be a pointer to double and the compiler does the necessary calculations to ensure that when using pointer arithmetic, appropriate allowance is made for the storage requirements of the different types. Essentially, if pt is a pointer to an element of an array, then pt+1 is a pointer to the next element in the array.

All of our operations on pointers are only valid provided we stay within the memory allocated to the array. In the following code fragment, we attempt to point to memory outside of that allocated to the array:

```
double *pt;
double a[10];
pt = &a[0];
pt += 100;   // WRONG: attempts to point outside of the array.
```

Pointing outside of the array is undefined, even if we don't attempt to access the memory location. One obvious way in which such operations may fail is if a small number of bytes are used to store pointers; going beyond the end of an array may cause the value of the pointer to wrap round.

There is one important case when it *is* permissible to point to memory outside of that allocated to an array. It is permissible to point the element one step beyond the high end of the array. However, it is still incorrect to dereference such a pointer, as demonstrated by the following code fragment:

```
double *pt;
double a[10];
pt = &a[0];
pt += 10;    // O.K.:  points to element one step beyond a[9].
*pt = 3.14;  // WRONG: cannot dereference this address.
```

The memory outside of an array may well be allocated for some specific purpose and the result would be the usual symptoms of a pointer error.

2. **Subtraction of two pointers of the same type**

The type of the result of a subtraction of pointers is dependent on the C++ compiler, but is defined as `ptrdiff_t` in the header file `<cstddef>`. The result is a signed integer representing the number of elements separating the two pointers, as illustrated by the following program:

```
#include <iostream>
#include <cstdlib>       // For EXIT_SUCCESS
#include <cstddef>       // For ptrdiff_t
using namespace std;

int main()
{
    double *pt_1, *pt_2;
    double a[10];
    ptrdiff_t diff;

    pt_1 = &a[1];
    pt_2 = &a[4];
    diff = pt_2 - pt_1;      // Assigns 3 to diff.
    cout << "Difference is " << diff << "\n";
    diff = pt_1 - pt_2;      // Assigns -3 to diff.
    cout << "Difference is " << diff << "\n";

    return(EXIT_SUCCESS);
}
```

The result of subtracting pointers is undefined if the pointers do not point to elements within the same array. It is also undefined if the result does not point to an element of the array, except that it is again permissible to point to the element one step beyond the high end of the array. These features are illustrated in the following program:

```cpp
#include <iostream>
#include <cstdlib>          // For EXIT_SUCCESS
#include <cstddef>          // For ptrdiff_t
using namespace std;

int main()
{
    double *pt_1, *pt_2;
    double a[10], b[20];
    ptrdiff_t diff;

    pt_1 = &a[0] + 10;
    pt_2 = &a[0] + 9;
    diff = pt_1 - pt_2;      // Correctly assigns 1.
    cout << "Difference is " << diff << "\n";
    pt_1 = &b[19];
    diff = pt_1 - pt_2;      // WRONG: undefined.
    cout << "Difference is " << diff << "\n";

    return(EXIT_SUCCESS);
}
```

In this example we attempt to find the difference between pointers to elements
of different arrays. The result is "undefined" and may therefore depends on the
compiler and the particular computer on which the program is run.

3. **Relational operations**

   It is possible to compare pointers, but the elements pointed to must again be in
   the same array. Once again, it is legal to point to the element one step beyond
   the high end of the array. Some examples of relational operations on pointers are
   given in the following program:

```cpp
#include <iostream>
#include <cstdlib>          // For EXIT_SUCCESS
using namespace std;

int main()
{
    double *pt_1, *pt_2;
    double a[10];

    pt_1 = &a[10];
    pt_2 = & a[5];
    cout << "pt_1: " << pt_1 << "\npt_2: " << pt_2 << "\n";
    if (pt_2 > pt_1)
        cout << "pt_2 > pt_1\n";
    else
        cout << "pt_2 <= pt_1\n";
```

```
        if (pt_1 != pt_2)
            cout << "pt_1 != pt_2\n";
        if (pt_1 == pt_2)
            cout << "pt_1 == pt_2\n";
        else
            cout << "pt_1 == pt_2 is false\n";

        return(EXIT_SUCCESS);
    }
```

All other arithmetic pointer operations are illegal; for instance multiplying, adding and dividing two pointers are not allowed. In any case, such operations would not be particularly useful!

One consequence of the relationship between pointers and arrays is that operations on arrays can always be rewritten directly in terms of pointer arithmetic. For instance, if we have an array of size ten and type `double`, then the following code fragment would provide the sum of the values stored by the array:[3]

```
double sum = 0.0, x[10];

// Assignments to x[i] would go in here.

for (int i = 0; i < 10; ++i)
    sum += x[i];
```

This summation can be rewritten in terms of pointers as:

```
double sum = 0.0, x[10];
double *pt, *pt_end;

// Assignments to x[i] would go in here.

pt = &x[0];
pt_end = pt + 10;    // Points one element beyond the array.
while (pt < pt_end)
    sum += *pt++;
```

which is more efficient, since incrementing `i` and performing an address calculation is replaced by the single pointer operation.[4] This code also demonstrates why the pointer one element beyond the high end of an array is defined, even though the pointer cannot be dereferenced. The pointer is needed for comparison in a loop termination condition.

Exercise

Using the above code, assign the values 10, 20, 30 ... to the elements of `x[]` and check that the correct value of `sum` is calculated.

---

[3]Notice once again that the terminating condition is `i < 10`. A common error is to write `i <= 10`, which uses the value held in memory after the high end of the array.

[4]In simple situations a good modern compiler will probably produce executable code for a program written in terms of array elements that is as fast as for a program written in terms of pointers. Since programming errors are more likely when using pointers, it is better to restrict their use to situations where it really makes a difference to the execution time.

For an array, x[ARRAY_SIZE], a notational convenience is that x is defined to be the base address of the array. So we could sum the elements of an array by using the program given below.

```cpp
#include <iostream>
#include <cstdlib>          // For EXIT_SUCCESS
using namespace std;

int main()
{
    double sum = 0.0, x[10];
    double *pt, *pt_end;

    for (int i = 0; i < 10; ++i)
        x[i] = 1.0 + i;
    pt = x;
    pt_end = pt + 10;
    while (pt < pt_end)
        sum += *pt++;
    cout << "Sum is: " << sum << "\n";

    return(EXIT_SUCCESS);
}
```

This notation is one that we employ frequently. Notice that, since x is an unmodifiable lvalue, we cannot do anything that attempts to change its value. Two examples of invalid attempts to modify x are given in the following code fragment:[5]

```cpp
double x[10], y[10];
double **pt;
x = y;      // WRONG: cannot assign to a constant.
++x;        // WRONG: cannot increment a constant.
```

There is an interesting curiosity which results from the fact that an array, such as

```cpp
double x[10];
```

can be accessed by x[i], which is directly equivalent to:

```cpp
*(x+i)
```

Since the order of x and i is irrelevant in this expression (addition commutes), x[i] can be validly written as i[x]. The latter looks like an array of integers that is indexed by the value of a double. This bizarre expression will be correctly interpreted by the compiler, but seems to serve no useful purpose for human readers!

---

[5]Recall that an lvalue is an expression referring to an object or function. An lvalue is unmodifiable if it is an array or function name or if it is declared const.

6.3 Type void*

Suppose we want to store a memory address, but we don't know what type of object will be stored at that address. In such cases the type void* should be used as a generic pointer type. As was emphasized in Chapter 5, an object cannot be declared to be of type void. (One reason is that the compiler would not know how many bytes to allocate for such an object.) However, a pointer can be declared to be of type void*, as illustrated by the following program:

```
#include <iostream>
#include <cstdlib>        // For EXIT_SUCCESS
using namespace std;
int main()
{
    int i, a[10];
    double v[20];
    void *pt;
    pt = &i;
    cout << "Address of i = " << pt << "\n";
    pt = a;
    cout << "Address of a[0] = " << pt << "\n";
    pt = v;
    cout << "Address of v[0] = " << pt << "\n";
    return(EXIT_SUCCESS);
}
```

The address of any type can be assigned to a pointer of type void*, but arithmetic operations are not allowed on such pointers, as demonstrated by the following code fragment:

```
void *pt;
double a[10];
pt = a;
++pt;         // WRONG: cannot increment a pointer to void.
```

This restriction is perfectly reasonable. Since pt simply holds an address, the compiler has no information on the type of object stored at that address and therefore has no way of calculating the address of the next element in the array.

6.4 Pointer Conversions

A pointer of any type can be assigned to a pointer of type void*, but the converse is not true, as illustrated by the following code fragment:

```
int a[10];
int *pt_i;
void *pt;
pt = a;                   // O.K.
pt_i = pt;                // WRONG!
```

Such prohibitions are for our own protection. If pt holds an address and we don't know what type of object may be stored there, then assigning that address to a pointer to a type other than void would be risky; we might, for instance, dereference that pointer. If it is necessary to get round these prohibitions, then we can use an explicit cast, but we had better understand what we are doing. The following program shows what can go wrong. Initially the address of a[1] is assigned to pt, which means we have lost any information about what sort of object is pointed to by pt. A static cast is then used to assign this address to pt_i. Since pt_i is an integer pointer we may reasonably expect to use the memory at this address is to store the integer 3. So far this is a valid if rather convoluted way of storing an integer. The next thing that happens is that the address of a[0] is assigned to pt. Again, at this point we have lost any information about what sort of object is pointed to by pt. A static cast is then used to assign this address to pt_d. Since pt_d is a pointer to a double we may reasonably expect to use the memory at this address to store 3.142, which is a double. This is not a good thing to do since it writes over the integer we have stored at a[1], as the output obtained from the program shows. Of course, this simple program is only meant to illustrate what can go wrong. In a larger, more realistic program it may be much harder to spot the error.

```cpp
#include <iostream>
#include <cstdlib>          // For EXIT_SUCCESS
using namespace std;
int main()
{
    int a[2];
    int *pt_i;
    double *pt_d;
    void *pt;
    pt = a + 1;
    pt_i = static_cast<int*>(pt);
    *pt_i = 3;                              // O.K.
    cout << "a[1] = " << a[1] << "\n";
    pt = a;
    pt_d = static_cast<double*>(pt);
    *pt_d = 3.142;                          // Very risky!
    cout << "*pt_d = " << *pt_d << "\n";
    cout << "a[1] = " << a[1] << "\n";
    return(EXIT_SUCCESS);
}
```

In general, it is better to avoid making casts from void* to other pointer types since such casts are inherently risky. Moreover, they are also often a consequence of poor program design.

6.5   Multi-dimensional Arrays

C++ also supports *multi-dimensional arrays*, which are defined by repeated square brackets. The following code fragment gives some examples of how to declare multi-dimensional arrays:

```
double x[3][5];      // Defines a 3 x 5 array of doubles.
float y[2][3][4];    // Defines a 2 x 3 x 4 array of floats.
```

Notations typical of some other languages, such as:

```
double x[3, 5];      // WRONG!
float y(2, 3, 6);    // WRONG!
```

are not valid.

The definition of a multi-dimensional array, such as:

```
int a[4][2];
```

allocates sufficient contiguous memory to store the array. In this example, eight values
of type `int` could be stored. Each array index runs from zero to one less than the
size specified for that index. The following code fragment demonstrates both valid and
invalid array access:

```
int a[4][5];
a[0][0] = 1;         // Low end of the array.
a[3][4] = 25;        // High end of the array.
a[4][5] = 100;       // WRONG: outside of the allocated memory.
```

For a two-dimensional array, it is conventional to regard the first index as labelling
rows and the second as labelling columns; this is consistent with the standard notation
for matrices.

Multi-dimensional arrays can be used to represent matrices and tensors, but arithmetic operations must be done element by element. For example, adding two 4×5
matrices A and B is expressed mathematically in terms of elements as:

$$C_{ij} = A_{ij} + B_{ij}$$

The equivalent C++ statements are given below.

```
double A[4][5], B[4][5], C[4][5];
for (int i = 0; i < 4; ++i)
    for (int j = 0; j < 5; ++j)
        C[i][j] = A[i][j] + B[i][j];
```

For another example, consider the multiplication of a 4×5 matrix, X, by a 5×6 matrix,
Y. This is expressed mathematically as:

$$Z_{ij} = \sum_{k=0}^{k=5} X_{ik} \times Y_{kj}$$

The equivalent C++ code fragment is as follows:

```
double X[4][5], Y[5][6], Z[4][6];
for (int i = 0; i < 4; ++i)
    for (int j = 0; j < 6; ++j) {
```

```
        double temp = 0.0;
        for (int k = 0; k < 5; ++k)
            temp += X[i][k] * Y[k][j];
        Z[i][j] = temp;
    }
```

In terms of matrices, the equivalent mathematical equations are much simpler. The addition is given by

$$C = A + B$$

and the multiplication by

$$Z = X * Y$$

However, it is only by introducing overloaded operators (in Chapter 9) that we can truly manipulate matrices, rather than matrix components, as objects.

Exercise

By assigning appropriately chosen integers to the elements of A[][], B[][], X[][] and Y[][], check the correctness of matrix addition and multiplication as implemented above. You should display the calculated matrices as two-dimensional arrays.

Since a memory location is specified by a single address, the two or more dimensions of a multi-dimensional array must be mapped into the linear address space of physical memory. This mapping is often known as a *storage map*. In C++, two-dimensional arrays are stored by rows and a typical storage map is shown in Figure 6.4, where the symbol x_0 represents the array element &x[0][0].

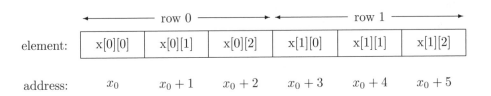

Figure 6.4: Storage map for x[2][3].

6.5.1   Pointers and Multi-dimensional Arrays[†]

An array defined by:

```
    int x[2][3];
```

is accessed via x[i][j], which is equivalent to:

```
    *(&x[0][0] + 3 * i + j)
```

In fact, two-dimensional arrays have no more significance than this equivalence. In this example, x[0][0] is the element at the low end of the array and therefore &x[0][0] is the base address of the array.

Great care should be exercised when using pointers to access multi-dimensional arrays. The constant, x, is also the base address of this array, but

```
*(x + 3 * i + j)
```

is not equivalent to x[i][j] since (x + i) is actually the base address of row i; for instance, (x + 1) is the base address of the row 1. A correct way of using x, rather than &x[0][0], to access an element of the two-dimensional array is to use:

```
(*(x + i))[j]
```

The outer parentheses are required because [] binds tighter than the dereferencing operator. In fact, by studying the above expression, we can see how the notation for two-dimensional arrays arises. The expression, *(x+i), is the same as x[i], the base address of row i, so the whole expression is directly equivalent to x[i][j]. There are two other, equally devious, ways of rewriting x[i][j]; these are

```
*((*(x + i)) + j)
```

and

```
*(x[i] + j)
```

It is worth convincing yourself that these expressions really are equivalent to x[i][j], although in practice it is best to stick to the more obvious notation.

Exercise

You have now encountered four different non-standard ways of accessing a two-dimensional array by using pointers. Write a program that uses each of these four techniques to write numbers to a 2×3 array of type int. Use the standard way of accessing the array to demonstrate that the correct values are assigned in each case.

For arrays of three dimensions and higher, the storage map is a straightforward extension of the two-dimensional case. For instance, if we make the definition:

```
float y[2][3][4];
```

then an element, y[i][j][k], can equivalently be accessed by:

```
*(&y[0][0][0] + 3 * 4 * i + 4 * j + k);
```

Needless to say, there are devious ways of rewriting this expression in terms of y rather than &y[0][0][0].

We don't need to know how arrays are stored in order to use them, but doing so can help us to understand (and even reduce) the overhead caused by indexing into multi-dimensional arrays. For instance, if a calculation involves going down columns, one step at a time, then it may be faster to rearrange the code so that the stepping is done along rows. In Section 7.3.2 we show how knowledge of the array storage map can speed up a typical numerical application.

It is worth emphasizing that in many situations it is better to keep to the standard array notation since a small increase in efficiency is not worth the risk of making a mistake with the pointer arithmetic. However, in situations where using pointer arithmetic would lead to a worthwhile increase in speed, then the more advanced techniques of C++ should be used to encapsulate the pointer arithmetic in isolated pieces of code that can be carefully checked.

6.6   Initializing Arrays

One-dimensional arrays can be initialized by a comma-separated list between a pair of braces, known as an *initialization list*, as in:

```
int x[] = {1, 2, 3};
float y[4] = {1.1, 2.2};
```

If the array size is not specified, then it is taken as the number of elements between the braces. In this example, x[] can store three elements and the initialization of x[] is equivalent to:

```
x[0] = 1;
x[1] = 2;
x[2] = 3;
```

If the size specified is greater than the number of values given, then the low end of the array is initialized with these values and the remaining elements are set to zero (of an appropriate type). The initialization of y[] is therefore equivalent to:

```
y[0] = 1.1;
y[1] = 2.2;
y[2] = 0.0;
y[3] = 0.0;
```

It is illegal to have more members in the list than the specified array size, so the following code fragment is incorrect:

```
double z[2] = {1.1, 2.2, 3.3};   // WRONG: size specified is too
                                 //              small.
```

Multi-dimensional arrays can be initialized by comma-separated, nested, braces, as demonstrated by:

```
int w[4][3] = {{1, 2, 3}, {4, 5, 6}};
```

This two-dimensional array is initialized row-wise, with the last two rows being filled with zeros; that is, the initialization is equivalent to:

```
w[0][0] = 1;   w[0][1] = 2;   w[0][2] = 3;
w[1][0] = 4;   w[1][1] = 5;   w[1][2] = 6;
w[2][0] = 0;   w[2][1] = 0;   w[2][2] = 0;
w[3][0] = 0;   w[3][1] = 0;   w[3][2] = 0;
```

As discussed in the previous section, two-dimensional arrays are actually stored row-wise, so the above initialization can also be written as:

```
int w[4][3] = {1, 2, 3, 4, 5, 6};
```

Incomplete rows of multi-dimensional arrays can also be initialized by means of coma-separated, nested, braces, as shown in the following code fragment:

```
int w[2][3] = {{1}, {2, 3}};
```

Since the low ends of the rows are initialized first (with any remaining elements being filled with zeros) this is equivalent to:

```
w[0][0] = 1;   w[0][1] = 0;   w[0][2] = 0;
w[1][0] = 2;   w[1][1] = 3;   w[1][2] = 0;
```

Nested, comma-separated braces can also be used to initialize arrays of more than two dimensions in a completely analogous way.

Exercise

Verify the initialization of the 2×3 array, `w[][]`. Achieve the same initialization by using a single, rather than nested, comma-separated list.

It should be noted that there is no array assignment analogous to an initialization list. So although

```
int x[2] = {4, 8};
```

is legal, the following is not:

```
int x[2];
x[2] = {4, 8};   // WRONG: an assignment list is illegal.
```

6.7 Size of Arrays[†]

So far we have used the `sizeof` operator to determine the size of fundamental types, such as `int`, `double`, `char` etc. This operator can also give the size of an array, but we must make a clear distinction between the size of one element in an array and the size of the entire array. If we specify a particular element, then, as expected, we obtain the size of just a single element, as demonstrated by the following program:

```
#include <iostream>
#include <cstdlib>        // For EXIT_SUCCESS
using namespace std;
int main()
{
    double x[10], y[5][10];
    cout << "The size of a single element of x[10] is " <<
        sizeof(x[0]) << "\n";
    cout << "The size of a single element of y[5][10] " <<
        "is " << sizeof(y[0][0]) << "\n";
    cout << "For comparison, the size of a double is " <<
        sizeof(double) << "\n";
    return(EXIT_SUCCESS);
}
```

Notice that in general the size of a type can be obtained by using the type name, rather than a specific variable.

Specifying the name of the array as the argument of the `sizeof` operator gives the size of the entire array. This is demonstrated by the following program. (Notice that there is a minor inconsistency here, since in most circumstances x and y are the (unmodifiable) base addresses of the arrays x[] and y[][].)

```
#include <iostream>
#include <cstdlib>          // For EXIT_SUCCESS
using namespace std;
int main()
{
    double x[10], y[5][10];
    cout << "The size of the array x[10] is " << sizeof(x) << "\n";
    cout << "The size of the array y[5][10] is " << sizeof(y) <<
        "\n";
    cout << "The size of an element of the array y[5] is " <<
        sizeof(y[0]) << "\n";
    cout << "For comparison:\n\tThe size of a pointer to " <<
        "a double is " << sizeof(double*) <<
        "\n\tThe size of a double is " << sizeof(double) << "\n";
    return(EXIT_SUCCESS);
}
```

Exercise

What sizes are given by running the above code on your system? Check that they are what you would expect.

On a particular computer we obtain the following output:

```
The size of the array x[10] is 80
The size of the array y[5][10] is 400
The size of an element of the array y[5] is 80
For comparison:
        The size of a pointer to a double is 4
        The size of a double is 8
```

Notice that sizeof(x) gives the size of the entire array, x[10], and not the size of a pointer to an element of x; the meaning of the array name when used as an argument to the sizeof operator is not the same as when the array name is used in pointer arithmetic. Similarly, sizeof(y) gives the size of the entire 5 × 10 array and sizeof(y[0]) gives the size of row 0, that is 10 × sizeof(double).

6.8   Arrays of Pointers

It is possible to define arrays of pointers. Two examples are given in the following code fragment:

```
double *pt_a[5];
float *pt_b[4][10];
```

Since the array index operator, [], has higher precedence than the indirection operator, *, the two statements do indeed each define an array of pointers, rather than a pointer

to an array (which will be discussed shortly). The first statement defines an array of
size 5, which can store the addresses of objects of type double. The second statement
defines a 4×10 array, which can store 40 addresses of objects of type float. It must be
clearly understood that these arrays can only store addresses. They do not themselves
define any memory for storing values of type double or float. Such memory must be
defined separately. In the following code fragment, the memory to store two objects of
type double is allocated by the first statement; the array of pointers is only used to
point to this memory.

```
double a, b;
double *pt[2];
pt[0] = &a;
pt[1] = &b;
```

The distinction between an array of pointers and a pointer to an array is worth
emphasizing. In the definitions:

```
double *p_d[3];
int (*p_i)[3];
```

p_d is an array of three pointers to type double and can therefore store three addresses,
as demonstrated by the following code fragment:

```
double x, y, z;
p_d[0] = &x;
p_d[1] = &y;
p_d[2] = &z;
```

However, p_i is a pointer to an array of three integers and can therefore only store one
address, as in the code fragment given below.

```
int (*p_i)[3];
int a[2][3];
p_i = a;
```

Defining a pointer to an array is sometimes useful because pointer arithmetic automatically allows for the size of the array. In this particular case, p_i points to the base
address of row 0 of the array and therefore (p_i+1) points to row 1.

The following program demonstrates the use of an array of pointers, in addition to
pointers to pointers:

```
#include <iostream>
#include <cstdlib>        // For EXIT_SUCCESS
using namespace std;

int main()
{
    double x, y, z;
    double *pt[3], **p, **p_end;
    pt[0] = &x;
    pt[1] = &y;
```

```
        pt[2] = &z;
        x = 1.11;
        y = 2.22;
        z = 3.33;
        p = pt;
        p_end = p + 3;
        while (p < p_end)
            cout << "data = " << **p++ << "\n";
        return(EXIT_SUCCESS);
    }
```

As shown in Figure 6.5, x, y, z are stored in some arbitrary memory locations and it should not be assumed that these locations are either contiguous or ordered in any particular way. The addresses of x, y and z are stored in pt[0], pt[1] and pt[2] respectively and these array elements must therefore be declared to be of type pointer to double. The address of the element one step beyond pt[2] is stored in p_end and p scans the addresses of the array, pt[]; both p_end and p are of type pointer to pointer to double. Dereferencing p twice gives the data stored in x, y or z. In this simple example, using pointers to pointers is not an improvement on manipulating the array elements directly, but such techniques are useful for manipulating large amounts of data, where copying the data would be very inefficient.

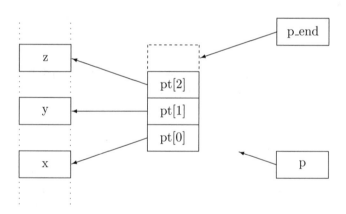

Figure 6.5: Array of pointers.

6.9   Using Pointers and Arrays

In this section we consider two examples that employ some of the techniques introduced in this chapter. The first example uses one-dimensional arrays, rather than explicit pointers, and highlights some of the deficiencies with the techniques learned so far. In the second example we use an array of pointers to construct a ragged two-dimensional array.

6.9.1 Fitting Data to a Straight Line

A common task in science and engineering is to fit a set of data points to a straight line. More specifically, if we are given data x_i and y_i, where $i = 1, 2, \ldots, n$, then we want to find a and b such that:

$$y = a + bx$$

We suppose that the x_i are known exactly, but that there is a error, σ_i, associated with each measurement, y_i. It can be shown that a and b are given by the following definitions and equations:[6]

$$S_x \equiv \sum_{i=1}^{N} \frac{x_i}{\sigma_i^2}$$

$$S_y \equiv \sum_{i=1}^{N} \frac{y_i}{\sigma_i^2}$$

$$S \equiv \sum_{i=1}^{N} \frac{1}{\sigma_i^2}$$

$$t_i \equiv \frac{1}{\sigma_i}\left(x_i - \frac{S_x}{S}\right)$$

$$S_{tt} \equiv \sum_{i=1}^{N} t_i^2$$

$$b = \frac{1}{S_{tt}} \sum_{i=1}^{N} \frac{t_i y_i}{\sigma_i}$$

$$a = \frac{S_y - S_x b}{S}$$

A program that prompts for data points and then computes the values of a and b is given below.

```
// source:      fit.cxx
// use:         Does straight line fitting.

#include <iostream>
#include <cstdlib>        // For EXIT_SUCCESS
using namespace std;

int main()
{
    const int max_points = 100;
    int points;
    double x[max_points], y[max_points], sigma[max_points];
    do {
```

---

[6]We use *exactly* the same notation as [14] so that background reading for this example is readily accessible.

```
        cout << "How many points (less than " << max_points <<
            ") do you want to fit?\n";
        cin >> points;
    } while (points < 1 || points > max_points);
    cout << "\nEnter " << points << " data points in the form:" <<
        "\n   x coordinate  y coordinate  error:\n\n";
    for (int i = 0; i < points; ++i) {
        cin >> x[i] >> y[i] >> sigma[i];
    }
    double s = 0.0, s_x = 0.0, s_y = 0.0;
    for (int i = 0; i < points ; ++i) {
        double temp = 1.0 / (sigma[i] * sigma[i]);
        s += temp;
        s_x += temp * x[i];
        s_y += temp * y[i];
    }
    double s_x_s = s_x / s;
    double s_tt = 0.0, b = 0.0;
    for (int i = 0; i < points ; ++i) {
        double t = (x[i] - s_x_s) / sigma[i];
        s_tt += t * t;
        b += t * y[i] / sigma[i];
    }
    b /= s_tt;
    double a = (s_y - s_x * b) / s;
    cout << "The " << points << " points fit the equation: " <<
        "y = a + b * x\nwith:    a = " << a << " b = " << b << "\n";
    return(EXIT_SUCCESS);
}
```

Exercise

Try this program with your own data.

This line-fitting program exposes a number of limitations with the techniques learned so far:

- The maximum points that the program can handle is fixed at compile-time and can only be altered by editing and recompiling. Techniques for removing this difficulty are given in the next chapter.

- The data must be entered directly in response to a program prompt, rather than read from a file. This is particularly tedious with large data sets and infuriating if you make a typing mistake! Chapter 18 explains how to read data from a file.

- Since we do not know how to pass arrays as function arguments, the entire program consists of the single function, main(). This makes it difficult to reuse our line-fitting code as part of a larger program. Passing arrays as function arguments is introduced in the next chapter.

6.9.2 Ragged Arrays

One use of an array of pointers is to define a *ragged array*, in which the rows have different lengths. For instance, suppose we make some astronomical observations on seven different nights. One night is very cloudy and we only manage 5 observations, but another is exceptionally clear and we carry out 2000 observations. On the remaining nights we make various numbers of observations between these two extremes. We could define a 7 × 2000 two-dimensional array to store the data. However, a more efficient approach is to use a ragged array as shown in the following code fragment:[7]

```
// Define arrays of just sufficient size to
// store the observations made each night:
double Sunday[500];
double Monday[100];
double Tuesday[10];
double Wednesday[1000];
double Thursday[20];
double Friday[5];
double Saturday[2000];

// Define an array of pointers to the data arrays:
double *data[7];
data[0] = Sunday;
data[1] = Monday;
data[2] = Tuesday;
data[3] = Wednesday;
data[4] = Thursday;
data[5] = Friday;
data[6] = Saturday;

// Suppose data are entered in each array here.

// Average element one of each array:
double average = 0.0;
for (int i = 0; i < 7; ++i)
    average += data[i][1];
average /= 7;
cout << "Seven day average of data element 1 = " << average <<
    "\n";
```

The observations for each night are stored in a one-dimensional array of the appropriate length, named Sunday, Monday etc. An array of pointers, data[], then stores the base address of each array holding the observations. The result is a ragged array as shown in Figure 6.6. With care, we can now use data as if it were the name of a two-dimensional array. As a simple example we have averaged the second observation (that is element one) for the seven nights. Of course, it is entirely our responsibility to avoid attempting to access elements beyond the end of a defined row. In practice, an integer

---

[7]Sunday, Monday ... mean the nights of Sunday–Monday, Monday–Tuesday ...

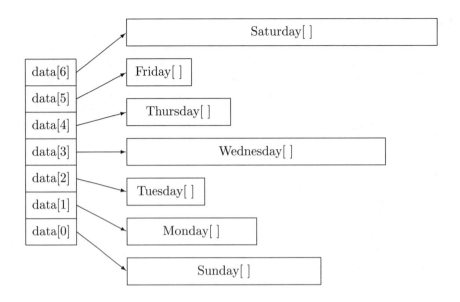

Figure 6.6: Ragged array.

array holding the length of each row would help to prevent such disasters. Notice that
in our interpretation of `data` as the name of a (ragged) array, the first index does indeed
label rows and the second index labels columns. The essential step in understanding
`data[i][j]` is to recall that the array subscripting operator, `[]`, associates left to right,
so that `data[i][j]` is equivalent to `(data[i])[j]`. For example, since `data[2]` is the
base address of the array `Tuesday[]`, `data[2][j]` reduces to the one-dimensional array,
`Tuesday[j]`, which is accessed in the usual way. There is, of course, no equivalence to
a storage map:

```
*(&data[0][0] + row_length * 2 + j)
```

as described in Section 6.5.1, since memory for the entire array is not necessarily
contiguous and the rows have different lengths.

Exercise

Make this ragged array example into a complete program. You should
define an array that stores the length of each row and test the program by
generating random data and finding the average for each night.

6.10   Summary

- A pointer is a variable that can be used to store an address:

  ```
  double *pt;
  ```

- The address of an object can be found by using the address-of operator, **&**. The data stored at a particular address is accessed by means of the dereferencing or indirection operator, *****, as in:

  ```
  double x;
  double *pt = &x;
  ```

- It is illegal to take the address of a constant:

  ```
  &10;                           // WRONG!
  ```

  or an expression:

  ```
  &(x + 3.142);                  // WRONG!
  ```

- Memory must be defined for a pointer to point to:

  ```
  double *pt;
  *pt = 3.1415926535897932;      // WRONG!
  ```

- The null pointer (obtained by assigning zero to any pointer) is guaranteed to be an invalid address.

- One-dimensional arrays are defined as in:

  ```
  double x[10];
  ```

  Individual elements are accessed by specifying an index:

  ```
  double y = x[5];
  ```

  Notice that this array starts at x[0], has 10 elements and that the element, x[10], is undefined.

- Pointer arithmetic is only valid for restricted operations, such as:

  ```
  ++pt_x;
  pt_z = pt_x - pt_y;
  if (pt_x < pt_y) { ... }
  ```

- For a one-dimensional array, an element a[i] is equivalent to *(&a[0]+i).

- The generic pointer type is **void***:

  ```
  void *pt;
  ```

  The address of any type can be assigned to this pointer:

  ```
  double x[100];
  pt = x;
  ```

but pointer arithmetic on `pt` is illegal:

```
++pt;                           // WRONG!
```

as is assigning `pt` to a pointer to a fundamental type:

```
double *pt_x = pt;      // WRONG!
```

- Multi-dimensional arrays are defined as in:

```
double x[2][4];
int i[2][5][4];
```

Indices are used to access such arrays:

```
y = x[0][1];
```

This means nothing more than successive applications of the subscripting operator and implies that two-dimensional arrays are stored by rows; an element, `x[i][j]`, is equivalent to:

```
*(&x[0][0] + 4 * i + j)
```

- Arrays can be initialized by comma-separated lists:

```
int i[] = {4, 5, 6};
double x[4] = {10.1, 10.5};
int w[3][2] = {{1, 2}, {0, -3}, {7, -5}};
```

- If an array is defined by:

```
double a[10];
```

then `sizeof(a)` gives the size of the entire array. However, the expression `sizeof(a[1])` gives the size of a single element.

- The statement:

```
double *pt_a[5];
```

defines an array of pointers, but:

```
double (*pt_a)[5];
```

defines a pointer to an array.

- The precedence and associativity of all C++ operators are given in Appendix B.

6.11 Exercises

1. Given a set of real numbers, $x_1, x_2, \ldots, x_N$, the mean, $\bar{x}$, is defined by:

$$\bar{x} = \frac{1}{N} \sum_{i=1}^{N} x_i$$

and the variance, Var, by:

$$\text{Var}(x_1 \ldots x_N) = \frac{1}{N-1} \sum_{i=1}^{N} (x_i - \bar{x})^2$$

and the standard deviation, σ, by:

$$\sigma(x_1 \ldots x_N) = \sqrt{\text{Var}(x_1 \ldots x_N)}$$

Write a program that calculates these three values from a list of numbers stored as a one-dimensional array. For test purposes you can use the `rand()` function, declared in `<cstdlib>`, to generate the list. (You should find that the mean is approximately 0.5.) Try lists of increasing length. Is there any point in generating more than `RAND_MAX` random numbers?

2. A χ^2 test is often used to check the validity of binned data and consists of evaluating:

$$\chi^2 = \sum_i \frac{(N_i - n_i)^2}{n_i}$$

where N_i is the number of observed items in bin i and n_i is the number of items expected in bin i. Write a program that has an integer array representing ten bins and uses the library function, `rand()`, to generate a sequence of random numbers in the range 0 to 999. If the generated number is in the range 0 to 99, then increment bin 1, if the number is in the range 100 to 199, then increment bin 2, and so on. Your program should send the value of χ^2 and the total number of items to the output stream.[8]

3. P and Q are polynomials in x and y, given by:

$$P = \sum_{i,j=0}^{3} p_{ij} x^i y^j \quad \text{and} \quad Q = \sum_{i,j=0}^{3} q_{ij} x^i y^j$$

where p_{ij} and q_{ij} are integers. Use two-dimensional arrays to add P and Q for various values of p_{ij} and q_{ij}. The output should be given in terms of a polynomial, rather than a list of coefficients.

4. Write a program that *initializes* the `int` array `a[2][3][4]`, with values corresponding to:

$$a_{ijk} = (1 - i)(2 - j)(3 - k)$$

by using nested, comma-separated lists. Elements that have the value zero should be initialized by default and *not* explicitly. Check your program by comparing the initialization actually achieved with the expected values of `a[i][j][k]`.

---

[8]An introduction to the χ^2 test is given in [14].

The following exercises require an understanding of Section 6.5.1:

5. The results of an experiment are stored in a five-dimensional array:

    `float data[SIZE_1][SIZE_2][SIZE_3][SIZE_4][SIZE_5];`

    One requirement is to find the average of all elements. Write two programs
    to calculate this average; one version should use the standard array notation,
    whereas the other should use a pointer. Compare the times taken by the two
    programs for various values of `SIZE_1`, ..., `SIZE_5`. Can you make the pointer
    version any faster? (See the comment on program timing given in Exercise 1 of
    Chapter 5.)

6. Use nested, comma-separated lists to initialize the array, `a[2][3][4]`, so that it
    corresponds to:

$$a_{0ij} = \begin{pmatrix} 1 & 2 & 3 & 4 \\ 5 & 6 & 7 & 8 \\ 9 & 10 & 11 & 12 \end{pmatrix}$$

    and

$$a_{1ij} = \begin{pmatrix} 13 & 14 & 15 & 16 \\ 17 & 18 & 19 & 20 \\ 21 & 22 & 23 & 24 \end{pmatrix}$$

    Using only pointer arithmetic and the array name (that is `a`, but *not* `a[0][0][0]`
    or `a[i][j][k]`), list the array elements and hence verify your method of access.

The following exercise requires an understanding of Sections 6.5.1 and 6.7, which you
may have decided to omit:

7. If the definition of an array is:

    `double y[2][3][4];`

    what is the type of `y[2]`? Verify your answer by assigning `y[2]` to a suitable
    pointer. What is the result of `sizeof(y[2])`?

Chapter 7

Further Pointer Techniques

Pointers are fundamental to many aspects of C++. In this chapter we introduce a number of more advanced topics concerning pointers. In particular, we consider strings, pointers as function arguments, pointers to functions, dynamic memory allocation and reference arguments. It should be emphasized that in this chapter we are *not* concerned with the more sophisticated string defined as a C++ template. We consider such "strings" in Chapter 17.

7.1 Strings

The basic idea of a *string* in C++ is the same as in C. A string is simply a one-dimensional array of type `char`, with the character constant `'\0'`, indicating the end of the string.[1] This string terminator is the crucial feature that distinguishes a string from a general `char` array; if the terminator is omitted, functions that manipulate strings usually fail disastrously. The disaster occurs through attempting to interpret a perfectly valid `char` array, such as `char c[3] = {'C', '+', '+'}`, as a string. The absence of a string terminator causes string manipulation functions to access memory beyond the end of the array.

The idea of a string constant (or string literal) was introduced in Section 2.5.5. A string constant consists of a sequence of characters enclosed by double quotes, as in:

```
"This is a string constant (or string literal)."
```

There is a special way of initializing a `char` array with a string constant. This takes the form:

```
char message[] = "Hello world";
```

and avoids the equivalent, but more tedious alternative:

```
char message[] = {'H', 'e', 'l', 'l', 'o', ' ',
                  'w', 'o', 'r', 'l', 'd', '\0'};
```

---

[1] `'\0'` is the null character and is represented by zero in the ASCII character set. The null character differs, of course, from the character `'0'`.

As can be seen, this array actually has 12 elements, since the last element is '\0'. The empty pair of square brackets tells the compiler to allocate an array of sufficient length to store the given string constant, together with the '\0' terminator. An alternative strategy is to define a **char** array of sufficient size to hold at least the required number of characters, as in:

```
char message[80] = "Hello world";
```

Notice that the array need not be initialized with the maximum number of characters.

Once again, you must be careful to distinguish between initialization and assignment. Suppose we now want to assign a new string to **message[]**. It is tempting to use:

```
message = "Hello";        // WRONG!
```

However, we cannot assign anything to **message** since it is an unmodifiable lvalue (the base address of the array). In order to assign a string to an array we have to assign characters to individual elements of the array, as demonstrated in the following program:

```
#include <iostream>
#include <cstdlib>        // For EXIT_SUCCESS
using namespace std;
int main()
{
    char message[20] = "0123456789123456789";
    message[0] = 'H';
    message[1] = 'e';
    message[2] = 'l';
    message[3] = 'l';
    message[4] = 'o';
    message[5] = '\0';
    cout << message << "\n";
    return(EXIT_SUCCESS);
}
```

Notice how **cout** recognizes the '\0' string terminator and simply sends the first five characters to the screen, even though the array can hold up to 80 characters.

Exercise

Try running the above program and find out what happens if:

(a) the array is initialized with 20, rather than 19, characters;

(b) the statement:

```
message[5] = '\0';
```

is omitted.

The role of the string terminator cannot be over-emphasized. As an example, suppose we didn't initialize the **message[20]** array and we accidentally omitted to assign the

'\0' string terminator to `message[5]`. Then `cout` would extract whatever characters happened to be stored in memory until it encountered something that could be interpreted as a string terminator. This could very well be outside the memory allocated to `message[20]`.

The above example shows us a lot about the structure of strings, but a much easier technique is to use an appropriate string manipulation function, declared in `<cstring>`. The function, `strcpy()`, has the declaration:

```
char *strcpy(char *string1, const char *string2);
```

and copies a string (including the terminator) from `string2` to `string1`. (Notice the direction in which the copy is made. It may not be what you might have expected!) The function, `strcpy()`, also returns the destination address of the string (that is `string1`). As usual, it is up to the programmer to ensure that `string1` is the address of sufficient allocated memory to hold the required characters, together with the '\0' terminator. The penalty for failure is to write over some arbitrary, but possibly important, area of memory.

We can now assign values to the array, `message[]`, as demonstrated in the following program:

```
#include <iostream>
#include <cstdlib>       // For EXIT_SUCCESS
#include <cstring>       // For strcpy()
using namespace std;
int main()
{
    char message[80];
    strcpy(message, "Hello");
    cout << message << "\n";
    return(EXIT_SUCCESS);
}
```

Again notice that we don't add the '\0' terminator, since it is part of the double quotes or string constant syntax.

There are many functions in the `<cstring>` group. For instance:

```
char *strcat(char *string1, const char *string2);
```

appends `string2` to the end of `string1`. The following program shows the `strcpy` and `strcat()` functions in use:

```
#include <iostream>
#include <cstdlib>       // For EXIT_SUCCESS
#include <cstring>       // For strcpy(), strcat()
using namespace std;
int main()
{
    char message[80];
    strcpy(message, "Hello");
    strcat(message, " world.");
```

```
        strcat(message, "\nHave a nice day.");
        cout << message << "\n";
        return(EXIT_SUCCESS);
    }
```

The contents of the `char` array are initially arbitrary and don't necessarily have the string terminator as the first element. For this reason it is important that `"Hello"` is copied to `message` rather than concatenated. Once `message` contains a string, the `strcat()` function can be used to append first the string `" world."` and then `"\nHave a nice day."`. Again `strcat()` takes care of any subtleties involving the `'\0'` terminators and the result is:

```
Hello world.
Have a nice day.
```

Now consider the following code fragment:

```
char *pt;
pt = "Hello world.";
```

or, equivalently:

```
char *pt = "Hello world.";
```

and compare it with:

```
char pt[] = "Hello world.";
```

It is tempting to think that the `*pt` and `pt[]` definitions are equivalent. However, in the first case memory is allocated to hold the string constant, a pointer is declared and the base address of this string is assigned to the pointer. The only new assignment that we can make is to change the address held by the pointer, since we cannot make an assignment to the string constant, `"Hello world."`. In the second case, we have an array of just sufficient length to hold the given string and this array is initialized to that string. It is possible to copy another string (provided it is not too long) to this array, just as we have done previously.

Any temptation to use a pointer assignment to a string constant as if it were an array definition should be resisted. Suppose we had the following:

```
char *string1 = "Hello world";
char *string2 = "world";
```

then a clever compiler could simply make `string2` point to the appropriate place in the `"Hello world"` string. A programmer making assignments to the memory pointed to by `string1` or `string2` would be rather surprised by the result.

Scientists, engineers and mathematicians don't usually spend a lot of their time performing complicated string manipulations; they are more interested in inverting matrices than writing word processing programs. The places where you are likely to want to use strings are creating filenames for various storage media and annotating output. Moreover, with more advanced C++ techniques it is possible to implement a string class with the advantages of safer and more natural string manipulations. (See Sections 9.6.2 and 17.2.4.)

7.2 Pointers as Function Arguments

As we discussed in Chapter 5, function arguments in C++ are normally passed by value; the function makes a copy of the value of each argument and then manipulates these copies. The return value of a function can be used to change a single variable in the calling environment, but often we want to change several or even many variables. For instance, suppose we need to write a function that swaps two integers. One way of doing this is to use function arguments to pass the addresses of the integers, since once a function has the addresses, it can directly manipulate the values stored in memory. Of course, this is just the kind of potentially disastrous situation that we discussed previously, so we must be careful to get such pointer manipulations exactly right. The implementation of a possible `swap()` function using pointers is contained in the following program:[2]

```cpp
#include <iostream>
#include <cstdlib>        // For EXIT_SUCCESS
using namespace std;

void swap(int *pt_x, int *pt_y)
{
    int temp;
    temp = *pt_x;
    *pt_x = *pt_y;
    *pt_y = temp;
}

int main()
{
    int i = 10, j = 20;
    swap(&i, &j);
    cout << "i = " << i << " j = " << j << "\n";
    return(EXIT_SUCCESS);
}
```

In this example `pt_x` and `pt_y` are of type pointer to `int`; in other words they can store the addresses of variables of type `int`. The two arguments of the `swap()` function in the calling environment must also be the addresses of objects of type `int`. The way in which the `swap()` function works is that the addresses of `i` and `j` are passed as arguments. The value stored at the address of `i` (that is 10) is assigned to `temp`. The value currently stored at the address of `j` (that is 20) is then assigned to the memory location of `i`. The final statement in the swap function assigns 10 (the value of `temp`) to the memory location of `j`. Hence the values stored in `i` and `j` are interchanged. Notice that the statement:

```cpp
*pt_x = *pt_y;
```

has a totally different effect from:

---

[2]Now that you understand more about strings, you may wish to use `'\n'` rather than `"\n"`. However, the space saved is very slight.

```
    pt_x = pt_y;
```

The latter is valid C++, but simply assigns one pointer to another. In this context, the address of j (which bears no relation to 10) would be assigned to `pt_x`. Since the scope of `pt_x` is limited to the function body and nothing further is done with `pt_x`, this syntactically correct statement would achieve nothing.

The technique of using pointer arguments to change values in the calling environment can clearly be extended to more complicated situations than simply swapping two numbers. The main disadvantage is that function modularity is eroded. Once the address of a variable is passed to a function, then that variable may be changed in a way that can only be discerned by examining the function body. For this reason, pointers should only be used as function arguments when it is really necessary. No purpose is served by using pointers as function arguments just to save trivial quantities of memory.

Exercise

Write a program that stores 10 different integers as an array of type `int`. Use the `swap()` function, as implemented in this section, to reverse the order of the numbers held in the array.

7.3   Passing Arrays as Function Arguments

As has already been pointed out, arrays are of vital importance in many numerical applications and functions are an essential technique for controlling the complexity of programs. It should therefore come as no surprise to learn that functions frequently need to access the elements of arrays.

7.3.1   One-dimensional Arrays

It is straightforward to pass one-dimensional arrays as function arguments. The basic point to keep in mind is that accessing an element of a one-dimensional array simply involves dereferencing a pointer. The array size is only needed for the array definition so that the compiler can allocate the appropriate memory. Once the memory has been allocated, it is entirely the programmer's responsibility to ensure that a pointer really does point to the allocated memory. Therefore, to pass a one-dimensional array as a function argument, it is only necessary to pass an address, since the array size is not relevant for accessing array elements, although it may be necessary in order to ensure that an element really is a member of the array.

Suppose we want to write a function to return the sum of a sequence of elements of an array. A suitable function is given below.

```
    double sum(double pt[], int n)
    {
        double temp = 0.0;
        for (int i = 0; i < n; ++i)
            temp += pt[i];
        return temp;
    }
```

The argument, n, in the function declaration is the number of elements that we want to sum. This does not need to be the same as the total number of elements in the array. The following illustrates using the sum() function to find the total value of all elements of an array:

```
double height[100], total_height;
// Assignment of values to height[] goes here.
total_height = sum(height, 100);
```

However, there is no need to start the summation at the low end of the array, as the following code fragment demonstrates:

```
double height[100], total_height;
// Assignment of values to height[] goes here.
total_height = sum(&height[10], 24);
```

This example sums 24 elements of the height[] array, starting with the element height[10]. Notice that we must pass an address as the first argument to sum; height[10] would merely pass a double precision floating point number, which would tell sum() nothing about where the array is stored (and would also be a compile-time error).

Exercise

Assign 1, 2, 3 . . . to the elements of the height[] array and then verify that total_height has the expected value. Do this for both of the previous code fragments.

In the function declaration for sum() we omitted the size of the one-dimensional array, and even if we did insert the size it would be ignored by the compiler. It is worth remarking that we can only omit the array size in a function header or declaration; elsewhere a compile-time error occurs. For example, the following declaration is not valid:

```
double height[];            // WRONG: array size not specified.
```

It is also important to understand that, although in the above example we can access pt as if it were an array, it is actually a pointer, rather than an array name. In fact we could simply use a pointer in the function declaration, as in:

```
double sum(double *pt, int n);
```

We can even assign an address to pt, as shown in the version of the sum() function given below.

```
double sum(double pt[], int n)
{
    double temp = 0.0, *pt_end;
    pt_end = pt + n;
    while (pt < pt_end)
        temp += *pt++;          // Increments pt.
    return temp;
}
```

Exercise

Replace the `sum()` function used in the previous exercise by this pointer version. Does your program give the same results for `total_height` as you obtained previously?

Using the array notation for function arguments can lead to confusion, as in:

```
double sum(double pt[1000])
{
    double temp = 0.0;
    int n = sizeof(pt) / sizeof(double); // WRONG!
    for (int i = 0; i < n; ++i)
        temp += pt[i];
    return temp;
}
```

In this attempt at a `sum()` function the idea was to work out the number of elements, n, in the array by dividing the size of the array by the size of the type `double`. What actually happens is that n is assigned the size of a pointer to a `double` divided by the size of a `double`. This is very different to what was intended!

Exercise

Replace the `sum()` function used in the previous exercise by this incorrect version. Does your program give the same results for `total_height` as obtained in the previous exercise? Find out what value is assigned to n.

7.3.2 Multi-dimensional Arrays

An Introduction

Passing multi-dimensional arrays as function arguments is fairly straightforward. The only feature to note is that a function needs access to the values of all sizes associated with the array indices, apart from the first. This is necessary so that the compiler can generate the correct storage map. Details of storage maps for multi-dimensional arrays were given in Section 6.5.1, which you may have decided to skip for the moment. The essential idea is that a multi-dimensional array is mapped by the compiler into a one-dimensional array corresponding to the computer memory. In order to generate this mapping, the compiler must have available the sizes corresponding to all array indices apart from the first. For instance, in the case of a two-dimensional array, a function needs the number of columns in order to index into the correct row. As an example, the following program shows a function that returns the trace of a 5×5 matrix (represented here as a two-dimensional array):

```
#include <iostream>
#include <cstdlib>        // For EXIT_SUCCESS
using namespace std;

double trace(double y[][5])
{
```

```
        double sum = 0.0;
        for (int i = 0; i < 5; ++i)
            sum += y[i][i];
        return sum;
    }

    int main()
    {
        double x[5][5];
        for (int i = 0; i < 5; ++i)
            for (int j = 0; j < 5; ++j)
                x[i][j] = (i + 1) * (j +1);
        cout << "Trace = " << trace(x) << "\n";
        return(EXIT_SUCCESS);
    }
```

Notice that it is the base address of the array that is passed as the argument to `trace()`. We cannot use:

```
    cout << "Trace = " << trace(&x[0][0]) << "\n";
```

since this would attempt to pass the address of a `double` to a function that has an argument type `double [][5]`. Neither can we use:

```
    cout << "Trace = " << trace(x[0][0]) << "\n";    // WRONG!
```

since `x[0][0]` is of type `double`.

More Advanced Features[†]

The declaration in the `trace()` function header given above could equivalently be written as:

```
    double trace(double (*y)[5]);
```

This declaration implies that `y` is a pointer to an array of 5 doubles but is probably more obscure and there is the danger of writing:

```
    double trace(double *y[5]);
```

Since the square brackets have a higher precedence than the dereferencing operator, this declares an array of 5 pointers to the type `double`, which is not what is required in the `trace()` function header.

We could alternatively define a `trace()` function with the header:

```
    double trace(double *pt);
```

rather than:

```
    double trace(double (*y)[5]);
```

We would then have to replace the function call by:

```
trace(&x[0][0]);
```

The advantage of the original function header is that within the function body we can use the standard array-subscripting notation. Since y points to an object with size 5 × sizeof(double), array subscripting automatically does the correct pointer arithmetic. For instance, (y+1) is the base address of row 1 of the array, rather than element 1 of row 0.

Sometimes, knowledge of how arrays are stored can considerably improve the speed of calculations. As an simple example, suppose we want to sum two matrices represented by 100 × 100 arrays. We could imagine having something like the following in our program:

```
double a[100][100], b[100][100], c[100][100];
// Assignments to a[i][j] and b[i][j] go in here.
sum(a, b, c);
```

A possible sum() function is given below.

```
void sum(double a[][100], double b[][100], double c[][100])
{
    for (int i = 0; i < 100; ++i)
        for(int j = 0; j < 100; ++j)
            c[i][j] = a[i][j] + b[i][j];
}
```

The disadvantage of this implementation is that a considerable amount of unnecessary arithmetic is done in calculating the address of each element since, for matrix addition, there is no advantage in scanning over rows and columns. A faster, but more devious, implementation is given by the following sum() function:

```
void sum(double *pt_a, double *pt_b, double *pt_c)
{
    double *pt_end = pt_c + 100 * 100;
    while (pt_c < pt_end)
        *pt_c++ = *pt_a++ + *pt_b++;
}
```

This version simply steps from one element in memory to the next. Notice that the argument types in the two versions are different. The second version of the function expects arguments that are pointers to double so it should be called as in the code fragment given below.

```
double a[100][100], b[100][100], c[100][100];
// Assignments to a[i][j] and b[i][j] go in here.
sum(&a[0][0], &b[0][0], &c[0][0]);
```

For very large matrices, the increase in speed obtainable through using such techniques can be significant.

Exercise

Write a program that assigns values to the elements of b[][] and c[][] and then verifies the result given by the sum() function.[3]

7.4 Arguments to main()

It is fairly common for a program to require the user to input values for parameters. For instance, we may want to write a program to list all primes up to some maximum, with the largest value to be tested being entered by the user. So far our programs have all prompted the user to enter any necessary values. However, an alternative would be to have a command line argument, so that we could enter the required values at the same time as typing the program name. For example:

```
prime 100
```

could generate all prime numbers less than 100. This technique can be achieved by using arguments to main().

It is part of the C++ language that the function main() can take the following form:

```
int main(int argc, char *argv[])
{
    // Code goes in here.
}
```

Since the two arguments are *formal* arguments, they could be given any names, but argc and argv[] are established conventions. The number of command line arguments, including the program name, is given by argc. For our prime example, argc would be 2. The second argument, *argv[], is an array of char pointers. These pointers actually point to strings, terminated in the usual way by '\0'. The first element of the array is special since argv[0] points to the program name as a string. For instance, in the above example the string would be "prime". Subsequent elements of the argv array point to the command line arguments; argv[1] is the first command line parameter as a string, argv[2] is the second and so on. In this example argv[1] points to the string "100". The following program does no more than print the command line arguments. Try it for different arguments.

```
#include <iostream>
#include <cstdlib>        // For EXIT_SUCCESS
using namespace std;

int main(int argc, char *argv[])
{
    cout << "The program name is: " << argv[0] << "\n\n";
    cout << "There are " << argc << " arguments\n\n";
```

---

[3]As for one-dimensional arrays, we place the appropriate pairs of empty square brackets after an array name appearing in the text, as in b[][]. This is to emphasize that we are dealing with an array, but does not imply that this is a valid syntax within code.

```
        for (int i = 0; i < argc; ++i)
            cout << "Argument " << i << " is " << argv[i] << "\n\n";
        return(EXIT_SUCCESS);
}
```

Notice that the arguments are strings rather than integers. As a consequence, our prime example must convert the string, "100", to the integer, 100. Fortunately, a library function exists that converts a string to an integer. The function declaration is in <cstdlib> and is:

```
    int atoi(const char *pt);
```

If we supply a string as the argument to this function, then the corresponding integer is returned.

Exercise

Try running the following program:

```
        #include <iostream>
        #include <cstdlib>        // For atoi()
        using namespace std;
        int main()
        {
            int i = atoi("12345");
            cout << "The integer is " << i << "\n";
            return(EXIT_SUCCESS);
        }
```

What happens if the string is replaced by "123x5"?.

Using the answer to Exercise 7 in Chapter 5, we can now write a prime program to take a command line argument. An example of such a program is given below.

```
    #include <iostream>
    #include <cmath>        // For sqrt()
    #include <cstdlib>      // For exit(), atoi()
    using namespace std;

    bool test_prime(long n);
    void list_primes(long n);

    bool test_prime(long n)
    {
        bool prime;
        if (n == 0L || n == 1L)
            prime = false;
        else if (n == 2L)
            prime = true;
        else if (!(n % 2L))
```

```
            prime = false;
    else {
        // add 0.5 to ensure round up:
        long i_end = static_cast<long>( 0.5 + sqrt(n));
        prime = true;
        for (long i = 3L; i <= i_end; i += 2L) {
            if (!(n % i)) {
                prime = false;
                break;
            }
        }
    }
    return prime;
}

void list_primes(long n)
{
    cout << "Primes up to " << n << " are:\n\n";
    if (n >= 2)
        cout << "2\n";
    for (long i = 3L; i <= n; i += 2L)
        if (test_prime(i))
            cout << i << "\n";
}

int main(int argc, char *argv[])
{
    if (argc != 2) {
        cout << "Usage:  prime <max>\n";
        exit(EXIT_FAILURE);
        }
    long n = atoi(argv[1]);
    list_primes(n);
    return(EXIT_SUCCESS);
}
```

If, after compiling and linking this program, the executable file is **prime** then, to list all primes up to 100 you would enter the command:

```
prime 100
```

Exercise

(a) Run the **prime** program with various values for the command line parameter. Does **test_prime()** deal appropriately with all possible arguments? If not, make suitable modifications.

(b) Modify the **prime** program so that it only lists the primes between two numbers specified by two command line arguments. For example, typing:

```
prime 100 300
```

would list all prime numbers between 100 and 300. You should include code that traps alternative inputs, such as:

```
prime 300 100
```

and takes appropriate action.

7.5   Pointers to Functions

It is fairly common to want to pass a function as an argument to another function. For instance, suppose we want to write a function, called `sum()`, that sums the first n values of another function. We could use something like the following:

```
double f(int m)
{
    // Definition of f() goes here.
}

double sum(int n)
{
    double temp = 0.0;
    for (int i = 0; i < n; ++i)
        temp += f(n);
    return temp;
}
```

This approach does work, but now suppose we want to do the summation for a number of different functions, `f()`, `g()`, `h()` .... We would have to write a different `sum()` function for each of the `f()`, `g()`, `h()` ... functions we wanted to sum. This could be avoided if we could specify the function as an argument to `sum()`, rather than have each `f()`, `g()`, `h()` ... function embedded in its own `sum()`. We now explain how to achieve this.

In C++ it is possible to have a pointer to a function. As an example, a pointer to a function that takes no arguments and returns a `double` is defined by:

```
double (*g)();
```

The slightly bizarre notation is necessary because the function operator, (), binds tighter than the indirection operator, *. (See Appendix B.) So if we wrote:

```
double *g();
```

this would actually be a declaration for a function that takes no arguments and returns a pointer to a `double`. Apart from the fact that this declaration is not our intention, the compiler would not allow us to do this inside a function body; we cannot declare a function inside another function.

So our modified `sum()` function could be written as:

```
double sum(double (*g)(int m), int n)
{
    double temp = 0.0;
    for (int = 0; i < n; ++i)
        temp += (*g)(i);
    return temp;
}
```

In the header for this function, g is declared to be a pointer to a function that takes an argument of type int. (The identifier, m, could be omitted.) Notice that since g is a pointer it must be dereferenced when it is used in the body of sum(). However, the usual function notation, as in:

```
        temp += g(i);
```

is also permitted in this situation as the two forms are defined to be equivalent. This is consistent with the name of a function being that function's address.

A complete program in which the sum() function invokes two different functions is given below.

```
#include <iostream>
#include <cmath>          // For pow()
#include <cstdlib>        // For EXIT_SUCCESS
using namespace std;

double f2(int i);
double f4(int i);
double sum(double (*g)(int m), int n);

double f2(int i)
{
    return 1.0 / (i * i);
}

double f4(int i)
{
    double temp = i * i;
    return 1.0 / (temp * temp);
}

double sum(double (*g)(int m), int n)
{
    double temp = 0.0;
    for (int i = 1; i < n; ++i)
        temp += (*g)(i);
    return temp;
}

int main()
```

```
{
    cout << "Approximations to the Riemann Zeta " <<
        "functions are:\n\n";
    cout << "Zeta(2) = " << sum(f2, 20) <<
        "\nA better approx. is " << M_PI * M_PI / 6.0 << "\n\n";
    cout << "Zeta(4) = " << sum(f4, 20) <<
        "\nA better approx. is " << pow(M_PI, 4) / 90.0 << "\n\n";
    return(EXIT_SUCCESS);
}
```

We have already come across the `pow()` library function in Section 5.9.2. It returns the value of `x` raised to the power `y`. Notice that in the above program we need to pass the address of the appropriate function as the first argument to `sum()`. We do this by using `f2` and `f4` without the usual function call operator, `()`. This syntax is analogous to `x` being the base address of an array, `x[]`.

Exercise

(a) Compile and run this program (which evaluates the Riemann Zeta function, $\zeta(n)$, for $n = 2$ and 4).[4]

(b) Replace `f2()` by a function that calculates $1/i!$, and `f4()` by a function that calculates $-(-1)^i/i$. Hence, use `sum()` to calculate the well-known constants e and $\ln 2$. Check these constants against `M_E` and `M_LN2` defined in `<cmath>`.

We can now use the pointer to a function technique to improve our root program, given in Section 5.9.2. The modified program is given below.

```
#include <iostream>
#include <cmath>          // For exp(), pow(), cos()
#include <cstdlib>        // For exit()
using namespace std;

// Function declarations:
double f(double x);
double root(double x1, double x2, double f1, double f2,
        int &depth);
double find_root(double (*g)(double), double x1, double x2,
        int &depth);

double f(double x)   // A solution of f(x) = 0 is required.
{
    return (exp(x) + pow(2.0, -x) + 2.0 * cos(x) - 6.0);
}

double root(double x1, double x2, double f1, double f2,
```

---

[4]The Riemann Zeta function is defined by $\zeta(s) = \sum_{k=1}^{\infty} k^{-s}$. See [11] for further details.

```
        int &depth)
// Finds a root of f(x) = 0 using bisection.
// Assumes that x2 >= x1 and f1 * f2 < 0.0
{
    const int max_depth = 50;
    const double x_limit = 1e-5;
    double estimated_root;
    double x_mid = 0.5 * (x1 + x2);
    if (x2 - x1 <= x_limit)
        estimated_root = x_mid;
    else if (++depth > max_depth) {
        cout << "WARNING: maximum limit of " << max_depth <<
            " bisections reached.\n";
        estimated_root = x_mid;
    }
    else {
        double f_mid = f(x_mid);
        if (f_mid == 0.0) {
            // Zero at x_mid.
            estimated_root = x_mid;
        }
        else if (f(x1) * f_mid < 0.0) {
            // Zero in first segment.
            estimated_root = root(x1, x_mid, f1, f_mid,
                depth);
        }
        else {
            // Zero in second segment.
            estimated_root = root(x_mid, x2, f_mid, f2,
                depth);
        }
    }
    return estimated_root;
}

double find_root(double (*g)(double), double x1, double x2,
    int &depth)
{
    double f1 = g(x1);
    double f2 = g(x2);
    if (f1 * f2 > 0.0) {
        cout << "Error in find_root():   " <<
            "end-points have same sign\n";
        exit(EXIT_FAILURE);
    }
    else if (x2 - x1 > 0.0)
        return root(x1, x2, f1, f2, depth);
```

```
    else
        return root(x2, x1, f2, f1, depth);
}

int main()
{
    int depth = 0;
    double x = find_root(f, 1.0, 2.0, depth);
    cout << "Root is " << x << " recursion depth = " <<
        depth << "\n";
    cout << "f(x) at root = " << f(x) << "\n";
    return(EXIT_SUCCESS);
}
```

The function for which we want to find a root is passed as an argument to `find_root()`, rather than being embedded in the body of `find_root()`. This means that it is now straightforward to implement root finding for a number of different functions. A further improvement on the original program is that since `depth` is defined in `main()` (rather than as a `static` within the body of `root()`) its value can be reset to zero. This would enable a single program to use the `root()` and `find_root()` functions to discover multiple roots.

Exercise

Compile and run this root-finding program for a variety of different functions. Some suggestions are:

(a) $f(x) = e^{2x} - x - 2$

(b) $g(x) = 1000x^3 - x - 1$

(c) $v(r) = (5.67 \times 10^6)e^{-21.5(r/\sigma)} - 1.08(\frac{\sigma}{r})^6$ where $\sigma = 4.64$.

   ($v(r)$ is an approximation to the potential energy between two helium atoms.)

Declarations involving pointers to functions often look quite complicated, as in:

```
double f(double (*p1)(double), double (*p2)(double),
        double (*p2)(double));
```

but a typedef can provide some simplification. For instance, we could declare a `typedef` for a pointer to a function taking a single `double` argument and returning a `double` by using the following statement:

```
typedef double (*PT_FUNC)(double);
```

This would enable us to write the declaration for `f()` as:

```
double f(PT_FUNC p1, PT_FUNC p2, PT_FUNC p2);
```

Exercise

Modify the root-finding program on page 168 by declaring a `typedef` for a pointer to a function taking a single `double` argument and returning a `double`. Use this `typedef` in place of `double (*p1)(double)` in the function arguments. Verify that your modified program gives the same result as the original version.

After working through this section you should understand our remark in Chapter 5, that to invoke a function, the syntax is:

```
f();
```

rather than:

```
f;
```

The latter statement, without any parentheses, achieves nothing since `f` is the address of the function; an analogous error for an array, `x[]`, is:

```
x;
```

7.6 Dynamic Memory Management

At the time of writing a program, the size of an array may not be known. For instance, the size might depend on how much data we manage to collect or how much memory is available on the particular computer running our program. One way of coping with these situations is to define array sizes by means of global constants, such as:

```
const int MAX_DATA_POINTS = 100;

double pressure[MAX_DATA_POINTS];
double height[MAX_DATA_POINTS];
double temperature[MAX_DATA_POINTS];
double humidity[MAX_DATA_POINTS];
```

If the complete program consists of more than one source file, then the array size definition is best placed in a header file so that modifying the header file automatically adjusts the array size throughout the files. However, each change necessitates re-compiling and re-linking the program, both of which can be very time-consuming processes. Also, it may not be clear whether an array is needed until the program is actually running. A further difficulty is that the useful lifetime of an array within a program may not be obvious at compile-time.

Fortunately, in C++ it is possible to allocate and deallocate memory while a program is running; such techniques are often known as *dynamic memory management*. A C++ program has access to an area of memory, commonly known as the *heap* or the *free store*. Requests can be made to allocate contiguous portions of this memory and, when no longer required, the memory can be returned to the free store. The operators that allocate and deallocate memory are called **new** and **delete** respectively. These operators hide the necessary book-keeping associated with dynamic memory management. For example, it is not necessary to calculate the number of bytes of memory required to store a particular object, nor it is necessary to keep a record of what areas of memory remain unallocated.

7.6.1   Allocating Memory

The new operator returns the address of sufficient memory to store the specified fundamental or derived type. If we want to allocate memory to store a single object, then the syntax is as demonstrated in the examples given below.

```
int *pt_i = new int;        // *pt_i can store an int.
float *pt_f = new float;    // *pt_f can store a float.
double *pt_d = new double;  // *pt_d can store a double.
```

We could, of course, define the pointer and invoke the new operator in two separate statements, as in:

```
double *pt;
pt = new double;
```

Accessing dynamically allocated objects is straightforward, except that since we have a pointer to an address, we need to dereference the pointer, as demonstrated by the following code fragment:

```
float *pt = new float;
*pt = 100.14;               // O.K. Assigns 100.14 to *pt.
cout << "*pt = " << *pt << "\n";
float x = pt;               // WRONG: pt is a pointer not a float;
```

In the above examples, the new operator allocates memory, but does no initialization. It is possible to allocate and initialize with the same statement by enclosing the initializing expression in parentheses, as illustrated below.

```
float *pt_pi = new float(3.14); // 3.14 assigned to *pt_pi.
```

Notice that pt_pi stores the address of the allocated memory, so it is the memory pointed to by pt_pi that is initialized, rather than pt_pi itself. It is possible for the initializing expression to be missing; this is equivalent to omitting the parentheses and consequently the memory stores an arbitrary initial value, as shown below.

```
float *sum = new float();   // O.K.
for (int i = 0; i < 10; ++i)
    *sum += i;              // WRONG: the initial value of
                            //        *sum may not be zero.
```

It is not meaningful to use the new operator to allocate memory for a function, but memory for a pointer to a function can be allocated. In Section 7.5 it was explained that:

```
double (*g)(void);
```

or

```
double (*g)();
```

defines a pointer to a function, having no arguments and returning a double. The following statement allocates memory for such a pointer:

```
double (**g)() = new(double (*)());
```

Notice that we have used a function-like syntax for **new**. The statement:

```
double (**g)() = new double (*)();        // WRONG!
```

is incorrect because operator precedence (see Appendix B) implies that this is actually equivalent to:

```
double (**g)() = (new double)(*)();        // WRONG!
```

which is not what was intended.

The need to dereference g twice, together with the multiple parentheses may be a bit hard to follow. However, the following program should clarify the ideas involved:

```
#include <iostream>
#include <cstdlib>        // For EXIT_SUCCESS
using namespace std;

double f()
{
    return 3.142;
}

int main()
{
    double (**g)() = new(double (*)());
    *g = f;
    cout << "pi = " << (**g)() << "\n";
    return(EXIT_SUCCESS);
}
```

The (trivial) function f() takes no arguments and returns a **double**. The first statement in **main()** allocates memory for a pointer to a function taking no arguments and returning a **double**. Consequently, g is a pointer to a pointer to a function. The second statement in **main()** then assigns the address of the function f() to the dynamically allocated memory. Finally, in order to invoke the function f() by means of the dynamically allocated memory, we need to dereference g twice; hence the (**g)() syntax.

If a program is doing a lot of dynamic memory management, it may run out of free store, in which case the **new** operator will not be able to allocate the required memory. Since this would be disastrous for any program, we need some way of detecting a memory allocation failure. The old method (in pre-ANSI C++) was for **new** to return a null pointer. Since it is guaranteed that the null pointer does not point to valid memory, a test for the null pointer is a test for memory allocation failure. The following code fragment demonstrates this idea:

```
float *pt = new float(3.1415926);
if (pt == 0) {
    cout << "Failed to allocate memory for pt.\n";
    exit(EXIT_FAILURE);
}
```

However, in ANSI C++ a memory allocation error "throws an exception", which is then "caught" by a Standard Library function. The result is that if a memory allocation fails, then the program terminates with a system-supplied error message.[5] From now on we will assume we are using an ANSI compliant C++ compiler and that any failure of dynamic memory allocation will throw an exception. If this is so, there is no point in testing for a null pointer because the program will terminate before the test can be made.

For fundamental types, such as `int`, `float`, `double` etc., there is usually little, if anything, to be gained by using the **new** operator to allocate memory for single objects. But derived types, such as arrays and class objects, may require a lot of memory and then the **new** operator can prove to be very useful. A class is a rather special derived type that will be described in Chapter 8. Here we consider dynamic memory allocation for arrays of the fundamental types.

The syntax for dynamically allocating memory for an array of objects is a bit different from that used for a single object. Instead of using **new**, the **new** [] operator is used to allocate memory for an array, as demonstrated by the following statements:

```
int *pt_i = new int[10];        // Array of 10 ints.
double *pt_d = new double[100]; // Array of 100 doubles.
char *pt_c = new char[10];      // Array of 10 chars.
int *pt = new(int[10]);         // Alternative syntax.
```

In each case, the base address of the array is assigned to a pointer. To access the array we can either explicitly dereference or, equivalently, use the standard array-subscripting notation, as illustrated by the following code fragment:

```
int *m = new int[10];
*(m+2) = 4;                     // Assigns 4 to element m[2].
m[9] = m[2];                    // Assigns 4 to element m[9].
```

Notice that square brackets are used to enclose the expression specifying the number of elements required in an array. For example, the following statement does not allocate an array of ten elements; instead it allocates memory for a single `int` and then initializes that memory by storing the value 10:

```
int *pt = new int(10); // Initializes *pt to 10, instead of
                       // requesting an array of 10 ints.
```

The **new** [] operator can also be used to allocate an array of pointers. In the following statement, the base address of memory, sufficient to store twenty pointers to type `double`, is assigned to `pt`:

```
double **pt = new(double *[20]);
```

Notice that `pt` is a pointer to a pointer to `double`. One advantage of dynamic memory allocation is that there is no need to know how much memory is required when the program is compiled. This is illustrated by the following code fragment:

```
int i;
// Assignment to i goes in here.
float **pf = new(float *[i]);
```

---

[5]A detailed description of exception handling is given in Chapter 15.

In this example, the `new` `[]` operator dynamically allocates memory to store i pointers to type `float`, using the run-time value of `i`.

Exercise

Modify the program that used ragged arrays (in Section 6.9.2) so that all arrays are allocated dynamically.

Multi-dimensional arrays can also be allocated dynamically. The following code fragment illustrates the syntax:

```
int i;
// Assignment to i goes here.
float (*x)[10][20] = new float[i][10][20];
```

This example allocates memory for an i × 10 × 20 three-dimensional array of type `float`. The sizes associated with all array indices, apart from the first, must be positive constants since the compiler needs to create a storage map, but the value of `i` need only be known at run-time. A dynamically allocated multi-dimensional array can be accessed by means of the standard notation (in this case `x[i][j][k]`), since `x` is actually a pointer to an array.[6]

Memory can also be dynamically allocated for an array of pointers to functions. The following statement assigns to `g` the base address of an array of ten pointers to functions, where each function returns a `float` and takes no arguments:

```
float (**g)() = new(float(*[10])());
```

Now that we have introduced the idea of dynamically allocated arrays, it is worth noting some important features:

- If, for example, we define an array by:

```
int m[10];
```

then `m` is an unmodifiable lvalue and we cannot assign to `m`. Consequently the following code is incorrect:

```
int m[10], *pt;
// Assignment to pt goes here.
m = pt;          // WRONG: cannot assign to a constant.
```

However, if the array is dynamically allocated, then `m` is of type pointer to `int` and can be used to hold any `int` address:

```
int *m, *pt;
m = new int[10];
// Assignment to pt goes here.
m = pt;          // O.K. (but probably unwise).
```

---

[6]Notice that x is *not* an array of pointers, but rather a pointer to an array. See Section 6.8.

Such assignments are often the result of a mistake since the base address of
a dynamically allocated array is a valuable piece of information; without this
information we cannot access or deallocate the array.

- Dynamically allocated memory remains allocated even when control passes out
  of the function where the **new** operator was invoked. To illustrate this, consider
  the following program:

```
#include <iostream>
#include <cstdlib>              // For EXIT_SUCCESS
using namespace std;

double *f(void)
{
    double *x;
    x = new double[20];    // Allocates memory.
    for (int i = 0; i < 20; ++i)
        x[i] = i;          // Assigns to elements.
    return x;              // Returns base address of array.
}                          //   x goes out of scope.

void g(void)
{
    double *pt;
    pt = f();                  // Assigns base address of allocated
                               // array to pt. (x is not in scope.)
    for (int i = 0; i < 20; ++i)
        cout << "x["<< i << "] = " << pt[i] << "\n";
}

int main()
{
    g();
    return(EXIT_SUCCESS);
}
```

The function f() dynamically allocates memory for an array of 20 elements of
type double. The base address for this memory is assigned to the pointer x. After
assigning numbers to the elements of x[], this function returns the value of the
base address of the dynamically allocated memory. At this point, the variable
x goes out of scope and the memory allocated to x may well be used to store
another object of a different type. However, the dynamically allocated memory
persists, as is demonstrated by the fact that we can successfully list the integers
stored in this memory.

- When the **new []** operator is used to allocate an array, initializers cannot be
  specified. Consequently, the following statement is not correct:

```
int *pt = new int[3](1, 7, 2);   // WRONG
```

Attempts at variations on the array initialization syntax of Section 6.6 are also invalid. For example, the following statement will not compile:

```
int *pt = new int[3]{1, 7, 2};  // WRONG
```

- The elements of dynamically allocated arrays have arbitrary initial values. Consequently, the following code fragment will compile but may give the wrong results:

```
float *velocity = new float[10];
for (int i = 0; i < 10; ++i)
    velocity[i] = 3.142 * i;
float *total_velocity = new float[10];
for (int i = 0; i < 10; ++i)
    total_velocity[i] +=
        velocity[i];          // WRONG: total_velocity[i] may
                              //        not be initially zero.
```

One minor restriction on the **new** and **new** [] operators is that the type cannot include the **const** specifier. Consequently, the following statement is not valid:

```
const int *pt_i = new const int[10];   // WRONG: const not allowed.
```

This is perfectly reasonable because `pt_i` simply stores whatever address is allocated at run-time.

7.6.2 Deallocating Memory

Once memory is allocated dynamically, then unless the memory is explicitly deallocated, the allocation is valid until the program terminates. The **delete** operator is used to perform this deallocation for single objects and the syntax is demonstrated by the following code fragment:

```
float *pt_f = new float;
int *pt_i = new int;
// Code accessing *pt_f and *pt_i.
delete pt_f;
delete pt_i;
```

A different notation is used for deleting arrays. Arrays are deallocated by using the `delete[]` operator, as shown below.

```
int *pt_i = new int[10];
double *pt_d = new double[100];
// Code accessing pt_i[i] and pt_d[i].
delete[] pt_i;
delete[] pt_d;
```

Notice that we don't specify the number of elements to be deleted since the the compiler keeps track of the array size.

It is important to use the correct version of the `delete` operator when deallocating memory. Using `delete` to deallocate an array, or using `delete[]` to deallocate a single object, will have unpredictable and possibly disastrous consequences. Examples of what not to do are given below.

```
int *pt = new int;
int *pt_array = new int[10];
// Code accessing pt and pt_array goes in here.
delete[] pt;             // WRONG: should use delete.
delete pt_array;         // WRONG: should use delete[].
```

In general, incorrect attempts to use the `delete` or `delete[]` operators may not be detected by the compiler. If you are lucky, you may get a run-time error message. At worst, the consequences of a bad deletion will only show up at some later point in the program execution and the cause of the failure will be difficult to locate. Typical errors, some of which will be trapped by a good compiler, are:

- Deleting memory that was not allocated by `new` is undefined, as in:

  ```
  int a[10];
  // Code accessing a[i] goes here.
  delete[] a;             // WRONG: a[10] was not allocated by new.
  ```

- Since a `const` object cannot be allocated by `new`, such objects cannot be deleted. An example is given below.

  ```
  const int a[4] = {100, 200, 300, 400};
  // Code accessing a[i] goes here.
  delete[] a;             // WRONG: a[] was not allocated by new.
  ```

- Deleting memory that has already been deallocated has undefined consequences, as in:

  ```
  int *p = new int[10];
  // Code accessing int[i].
  delete[] p;             // O.K.
  // More code.
  delete[] p;             // WRONG: memory already deallocated.
  ```

However, applying either the `delete` or `delete[]` operator to the null pointer does nothing since the null pointer is guaranteed not to point to a valid address. For this reason, some programmers always set a pointer to null after the memory associated with that pointer has been deallocated. This ensures that any subsequent accidental attempt to deallocate the same memory is harmless, as illustrated by the following:

```
int *p = new int[10];
// Code accessing int[i].
delete[] p;             // O.K.
```

```
p = 0;
// More code.
delete[] p;          // Harmless attempt to deallocate
                     // memory already deallocated.
```

- Once the dynamically allocated memory has been deallocated, that same memory may get allocated for some other object or the values stored in memory may even have been changed by the deletion. Consequently, memory must not be accessed once it has been deallocated and the following is incorrect:

```
int *p = new int[20];
// Code accessing p[i].
delete[] p;
cout << "p[0] = " <<
    *p << "\n";        // WRONG: memory deallocated.
```

Exercise

Use the `delete` operator to deallocate memory in the ragged array program developed for the exercise on page 175.

7.7   Pass by Reference and Reference Variables

7.7.1   Reference Arguments

In Chapter 5 it was emphasized that function arguments in C++ are normally passed by value; that is the function makes a copy of the value of each argument and then manipulates these copies. However, sometimes a function needs to make changes in the calling environment. One way of doing this was given in Section 7.2 where the `swap()` function used the dereferenced variables `*pt_x` and `*pt_y`. This use of pointer arguments to change values in the calling environment is notationally rather cumbersome and it is possible to overcome this inconvenience by specifying that some or all of the function arguments are to be passed by reference. The syntax used to denote *reference arguments*, as they are known, is demonstrated below.

```
void swap(int &x, int &y)
{
    int temp;
    temp = x;
    x = y;
    y = temp;
}
```

In this context the token, `&`, is known as the *reference declarator* and simply means that the function arguments are to be passed by reference.[7] There is no direct connection with the address-of operator and in no sense does taking the address of `x` give the `int`

---

[7]Notice that in this context the `&` token is *not* an operator but rather is used in the declaration of a type.

type. The way in which our new `swap()` function is invoked is rather different from and certainly more convenient than using the pointer notation. As the following program demonstrates, a call by reference looks just like an ordinary function call:

```
#include <iostream>
#include <cstdlib>         // For EXIT_SUCCESS
using namespace std;

void swap(int &x, int &y)
{
    int temp = x;
    x = y;
    y = temp;
}

int main()
{
    int i = 10, j = 20;
    swap(i, j);
    cout << "i = " << i <<" j = " << j << "\n";
    return(EXIT_SUCCESS);
}
```

As with pointers, pass by reference erodes the modularity of functions since in order to find out how variables are changed in the calling environment, we need to examine the function body. There is even more chance of unexpected changes because in the calling environment there is no indication that the arguments are actually passed by reference. Consider the code fragment given below.

```
int sum = 0;
for (int i = 0; i < 10; ++i)
    sum += dark_sheep_function(i);
```

The `dark_sheep_function()` may actually have very unpleasant side-effects. For example, the following definition of the function would cause the above loop to continue for ever:

```
int dark_sheep_function(int &i)
{
    --i;
    return i * i;
}
```

Due to the possibility of such effects, it is only worth using pass by reference when it is really necessary, such as when a function needs to access a large quantity of memory or to change variables in the calling environment.

One restriction on functions that use pass by reference is that they cannot be invoked with a temporary as an argument. For example, the following statement is illegal for our `swap()` function:

```
swap(10, 20);
```

In this case swapping 10 and 20 is clearly futile, but even a function that does not modify the value of a reference argument cannot be invoked with a temporary as an argument.

If a function argument that is passed by reference does not get modified by that function, then it is worth using the `const` specifier. For example, the following declares a function that uses pass by reference but doesn't change its argument:[8]

```
void f(const float &x);
```

You will find that you are forced to be consistent and to introduce the specifier at a lower level. For instance, the following code segment does not compile:

```
void g(float &x);
void h(float *pt);

void f(const float &x)
{
    g(x);                // WRONG!
    h(&x);               // WRONG!
    float *pt = &x;      // WRONG!
    // Use pt here.
}
```

This is because the function `g()` uses pass by reference and the function `h()` has a pointer as an argument. Either of these functions could attempt to modify `x`. Moreover, assigning the address of `x` to the pointer `pt` would open up the possibility of changing the value of `x` by subsequently dereferencing `pt`. However, we can tell the compiler that none of these possibilities can happen by using the following code:[9]

```
void g(const float &x);
void h(const float *pt);

void f(const float &x)
{
    g(x);
    h(&x);
    const float *pt = &x;
    // Use pt here.
}
```

One advantage of the `const` specifier for functions using pass by reference is that invoking such functions with a temporary as an argument *is* legal; for instance:

```
f(1.414);
```

---

[8]This example is only for illustration since it is not actually worth using pass by reference for a `const` fundamental type. More realistic examples involve class objects, which are introduced in Chapter 8.

[9]You would of course need to supply function definitions for `g()` and `h()` somewhere.

is valid with the above declaration for f().

Exercise

Use pass by reference to implement and test a function that interchanges the values of three variables, x, y and z, so that $x \leq y \leq z$. The three variables should have the type double.

7.7.2   Reference Return Values

It is also possible for functions to return a reference. For instance:

```
double &component(double *vector, int i)
{
    return vector[i];
}
```

returns the reference, vector[i]. The fact that the function returns a reference is specified by the & in the function header. However, the way in which such a function is invoked is no different from our previous use of functions. For instance, an example of invoking the component() function is given below.

```
double x[10], y;
x[4] = 72.3;
y = component(x, 4);            // Assigns 72.3 to y.
```

What is new is that we can also put the component() function on the left-hand side of an assignment, as demonstrated by the following statement:

```
component(x, 5) = 34.7;        // Assigns 34.7 to x[5]
```

This is something we can only do by returning a reference.

So far, our component() function is too much like the C++ array element to be useful, but suppose we define the function as follows.

```
double &component(double *vector, int i)
{
    return vector[i-1];
}
```

We can now access the components by indexing from 1 rather than 0, as demonstrated below.

```
double x[10], y;
component(x, 10) = 72.3;       // Assigns 72.3 to x[9].
y = component(x, 10);          // Assigns value of x[9] to y.
```

By using more advanced techniques, it is possible to write a version of component() that uses a less cumbersome notation. Such techniques are introduced in Chapter 9.

Exercise

Write a simple program to verify that it really is x[9] that gets assigned the value 72.3 in the above code fragment.

You must be careful when returning a reference, since effectively an address is passed back to the calling environment. This means that you must return something that has scope outside the function. It is no good returning a reference to an object that is defined inside the function, since that memory may well be reused when control leaves the function. For instance, consider the two functions given below.

```
double &square(double x)
{
    double temp = x * x;
    return temp;          // WRONG: temp goes out of scope outside
                          //        the function body.
}

double &new_component(double *p, int i)
{
    return(2 * p[i]);     // WRONG: cannot return a reference
                          //        to an expression.
}
```

In the `square()` function, the `temp` variable goes out of scope when control is passed back to the calling environment. The memory used by `temp` may then be used for something completely different. The error in the `new_component()` function is very similar since an implicit temporary variable is created for the expression. Such mistakes in using reference returns are similar to returning a pointer to a local array, as in the following code fragment:

```
double *array(int i)
{
    double x[10];
    // Assignments to x[] go here.
    return &x[i];         // WRONG! x[] goes out of scope outside
                          //        the function body.
}
```

However, this type of error is probably more obvious.

7.7.3 Reference Variables

Use of the reference declarator is not restricted to functions. Consider the following code fragment:

```
int x;
int &y = x;
x = 10;
cout << "y = " << y << "\n";
```

In this example, y is known as a reference variable. The statement:

```
int &y = x;
```

means, in effect, that y is an alias for x. There is a single memory location that is allocated by the statement:

```
int x;
```

The value for x and y is held in this same memory location and changing the value of one changes the other, as demonstrated in the following program. (Try it!)

```
#include <iostream>
#include <cstdlib>        // For EXIT_SUCCESS
using namespace std;

int main()
{
    int x;
    int &y = x;
    x = 10;
    cout << "y = " << y << "\n";
    y = 200;
    cout << "x = " << x << "\n";
    return(EXIT_SUCCESS);
}
```

This use of the reference declarator is consistent with its use in function arguments. In our swap() function example on page 180, x and y are reference variables or aliases for the actual variables supplied when the function is called (i and j in the example given). However, in the above example, the way in which changing the value of x changes y (and vice versa) is confusing; it would be far better to remove the redundant variable. Indeed, there does not seem to be much purpose in having reference variables, except in the context of function arguments and **return** statements.

Two restrictions on reference variables are that they must be initialized at the time of declaration and that a reference variable cannot subsequently be changed to refer to a different object. The following code fragment demonstrates both of these errors:

```
int x, y;
int &a = x;     // O.K.
int &b;         // WRONG: no initializer.
&a = y;         // WRONG: attempt to re-initialize reference.
```

7.8   Using Pointers, Arrays and Strings

In this section, we develop two programs that demonstrate most of the ideas introduced in this chapter. The first program implements matrix addition and the second implements an alphabetic sort.

7.8.1 Matrix Addition

Suppose we want to add two $m \times n$ matrices, A and B, whose elements are real numbers. Mathematically we would write this as:

$$C = A + B$$

where the elements of C are given by:

$$c_{ij} = a_{ij} + b_{ij}$$

In C++ it would be natural to store the matrices as two-dimensional arrays, as in the following declarations:

```
double a[MAX_ROWS][MAX_COLS];
double b[MAX_ROWS][MAX_COLS];
double c[MAX_ROWS][MAX_COLS];
```

However, with this approach we are forced to decide on the maximum sizes of the matrices before the program is compiled.

Another point worth bearing in mind, it that we might be tempted to implement matrix addition by something like the following:

```
for (int i = 0; i < m; ++i)
    for (int j = 0; j < n; ++j)
        c[i][j] = a[i][j] + b[i][j];
```

However, this is an inefficient implementation since each access of a two-dimensional array, a[i][j], implicitly uses pointer arithmetic of the form:

```
*(&a[0][0] + i * MAX_COLS + j)
```

It would be more efficient to use pointers to scan directly through the memory used to store the matrices.

Finally, it is also worth noting that array indices in C++ take the values 0, 1, 2 ..., rather than 1, 2, 3 ..., as is often more natural for scientific applications.

So our project is to write a program to try out matrix addition. A possible program is given below.

```
#include <iostream>
#include <cstdlib>        // For exit()
using namespace std;

// Function declarations:
double *create_matrix(int rows, int columns);
double &element(double *pt_matrix, int columns, int row,
    int column);
void add_matrices(double *pt_result_matrix, double *pt_matrix_a,
    double *pt_matrix_b, int rows, int columns);
void print_matrix(double *pt_matrix, int rows, int columns);
```

```
double *create_matrix(int rows, int columns)
// Dynamically allocates matrix.
{
    double *pt = new double[rows * columns];
    return pt;
}


double &element(double *pt_matrix, int columns, int row,
    int column)
{
    return *(pt_matrix + (row - 1) * columns + column - 1);
}


void add_matrices(double *pt_result_matrix, double
    *pt_matrix_a, double *pt_matrix_b, int rows, int columns)
{
    double *pt_end = pt_result_matrix + rows * columns;
    while (pt_result_matrix < pt_end)
        *pt_result_matrix++ = *pt_matrix_a++ + *pt_matrix_b++;
}


void print_matrix(double *pt_matrix, int rows, int columns)
{
    for (int i = 1; i <= rows; ++i) {
        for (int j = 1; j < columns; ++j) {
            cout << element(pt_matrix, columns, i, j) << " ";
        }
        if (columns != 1)
            cout << element(pt_matrix, columns, i, columns)
                << "\n";
    }
    cout << "\n";
}


int main(int argc, char *argv[])
{
    if (argc != 3) {
        cout << "Usage:  mult <rows> <columns>\n";
        exit(EXIT_FAILURE);
    }
    int rows = atoi(argv[1]);
    if (rows < 1) {
        cout << "Rows in matrix = " << rows <<
            ".\nMust have:  0 < rows.\n";
        exit(EXIT_FAILURE);
    }
    int columns = atoi(argv[2]);
```

```
            if (columns < 1) {
                cout << "Columns in matrix A = " << columns <<
                    ".\nMust have:  0 < columns.\n";
                exit(EXIT_FAILURE);
            }

            // Create matrices:
            double *pt_a = create_matrix(rows, columns);
            double *pt_b = create_matrix(rows, columns);
            double *pt_c = create_matrix(rows, columns);

            // Fill matrices:
            for (int i = 1; i <= rows; ++i) {
                for (int j = 1; j <= columns; ++j) {
                    element(pt_a, columns, i, j) = i * (j + 1);
                    element(pt_b, columns, i, j) = (i + 2) * (j + 3);
                }
            }

            // Print matrices:
            cout << "Matrix A:\n";
            print_matrix(pt_a, rows, columns);
            cout << "Matrix B:\n";
            print_matrix(pt_b, rows, columns);

            // Add matrices:
            add_matrices(pt_c, pt_a, pt_b, rows, columns);

            // Print results:
            cout << "Result of adding matrices:\n";
            print_matrix(pt_c, rows, columns);

            return(EXIT_SUCCESS);
        }
```

The way this program works is that command line parameters are passed to `main()` by using the function arguments. If the compiled program is in a file called `add` (or whatever is required on your system) then:[10]

    add <rows> <columns>

causes the program to start executing. For example, if we entered the following at the command line prompt, then the program would test addition for 4×5 matrices:

    add 4 5

The first thing that `main()` does is to check that `argc` is equal to three; if it is not, then the correct syntax has not been used and there is no point in proceeding further.

---

[10]Throughout this book, command parameters are delimited by <>. This should not be confused with the notation for templates, introduced in Chapter 16.

If `argc` is correct, the C++ library function, `atoi()`, is used to convert the `argv[]` strings to integers, which are then checked to ensure they have appropriate values.[11] Next, we dynamically allocate three contiguous areas of memory with sufficient size to store data for the three matrices. In order to test addition, two matrices are filled with data corresponding to:

$$a_{ij} = i(j+1)$$

and:

$$b_{ij} = (i+2)(j+3)$$

by using the `element()` function. The elements of these matrices are then sent to the output stream. Finally, the matrices are added and the result is also sent to the output stream.

Having given an overall picture of what the program actually does, we can now look at the remaining functions. The `create_matrix()` function returns the address of dynamically allocated memory for a `rows` × `columns` matrix. The `element()` function uses a reference return statement so that assignments can be made to matrix data. Notice that indices (called `row` and `column` in the `element()` function) take the values 1, 2, 3 ..., rather than 0, 1, 2 .... Next, have a look at the `add_matrices()` function. Notice that it uses pointers rather than array indices and that it scans the one-dimensional memory space rather than the two-dimensional space of matrix rows and columns. Finally, the `print_matrix()` function makes straightforward use of the `element()` function to print the matrix data.

There are many improvements that could be made to this program by using techniques that are introduced in subsequent chapters. Such improvements include:

- defining a matrix by the statement:

    ```
    matrix a(rows, columns);
    ```

- assignment to the ijth matrix element by statements such as:

    ```
    a(i, j) = i * (j + 1):
    ```

- printing a matrix, a, by:

    ```
    cout << a;
    ```

- adding matrices by:

    ```
    c = a + b;
    ```

Exercise

Compile and link the program described in this section. Check that the correct answers are given for various rows and columns. Also check the answers when other values are assigned to the matrix elements.

---

[11] We have already met the `atoi()` function in Section 7.4.

7.8.2 An Alphabetic Sort

Earlier in this chapter, we remarked that engineers and scientists don't usually spend much time manipulating strings. However, just as an exercise, let's assume we want to write a program that will produce an ordered list of words. The input to our program is to be a list of words, each entered on a separate line. We will assume that the words must be converted to lower case and listed alphabetically. We will also impose the condition that we don't want the program to be restricted in either the number or length of the words. The number of words in the list is to be entered as a command line argument.

As an example, of what we expect the program to do, if we enter:

```
Inheritance
new
delete
class
object
overload
address
Constructor
inline
private
```

then the result should be:

```
address
class
constructor
delete
inheritance
inline
new
object
overload
private
```

The complete program for this project is given below.

```cpp
#include <iostream>
#include <cstdlib>      // For exit(), atoi()
#include <cstring>      // For strlen(), strcmp(), memcpy()
#include <cctype>       // For tolower()
using namespace std;

// Function declarations:
void get_words(char **pt_words, int words);
void list_words(char **pt_words, int words);
void words_to_lower_case(char **pt_words, int words);
void word_to_lower_case(char *word);
void bubble_sort(char **pt_data, int elements,
```

```cpp
        void (*order)(char**));
void order_strings(char **pt);

void get_words(char **pt_words, int words)
{
    char buffer[128];
    for (int entry = 0; entry < words; ++entry) {
        cin >> buffer;
        int length = 1 + strlen(buffer);
        pt_words[entry] = new char[length];
        memcpy(pt_words[entry], buffer, length);
    }
}

void list_words(char **pt_words, int words)
{
    cout << "\n";
    for (int i = 0; i < words; ++i)
        cout << pt_words[i] << "\n";
}

void words_to_lower_case(char **pt_words, int words)
{
    for (int i = 0; i < words; ++i)
        word_to_lower_case(pt_words[i]);
}

void word_to_lower_case(char *pt)
{
    while (*pt) {
        *pt = tolower(*pt);
        ++pt;
    }
}

void bubble_sort(char **pt_data, int elements,
    void (*order)(char**))
{
    int n = elements - 1;
    for (int i = 0; i < n; ++i)
        for (int j = n; j > i; --j)
            order(pt_data + j - 1);
}

void order_strings(char **pt)
{
    char *temp;
```

```
        if (strcmp(pt[0], pt[1]) > 0) {
            temp = pt[0];
            pt[0] = pt[1];
            pt[1] = temp;
        }
    }
}

int main(int argc, char *argv[])
{
    if (argc != 2) {
        cout << "Usage:  my_sort <number of words>\n";
        exit(EXIT_FAILURE);
    }
    int words = atoi(argv[1]);
    if (words < 1) {
        cout << "Cannot enter " << words << " words\n";
        exit(EXIT_FAILURE);
    }
    char **pt_words = new char *[words];

    get_words(pt_words, words);
    cout << "\nOriginal list:\n";
    list_words(pt_words, words);

    words_to_lower_case(pt_words, words);
    cout << "\nLower case list:\n";
    list_words(pt_words, words);

    bubble_sort(pt_words, words, order_strings);
    cout << "\nOrdered lower case list:\n";
    list_words(pt_words, words);

    return(EXIT_SUCCESS);
}
```

As a first step in understanding this program, have a look at the function main().
After checking the command line parameters, main() finds the number of words to be
sorted by converting argv[1] from a string to an integer by using the library function
atoi().[12] The function main() next gets the words from the input stream, lists the
words, converts them to lower case and lists them again. Then the list is sorted using
a *bubble sort* and the sorted list is displayed.

Next consider the get_words() function.[13] In order to allow for an arbitrary number of arbitrary length words, we store the characters in a ragged array. (See Section 6.9.2.) The new operator is used to allocate an array of pointers, just sufficient to

---

[12]See Section 7.4.

[13]This implementation is not very robust since it does not properly handle cases where the value of
words is inconsistent with the number of words actually in the list. Techniques for overcoming such
limitations are considered in Chapter 18.

store all of the base addresses of strings, which in turn store the entered words. Each word is first copied into a temporary buffer, which can store a word of up to 127 characters. (Although this wouldn't cope with arbitrary length words, 127 characters should be more than sufficient for any genuine word.) The C++ library function, `strlen()`, declared in <cstring> as:

```
size_t strlen(const char *string);
```

returns the length of a string, excluding the '\0' terminator, and knowing this length enables us to dynamically allocate sufficient memory for the string. We then copy the string from the buffer into this memory, using the library function, `memcpy()`. This function is declared as:

```
void *memcpy(void *pt_1, const void *pt_2, size_t count);
```

and copies the number of characters specified by `count` from the memory pointed to by `pt_2` to that pointed to by `pt_1`. The function returns `pt_1`.

The `list_words()` function is very straightforward and shouldn't need any further explanation. The `words_to_lower_case()` function simply invokes the function, `word_to_lower_case()`. This function uses the library function, `tolower()`, to convert any upper case character to lower case. This library function is declared in <cctype> as:

```
int tolower(int c);
```

and returns the value corresponding to a lower case character for a value of c corresponding to a character of any case. Notice how `word_to_lower_case()` uses a pointer instead of an index to step through the characters of a string. Also, notice that the `while` loop terminates when the end-of-string character, '\0', is reached.

We use a bubble sort to arrange the words in alphabetical order. A bubble sort sweeps through an array, comparing successive pairs of elements. If a pair is out of order, then the elements are swapped. Here we use a pointer to a function (called `order`) to do the comparison and reordering. This would allow us to invoke different functions for different data types, ensuring that we could reuse the `bubble_sort()` function in other programs. To be sure of getting all of the elements in the correct order, we must sweep through the array n times, where n is one less than the number of elements. As a result, a bubble sort is very inefficient, the time taken being $\mathcal{O}(n^2)$.

In the `main()` function, `bubble_sort()` is invoked with the address of a function called `order_strings()` as the third argument. Notice how `order_strings()` uses the library function, `strcmp()`, to compare two adjacent words and returns a value greater than zero if the words are out of order. The amount of copying is reduced by swapping the addresses of the strings within the `pt_words[]` array, rather than the strings themselves. The argument of `order_strings()` is the address of the element in `pt_words[]` that gives the address of the first word of the pair to be sorted. The `strcmp()` function is declared in <cstring> as:

```
int strcmp(const char *string_1, const char *string_2)
```

and compares two strings, character by character. The function returns zero if the two strings are identical.

If the compiled program is in a file called `my_sort`, then we can run the program by typing:[14]

```
my_sort 10
```

at the system prompt, and then entering the ten words that we want to sort (one per line). Having to type in the word list every time we want to run the program isn't very convenient. It would be much better if we could read the data from a file, and we will learn how to do this in Chapter 18. In the meantime, depending on your system, you may be able to use input and output indirection. Using an editor (*not* a word processor) put your ten words (one per line) in a file called `unsorted`. Then try typing the following at the system prompt:

```
my_sort 10 < unsorted > sorted
```

The `<` symbol implies that the input is to be taken from the `unsorted` file and the `>` symbol similarly means that the output is to be placed in the `sorted` file. You can then look at the contents of `sorted` by using a text editor. Of course, you are not limited to ten words, so try the program with other inputs.

Exercise

A bubble sort is so-called because elements of an array are "bubbled" into the correct order. Modify the above program so that you can observe this happening as the sort proceeds, by sending the partially sorted lists to the output stream.

7.9 Summary

- A string is a one-dimensional array of type `char` and is terminated by the `'\0'` character. A `char` array can be initialized by a string constant (a sequence of characters inside double quotes), as in:

```
char message[] = "Hello world";
char another_message[80] = "Hello";
```

Notice the difference between a string constant and an array of type `char`.

- Use the string functions, declared in `<cstring>`, to carry out manipulations such as string copying:

```
strcpy(another_message, "Bye");
```

but make sure that you allocate enough memory. In particular, don't forget to allow for the `'\0'` end-of-string terminator.

- Declaring functions with pointer arguments, as in:

---

[14]It isn't a good idea to call your program "sort". This is because for some systems, such as UNIX, there already exists a command with this name.

```
void swap(int *pt_x, int *pt_y);
```

enables a function to change values in the calling environment. This is both useful and potentially dangerous.

- Arrays can be passed as function arguments. Examples are:

```
double sum(double a[], int n);
double trace(double a[][5]);
```

All array sizes, apart from the first, must be known at compile-time.

- Command line arguments are implemented by using arguments to the function main(), as in:

```
int main(int argc, char *argv[]) { // ... }
```

- For a function, f(), f is the address of the function. It is possible to define a pointer to a function. For example, a pointer to a function that takes no arguments and returns a double is defined by:

```
int (*pt)();
```

- Memory for a single object is dynamically allocated by using the new operator, as in:

```
int *pt_i = new int;
```

and deallocated by the delete operator:

```
delete pt_i;
```

- Memory for an array of objects is dynamically allocated by using the new [] operator, as in:

```
int *pt_i = new int[10];
```

and deallocated by the delete[] operator, as in:

```
delete[] pt_i;
```

- The reference declarator, &, has three related uses:

  - A function can change variables in the calling environment by using pass by reference, as in:

    ```
    void swap(int &x, int &y);
    ```

  - Returning a reference, such as:

    ```
    double &component(double *vector, int i);
    ```

enables a function to be the *left* operand of an assignment operator. For example:

```
component(x, 5) = 550.0;
```

– Reference variables create aliases to the same storage location:

```
double x;
double &y = x;
```

but such aliases are confusing and it is better to avoid them.

7.10 Exercises

1. Modify the program developed in Section 7.8.1 so that it implements and tests a matrix multiplication function.

2. Implement a function to transpose a matrix. The transposition should overwrite the original matrix.

3. Modify the sort program so that it sorts a list of integers into ascending order.

4. The bubble sort in Section 7.8.2 continues comparing values even if an array is in order after the first few passes. Write a version without this inefficiency.

5. Write a program that accepts a single word as a command line parameter and then searches a list of lower case words. The program should print:

```
<n> OCCURRENCES OF <word>
```

or

```
NO OCCURRENCES OF <word>
```

depending on the result of the search. (You may wish to write a function that generates a list of words for test purposes. In practice, the list would be read from a file.)

Modify your program so that the list is sorted before being searched and the search is terminated as soon as possible.

6. A bubble sort is *very* inefficient. A much better method, invented by C. A. R. Hoare, is known as *Quicksort*, the time taken being typically $\mathcal{O}(n \log n)$ rather than $\mathcal{O}(n^2)$.[15] Quicksort is part of the ANSI C++ Library, where it is implemented by a function called qsort(). (This function is declared in <cstdlib>.) Use qsort() to sort a list of random integers. Compare the time taken to sort the same list, using both a bubble sort and Quicksort. Do the times have the expected dependency on the list size? (In order to be able to measure the times on a modern computer you will either have to use very long lists or do the sort many times.)

---

[15]The method is too complicated to describe in an exercise, but an explanation is given in [14].

7. The Newton–Raphson technique was described in Exercise 4 of Chapter 4, but now that we have introduced pointers to functions, we can implement a more general version. Recall that given a function, $f(x)$, the technique attempts to find a zero of the function by successive iterations of the form:

$$x_{i+1} = x_i - \frac{f(x_i)}{f'(x_i)}.$$

Write a function to implement this technique. The function should have the declaration:

```
double newton(double x_lower, double x_upper,
    double accuracy, void (*f_pt)(double *f_value,
    double *f_derivative, double x));
```

where x_lower and x_upper are the lower and upper limits of an interval within which a root is known to occur. f_pt is a pointer to a function that, given x, calculates $f(x)$ and $f'(x)$, corresponding to *f_value and *f_derivative respectively. The function, newton(), returns a root of $f(x)$ when $|f(x)|$ is within the specified accuracy.

For some functions, the Newton–Raphson technique is very powerful, since it converges quadratically to the root. However, because the method can go badly wrong, you should include code to check whether the iteration has jumped out of the specified interval and also put an upper bound on the number of iterations.

Try your Newton–Raphson implementation for the functions:

(a) $f(x) = e^x + 2^{-x} + 2\cos x - 6$ in the range $1 \le x \le 2$.

(b) $f(x) = x - \cos x$ in the range $0 \le x \le \pi/2$.

In each case, try various values for the accuracy parameter and list successive approximations to the roots. How does the speed of convergence compare with the bisection method described in Section 5.9.2? Try both methods for other functions.

Chapter 8

Classes

Classes are an essential feature of object-oriented programming techniques and can help to control the complexity of application programs. A class is a user-defined type (sometimes called an *abstract data type*) that has its own collection of data, functions and operators.[1] Various levels of data hiding are provided and these help to create an interface to the class, hiding its implementation.

8.1 Declaring Classes

The basic syntax for a class declaration is:

```
class class_name {
    // Class body goes here.
};
```

Notice the terminating semicolon. The `class` keyword introduces the class declaration and `class_name` is the user-defined name for the class. The class body within the braces can consist of variable declarations, together with function definitions and declarations. Variables, functions etc. declared within a class body are known as *members* of the class. More specifically, variables declared within a class are often called *data members* and functions are known as *member functions*. It is also possible, by means of overloading, to give operators a special meaning within the context of a class. We leave any discussion of operator overloading until the following chapter.

As a particular example, let us suppose that we are working on a project that uses spheres of various sizes and radii. The project may well use a variety of other shapes but, in order to concentrate on the C++ language aspects of our project, we will just consider spheres for most of this chapter. We start with the following declaration of a `sphere` class:

```
class sphere {
    double x_centre, y_centre, z_centre, radius;
};
```

---

[1]More accurately, a class is an implementation of an abstract data type. A good discussion of this distinction is given in [7].

This gives us a user-defined type, analogous to the fundamental types, such as `int`, `double` etc. However, we know that the type `int` does not itself allocate memory for any `int` objects; this is done by defining instances of the type, as in the following statement:

```
int i, j, k;
```

In exactly the same way, the `sphere` class does not allocate memory, but rather it declares a type. However, we can define instances of the class and these do allocate memory. For example, the following statement defines `sphere1`, `sphere2` and `sphere3`:

```
sphere sphere1, sphere2, sphere3;
```

Such instances of a class are known as *objects*. These are *the* objects of object-oriented programming. In this example, each object stores its own radius, together with the Cartesian coordinates of its centre.

It is possible to define objects by means of the same statement as the class declaration, as in:

```
class sphere {
    double x_centre, y_centre, z_centre, radius;
} sphere1, sphere2, sphere2;
```

However, it is usually worth keeping all the class declarations together (in a header file) and separate from objects. The class declaration is an extremely valuable self-documenting feature of C++ and, as such, is best kept as uncluttered as possible.

As well as defining single objects, it is straightforward to define arrays of objects. Two examples of arrays of objects are given below.

```
sphere small_array[10], big_array[100][100];
```

Of course, the information contained in each of these arrays of objects could be stored without defining a `sphere` class. For instance, the `big_array[][]` data could be stored in four separate arrays defined by the statements:

```
double x_centre[100][100];
double y_centre[100][100];
double z_centre[100][100];
double radius[100][100];
```

This approach has two disadvantages compared with using classes. Firstly, we have to invent different names for all four variables associated with `sphere1`, `sphere2`, `sphere3`, `small_array[][]`, `big_array[][]` .... This point may seem trivial, but choosing good names helps to make programs more readable and therefore helps with program maintenance. Secondly, this approach forces us to think in terms of the details of the particular implementation, rather than in terms of broad concepts. Instead of thinking in terms of an array of spheres, we would be considering a collection of arrays of coordinates and radii.

8.2 Class Access Specifiers

Our current definition of the `sphere` class has one fundamental problem; we cannot assign values to members of this class. By default, all members of a class are private and private members can only be accessed by member functions of the same class. Our declaration of the `sphere` class is equivalent to using the `private` keyword, as in:

```
class sphere {
private:
    double x_centre, y_centre, z_centre, radius;
};
```

However, if all members of the class are private, the class forms an isolated system. This would serve no useful purpose since there would be no way to access the members.

The keyword, `public`, designates all subsequent members of the class to be public. In the following declaration, the four data members can all be freely accessed:

```
class sphere {
public:
    double x_centre, y_centre, z_centre, radius;
};
```

Members that are declared `public` can be accessed by both member and non-member functions. The keywords, `public` and `private`, are known as *class access specifiers*. A class access specifier is valid until either another class access specifier, or the end of the class, is reached. For instance, a class could have the following sequence of access specifiers:

```
class my_class {
public:
    // Public members go here.
private:
    // Private members go here.
public:
    // More public members could go here, but it is unusual
    // to repeat access specifiers like this.
};
```

The `public` and `private` keywords can be given in any order within the class body. Some C++ programmers consistently put the `public` members first and there is some motivation for doing this since it is the `public` members that constitute the class interface. It is certainly normal practice to group together all members with the same access specifier. For this reason you will rarely see the access specifier repeated in a single class declaration.

8.3 Accessing Members

With our present definition of the `sphere class` (including the `public` keyword) we can directly access individual members of the class by using the *class member access operator*, which is a single dot, as demonstrated below.

```
sphere1.x_centre = 2.2;        // Assigns 2.2 to the x_centre
                               // member of the sphere1 object.
sphere1.radius = 10.4;         // Assigns 10.4 to the radius
                               // member of the sphere1 object.
double x = sphere1.x_centre;   // 2.2 is assigned to x.
```

Members of arrays of objects can also be accessed. The following assigns a value to x_centre for element 5 in a sphere array:

```
sphere many_spheres[10];
many_spheres[5].x_centre = 4.5;
```

Notice that we first find the sphere and then find the required member; that is the syntax is:

```
many_spheres[5].x_centre = 4.5; // O.K.
```

rather than:

```
many_spheres.x_centre[5] = 4.5; // WRONG!
```

A complete, but very simple, program accessing data members is given below.

```
#include <iostream>
#include <cstdlib>        // For exit()
using namespace std;

class sphere {
public:
    double x_centre, y_centre, z_centre, radius;
};

int main()
{
    sphere s;

    s.x_centre = 1.1;
    s.y_centre = 2.2;
    s.z_centre = 3.3;
    s.radius = 5.5;

    ++s.radius;
    s.x_centre += 3.4;

    cout << "x_centre = " << s.x_centre << "\ny_centre = " <<
        s.y_centre << "\nz_centre = " << s.z_centre << "\n" <<
        "radius = " << s.radius << "\n";

    return(EXIT_SUCCESS);
}
```

This program defines a single sphere object, known as **s**, and assigns numbers to all four data members. The radius of **s** is incremented by one and the value of **x_centre** is increased by 3.4. The data for this sphere object is then sent to the output stream. The program demonstrates that, by using the class member access operator, we can manipulate public data members in a similar way to instances of the fundamental types.

It is also possible to dynamically allocate objects. The following code fragment allocates memory for a single **sphere** object and memory for an array of 100 **sphere** objects:

```
sphere *a_sphere = new sphere;
sphere *array_of_spheres = new sphere[100];
```

Class objects are deallocated in the same way as for the fundamental types. So the objects allocated by the above statements are deallocated by using:

```
delete a_sphere;
delete[] array_of_spheres;
```

Once again, it is essential that objects allocated by **new** are only deallocated by **delete**, and objects allocated by **new []** are only deallocated by **delete[]**. The results of using the wrong version of the **delete** operator are undefined.

Exercise

Compile and run the program (on page 200) accessing data members of a **sphere** object. Modify the program to dynamically allocate an array of three spheres and assign different values to all data members of these objects. You should check the assignments by listing the data members and also delete the array before the program terminates.

8.4 Assigning Objects

We can assign an object to another object in the same way as with instances of the fundamental types, as demonstrated by the following code fragment:

```
sphere sphere1, sphere2;

sphere1.x_centre = 1.1;
sphere1.y_centre = 2.2;
sphere1.z_centre = 3.3;
sphere1.radius = 5.5;

sphere2 = sphere1;
```

This assignment copies whatever is stored by the data members of **sphere1** to the members of **sphere2**. However, the default assignment operator may not be what is actually needed. For instance, the data members could include a pointer to an area of memory used to store data associated with **sphere1**, as shown in Figure 8.1.[2] In

---

[2]In Figure 8.1 the numbers outside boxes denote memory addresses and the numbers within boxes represent data stored in memory. (Also see Figure 6.1.)

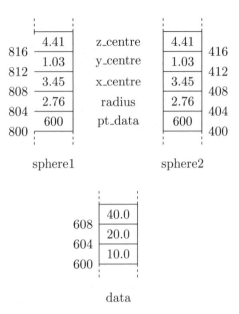

Figure 8.1: Default assignment.

this case the **pt** member of each **sphere** object stores the address of the same **data** object. Consequently, changing the values stored in **data** changes the values for both **sphere** objects. It most cases it would be better if the pointer for **sphere2** pointed to a different area of memory, containing a separate copy of the data values stored in the **data** object. As we will see later, it is possible to overload operators, such as the assignment operator, in order to meet our specific requirements.

Exercise

Verify that the data stored by **sphere1** is copied to **sphere2** by the default assignment operator.

8.5   Functions and Classes

Functions can access the **public** members of a class by using the class access operator. For instance, a function to return the volume of a sphere could be implemented as follows. (You would need to #include <cmath> in order to have the declaration of pow() and the definition of M_PI.)

```
double volume(sphere s)
{
    return 4.0 * M_PI * pow(s.radius, 3) / 3.0;
}
```

This function accesses the **radius** member of **s** as **s.radius**.

The following short program demonstrates how to use the **volume** function:

```
#include <iostream>
#include <cmath>        // For pow(), M_PI
#include <cstdlib>      // For EXIT_SUCCESS
using namespace std;

class sphere {
public:
    double x_centre, y_centre, z_centre, radius;
};

double volume(sphere s)
{
    return 4.0 * M_PI * pow(s.radius, 3) / 3.0;
}

int main()
{
    sphere s;
    s.radius = 2.0;
    cout << "Volume of sphere is " << volume(s) << "\n";
    return(EXIT_SUCCESS);
}
```

Since the volume() function is intimately connected with objects that are spheres, it would be sensible to make the function part of the sphere class. To do this, we include the function in the class declaration. This modified class declaration is given below.

```
class sphere {
public:
    double x_centre, y_centre, z_centre, radius;
    double volume(void) {return 4.0 * M_PI * pow(radius, 3) / 3.0;}
};
```

The function, volume(), is now a member function, with direct access to data members of the object for which it is invoked. For this reason the function does not take any argument and the function body uses the variable radius, rather than s.radius. Indeed, s.radius would be incorrect because, within the class definition, the volume() function has no access to an object named s. Notice that, in order to avoid over-emphasizing member function definitions, we allow a condensed layout style within the class body.

Including complicated function definitions within a class definition would drastically reduce the usefulness of a class as a self-documenting interface. However, the function declaration can be given within the class body and the function definition outside. This is demonstrated below.

```
class sphere {
public:
    double x_centre, y_centre, z_centre, radius;
```

```
        double volume(void);
    };

    double sphere::volume(void)
    {
        return 4.0 * M_PI * pow(radius, 3) / 3.0;
    }
```

The double colon in the function header is a single token, known as the *scope resolution operator*. This operator signifies that `volume()` is a member of the `sphere` class. Outside the class body, the function header must use the `class_name::`function syntax. Now that `volume()` is a member function, the way it is invoked is different. A function that is a class member, is accessed in the same manner as a data member; by using the class member access operator. If we have a `sphere` object called `s`, then instead of using `volume(s)` we use `s.volume()`. This is demonstrated in the following program:

```
    #include <iostream>
    #include <cmath>          // For pow(), M_PI
    #include <cstdlib>        // For EXIT_SUCCESS

    using namespace std;

    class sphere {
    public:
        double x_centre, y_centre, z_centre, radius;
        double volume(void) {return 4.0 * M_PI * pow(radius, 3) / 3.0;}
    };

    int main()
    {
        sphere s;
        s.radius = 2.0;
        cout << "volume of sphere is " << s.volume() << "\n";
        return(EXIT_SUCCESS);
    }
```

All functions defined (in contrast to declared) within a class are implicitly `inline`. A member function defined outside of the class body is not implicitly `inline`, but the `inline` keyword can be added if we want to suggest that the function is made `inline`. This is illustrated below.[3]

```
    class sphere {
    public:
        double x_centre, y_centre, z_centre, radius;
        double volume(void);
    };
```

---

[3]For the `volume()` function given here there may not be a lot to be gained by making it `inline` due to the amount of computation involved.

```
inline double sphere::volume(void)
{
    return 4.0 * M_PI * pow(radius, 3) / 3.0;
}
```

It is optional as to whether or not the `inline` specification is included in the function declaration within the class declaration. Although safer, it could be argued that the `inline` specifier is an implementation detail and so shouldn't be in a class declaration. This approach, which we adopt, is valid provided that the `inline` function implementation occurs before the function is invoked. As always, an `inline` function implementation belongs in a header (`.h`) rather than source (`.cxx`) file. (See Section 5.3.)

A function defined outside any class, such as:

```
double volume(sphere s)
{
    return 4.0 * M_PI * pow(s.radius, 3) / 3.0;
}
```

is a valid function definition, but not a class member. It is possible to have both a class definition of `volume()` (using the scope resolution operator) and a non-class definition. Indeed, it is possible to have functions with the same name belonging to a number of different classes, as in:

```
double sphere::volume(void)
{
    return 4.0 * M_PI * pow(radius, 3) / 3.0;
}

double cube::volume(void)
{
    return pow(side, 3);
}
```

Such functions are *not* overloaded since class members have *class* scope. However, if a class has member functions with the same name then the name *is* overloaded, just as with non-member functions. In the following code fragment, the `sphere` class has both member and non-member `volume` functions, and the `cube` class also has a `volume` function:

```
class sphere {
public:
    double x_centre, y_centre, z_centre, radius;
    double volume(void);
};

double sphere::volume(void)
{
    return 4.0 * M_PI * pow(radius, 3) / 3.0;
}
```

```
    double volume(sphere s)
    {
        return 4.0 * M_PI * pow(s.radius, 3) / 3.0;
    }

    class cube {
    public:
        double x_centre, y_centre, z_centre, side;
        double volume(void);
    };

    double cube::volume(void)
    {
        return pow(side, 3);
    }
```

The following code fragment demonstrates how to invoke the three different volume()
functions:

```
    sphere sphere1, sphere2;
    cube cube1;

    // Assignments to data members of sphere1, sphere2, cube1 go here.

    double vol1 = sphere1.volume();
    double vol2 = volume(sphere2);
    double vol3 = cube1.volume();
```

Exercise

Make the code fragment given above into a complete program that implements the cube and sphere classes. Assign values to the data members of
sphere1, sphere2 and cube1, and check that vol1, vol2 and vol3 give
the expected results.

8.6   Data Hiding

So far, using our sphere class has just been a convenient way of collecting related
data. However, we now introduce data hiding, which is one of the central concepts of
object-oriented programming. The idea is that we should deal with objects, such as
spheres, rather than their explicit representation, such as Cartesian coordinates; the
data associated with an object should only be accessible by means of function calls
that hide details of the class implementation.

In fact we already know how to hide the sphere data; what we need are some functions to access that data. The following code fragment gives a sphere class declaration
that involves data hiding, together with member functions accessing that data:

```
class sphere {
public:
    void assign_centre(double x, double y, double z);
    double get_x_coordinate(void);
    double get_y_coordinate(void);
    double get_z_coordinate(voi);
    void assign_radius(double r);
    double get_radius(void);
    double volume(void);
private:
    double x_centre, y_centre, z_centre, radius;
};
```

Notice how the public functions provide the class interface and hide details of the class implementation. (In fact, we could store the centre in terms of polar coordinates, r, θ and ϕ, without this change being discernible through the interface provided by the public functions.) A program with implementations for the sphere class member functions is given below. The program assigns data to an instance of the sphere class, and then uses the member functions to give the coordinates and volume of the object.

```
#include <iostream>
#include <cmath>         // For pow()
#include <cstdlib>       // For EXIT_SUCCESS
using namespace std;

class sphere {
public:
    void assign_centre(double x, double y, double z);
    double get_x_coordinate(void);
    double get_y_coordinate(void);
    double get_z_coordinate(void);
    void assign_radius(double r);
    double get_radius(void);
    double volume(void);
private:
    double x_centre, y_centre, z_centre, radius;
};

inline void sphere::assign_centre(double x, double y,
    double z)
{
    x_centre = x;
    y_centre = y;
    z_centre = z;
}

inline double sphere::get_x_coordinate(void)
{
```

```
        return x_centre;
    }

    inline double sphere::get_y_coordinate(void)
    {
        return y_centre;
    }

    inline double sphere::get_z_coordinate(void)
    {
        return z_centre;
    }

    inline void sphere::assign_radius(double r)
    {
        radius = r;
    }

    inline double sphere::get_radius(void)
    {
        return radius;
    }

    double sphere::volume(void)
    {
        return 4.0 * M_PI * pow(radius, 3) / 3.0;
    }

    int main()
    {
        sphere s;
        s.assign_centre(10, 24, 36);
        s.assign_radius(2.0);
        cout << "Volume is " << s.volume() << "\nCentre is " <<
            s.get_x_coordinate() << ", " << s.get_y_coordinate() <<
            ", " << s.get_z_coordinate() << "\n";
        return(EXIT_SUCCESS);
    }
```

In the context of some object-oriented languages (such as Smalltalk), member functions are referred to as *methods*, which send *messages* to objects. An object has a state and the effect of the methods is to access this state. Such terminology is not emphasized in this book, although it is used by some C++ programmers.

It is worth emphasizing again that memory for an object is not defined by the class declaration, but by the object definition itself. For example, the statement:

```
    sphere s;
```

not only declares s to be of type **sphere**, but also assigns sufficient memory to s for storing the values of `x_centre` etc. When we invoke a member function, as in:

```
double vol1 = s.volume();
s.assign_centre(10, 24, 36);
```

it is the specific memory associated with `s` that is accessed. Moreover, there may be many `sphere` objects, each with its own memory for storing data and access to the *single* copy of the member functions.

Exercise

Implement a `cube` class, analogous to this new `sphere` class. Declare an instance of the `cube` class, and assign data to it for the coordinates of the centre and the side length. Use member functions to list the coordinates of the centre, the side length, and the volume.

8.7 Returning an Object

Previously, we have made frequent use of functions that return a fundamental type; for instance, the `volume()` functions defined in this chapter all return the type `double`. It is also straightforward for a function to return an instance of a derived type. Suppose we need a function to return a new sphere, with the same radius as an existing sphere, but a translated centre. A suitable implementation of such a function, which should be made a `public` member of the `sphere` class, is given below.

```
sphere sphere::translated_sphere(double d_x, double d_y,
    double d_z)
{
    sphere new_sphere;
    new_sphere.x_centre = x_centre + d_x;
    new_sphere.y_centre = y_centre + d_y;
    new_sphere.z_centre = z_centre + d_z;
    new_sphere.radius = radius;
    return new_sphere;
}
```

We can invoke this `translated_sphere()` function as follows:

```
sphere s1;
// Assign values to s1 here.
sphere s2 = s1.translated_sphere(1.1, 2.2, 3.3);
```

Notice that the `translated_sphere()` function is used to assign values to another `sphere` object; the function does not translate the original object. We could achieve translation of the origin of a sphere by using the `sphere` class member function given below.

```
void sphere::translate(double d_x, double d_y, double d_z)
{
    x_centre += d_x;
```

```
        y_centre += d_y;
        z_centre += d_z;
}
```

An example of how this function would be invoked for a **sphere** object **s** is given in the following statement:

```
s.translate(1.1, 2.2, 3.3);
```

Exercise

(a) Modify the program implementing the **sphere** class given on page 207 so that it includes the **translated_sphere()** and the **translate()** member functions.

(b) In addition to the object **s**, define a second **sphere** object, **ss**, and use the **translated_sphere()** function with suitable arguments so that **ss** is a copy of **s**, but with its centre shifted by $(1, 2, 3)$. By using appropriate member functions, send the data for **ss** to the output stream.

(c) Use the **translate()** member function to translate the coordinates of the **s** object by $(2, 4, 6)$, and then send the data for **s** to the output stream.

It is worth having a look at the **translated_sphere()** function in more detail. The object called **new_sphere** is defined within the function and does not exist outside of the function body. This means a statement such as:

```
s2 = s1.translated_sphere(1.1, 2.2, 3.3);
```

creates at least one temporary copy of the **new_sphere** object. Creating these temporary objects is inefficient since only the **s1** and **s2** objects need exist. This would be particularly important for objects with large data members.

In order to overcome this problem of temporary copies, we might be tempted to return a reference object, as in the following function definition:

```
sphere &sphere::translated_sphere(double d_x, double d_y,
    double d_z)
{
    sphere new_sphere;
    new_sphere.x_centre = x_centre + d_x;
    new_sphere.y_centre = y_centre + d_y;
    new_sphere.z_centre = z_centre + d_z;
    new_sphere.radius = radius;
    return new_sphere;   // WRONG: cannot return a reference
                         //        to a local object.
}
```

However, any variation on this theme will fail since **new_sphere** only exists within the scope of the function body and so we cannot return a reference to this object.

Exercise

In the previous exercise on page 210 you included an implementation of the `translated_sphere()` function. What happens if you replace the correct version of that function by this invalid version that attempts to return a local object?

8.8 Reference Arguments

A valid way of defining a `translated_sphere()` function that does not introduce spurious copies is by means of a `sphere` reference argument. This uses the same `&` token (the reference declarator) as in Section 7.7.1. An implementation of a function using this technique is given below.

```
void sphere::translated_sphere(sphere &s, double d_x,
    double d_y, double d_z)
{
    s.x_centre = x_centre + d_x;
    s.y_centre = y_centre + d_y;
    s.z_centre = z_centre + d_z;
    s.radius = radius;
}
```

Notice that a member of a reference to an object is accessed in the same way as a member of an object. The formal argument, `s`, is a reference to a `sphere` object and hence no copy of this object is made by the `translated_sphere()` function. An example of how to invoke this function is given below.

```
sphere s1, s2;
// Assignments to s1.
s1.translated_sphere(s2, 1.1, 2.2, 3.3);
```

The only `sphere` objects are `s1` and `s2`. No temporary copies are made since the function works directly with `s2`.

Exercise

Modify the program you wrote for the exercise on page 210 so that it uses reference arguments to avoid unnecessary temporary copies of a `sphere` object.

8.9 Pointers to Members

An alternative to the reference argument technique, given in the previous section, is to use a pointer to a `sphere` object, as demonstrated by the following function:

```
void sphere::translated_sphere(sphere *pt, double d_x,
    double d_y, double d_z)
{
    (*pt).x_centre = x_centre + d_x;
    (*pt).y_centre = y_centre + d_y;
    (*pt).z_centre = z_centre + d_z;
    (*pt).radius = radius;
}
```

In this implementation, `pt` is a pointer to the `sphere` type; in other words it is a variable that can store the address of any object of type `sphere`. When the function is invoked, the address of a `sphere` object is passed. An example of how the `translated_sphere()` function is invoked is given below.

```
sphere s1, s2;
// Assignments to s1.
s1.translated_sphere(&s2, 1.1, 2.2, 3.3);
```

Within the function body, `*pt` is the dereferenced address and hence the object, `s2`. The class member access operator, with `*pt` and `x_centre` as operands, therefore modifies the x coordinate of the centre of the sphere, `s2`. Notice that the parentheses enclosing `*pt` are necessary because the class member access operator binds tighter than the dereferencing operator.[4]

Since accessing a member of an object pointed to by a pointer is a very common requirement, there is a special operator for this purpose. This operator is represented by a right-pointing arrow and, rather confusingly, it is also known as the *class member access operator*. The following statement illustrates how the class member access operator is used:

```
pt->x_centre = x_centre + d_x;
```

which is exactly equivalent to:

```
(*pt).x_centre = x_centre + d_x;
```

The arrow, which is a single token, consists of the minus sign followed by the greater than sign. The `translated_sphere()` function could therefore be written as shown below.

```
void sphere::translated_sphere(sphere *pt, double d_x,
    double d_y, double d_z)
{
    pt->x_centre = x_centre + d_x;
    pt->y_centre = y_centre + d_y;
    pt->z_centre = z_centre + d_z;
    pt->radius = radius;
}
```

---

[4]Don't forget that Appendix B contains a list giving the precedence and associativity of all operators, including those introduced in this chapter.

Using the -> operator, rather than the original pointer dereference and first class member access operator, is simply a notational convenience and has no effect on the way the `translated_sphere()` function is invoked. It is worth emphasizing that, since the class member access operator is a single token, the following statement is illegal:

```
pt - > x_centre = x_centre + d_x;    // WRONG!
```

Exercise

Change the program for the previous exercise on page 211 so that instead of using pass by reference, pointers are used for the function arguments.

8.10   Pointer-to-Member Operators[††]

Throughout this section and purely for the purpose of demonstration, we consider a simplified `sphere` class in which the only data member is `public`. The declaration for this `sphere` class is given below.

```
class sphere {
public:
    void assign_radius(double r) { radius = r; }
    double radius;
};
```

There are two fundamental ways in which we can specify an object; we can either use the object itself or else dereference a pointer to an object. For example, in our `sphere` class we can define a `sphere` object, `s`, and a pointer to a `sphere` object, `pt_s`, as shown below.

```
sphere s;
sphere *pt_s;
pt_s = &s;              // Address of object assigned to pt_s.
```

There are also two ways in which we can refer to members of a class. The obvious method is to use the member directly, such as `assign_radius(10.0)` (for a member function) or `radius` (for a data member). However, we can also dereference a pointer to class members. Such a pointer must be restricted to a particular class and can only point to a particular type of member function or data member. The scope resolution operator, `::`, is used to specify the class. For instance:

```
double sphere::*pt;
```

defines a pointer, `pt`, that can point to a data member, of type `double`, belonging to the `sphere` class. In the class defined above, `pt` can only point to the `radius` data member, as in:

```
pt = &sphere::radius;
```

However, in some classes there may be many different members that could be pointed to by the same pointer. The combination, `::*`, is sometimes known as a *pointer-to-member declarator*.

Don't forget that, since defining a class does not allocate memory, `pt` does not point to any memory suitable for storing a value for the radius; `pt` is a pointer to a member of a class and *not* to a member of a class object. Statements such as the following are illegal:

```
*pt = 10.0;                 // WRONG: incompatible types.
```

It is also possible to point to member functions. The following statement defines a `sphere` class pointer, `pt_f`, that returns the type, `void`, and has an argument of type, `double`:

```
void (sphere::*pt_f)(double);
```

The first parentheses are necessary because the function call operator, `()`, has a higher precedence than the dereferencing operator, `*`. Once again, since our current `sphere` class is rather limited, `pt_f` can only point to the `assign_radius()` function, as illustrated by the following statement:

```
pt_f = &sphere::assign_radius;
```

This assignment can also be written without using the address operator, as in:

```
pt_f = sphere::assign_radius;
```

This is consistent with the idea that if `f()` is a function then `f` is the address of that function. (See page 168.) Some compilers (such as the GNU `g++` compiler) may issue a warning that they are assuming that the address operator acts on the function name.

We now have two ways of specifying a `sphere` object (`s` and `*pt_s`). We also have two ways of specifying a class member (typically `radius` and `*pt` for data, or `assign_radius()` and `*pt_f` for functions). There are therefore four different ways of accessing object members and each of these methods has its own special binary operator. If we make the definitions:

```
sphere s;
sphere *pt_s;
double sphere::*pt;
void (sphere::*pt_f)(double);

pt_s = &s;
pt = &sphere::radius;
pt_f = &sphere::assign_radius;
```

then the four methods that can be used to access members of the object, `s`, are as given below.

1. The *class member access operator* (a single dot) is used when the operands are an object and a member, as in:

   ```
   s.radius = 100.0;        // Access data member.
   s.assign_radius(10.0);   // Access member function.
   ```

This operator was introduced in Section 8.3.

2. The second *class member access operator* (a right pointing arrow, ->) is used when the operands are a pointer to an object and an object member, as in:

```
pt_s->radius = 200.0;        // Access data member.
pt_s->assign_radius(20.0);   // Access member function.
```

We introduced this operator in the previous section.

3. A *pointer-to-member operator*, denoted, .*, is used when the operands are an object and a pointer to a class member, as in:

```
s.*pt = 300.0;               // Access data member.
(s.*pt_f)(30.0);             // Access member function.
```

Notice that the first parentheses in the second statement are necessary because the function call operator, (), has a higher precedence than the .* operator.

This is the first place that we have used the pointer-to-member operator. It is a single token, consisting of a dot followed by an asterisk. As always, white space is not allowed within the token, as illustrated by the following code fragment:

```
s .* pt = 30.0;        // O.K.
s.*pt = 30.0;          // O.K.    Probably the usual style.
s. *pt = 30.0;         // WRONG: White space is not allowed
                       //        within a token.
```

The same notation is used to access members of a reference to an object, as shown below.

```
sphere &ss = s;
ss.*pt = 400.0;        // Access data member.
(ss.*pt_f)(40.0);      // Access member function.
```

4. A second *pointer-to-member operator*, denoted ->*, is used if the operands are a pointer to an object and a pointer to a class member. Use of this operator is illustrated below.

```
pt_s->*pt = 40.0;        // Access data member.
(pt_s->*pt_f)(40.0);     // Access member function.
```

Again, the parentheses are necessary because the function call operator, (), has a higher precedence than the pointer-to-member operator, ->*. This is the first place that we have used this pointer-to-member operator; it is a single token, consisting of a minus sign, a greater than sign and an asterisk.[5]

---

[5]Notice that the usual names given to the four operators considered in this section are a bit confusing. There are two *class member access* operators and two *pointer-to-member* operators.

The following program shows the use of all four operators considered above. The program assigns different values to `radius` and then prints the values stored:

```cpp
#include <iostream>
#include <cstdlib>          // For EXIT_SUCCESS
using namespace std;

class sphere {
public:
    void assign_radius(double r) { radius = r; }
    double radius;
};

int main()
{
    sphere s;
    sphere *pt_s;
    double sphere::*pt;
    void (sphere::*pt_f)(double);

    pt_s = &s;
    pt = &sphere::radius;
    pt_f = &sphere::assign_radius;

    s.radius = 100.0;           // Access data member.
    cout << "Radius is " << s.radius << "\n";
    s.assign_radius(10.0);      // Access member function.
    cout << "Radius is " << s.radius << "\n";

    pt_s->radius = 200.0;       // Access data member.
    cout << "Radius is " << s.radius << "\n";
    pt_s->assign_radius(20.0);  // Access member function.
    cout << "Radius is " << s.radius << "\n";

    s.*pt = 300.0;              // Access data member.
    cout << "Radius is " << s.radius << "\n";
    (s.*pt_f)(30.0);            // Access member function.
    cout << "Radius is " << s.radius << "\n";

    sphere &r = s;
    r.*pt = 400.0;              // Access data member.
    cout << "Radius is " << s.radius << "\n";
    (r.*pt_f)(40.0);            // Access member function.
    cout << "Radius is " << s.radius << "\n";

    pt_s->*pt = 500.0;          // Access data member.
    cout << "Radius is " << s.radius << "\n";
    (pt_s->*pt_f)(50.0);        // Access member function.
```

```
    cout << "Radius is " << s.radius << "\n";
    return(EXIT_SUCCESS);
}
```

Whereas the two class member access operators, . and ->, are used extensively in typical C++ applications, the two pointer-to-member operators, .* and ->*, are less common. It is worth noting that these four operators do *not* override class protection, which is why we have made the `radius` data member `public` for this section.

8.11   Scope and Data Protection

We have already seen in Section 8.5 that functions declared within a class are not visible outside that class, unless the scope resolution operator is used. The same is true for both `public` and `private` data members. Consider the following modified version of the program on page 207:

```
#include <iostream>
#include <cstdlib>        // For EXIT_SUCCESS
using namespace std;

class sphere {
public:
    void assign_centre(double x, double y, double z);
    double get_x_coordinate(void);
    double get_y_coordinate(void);
    double get_z_coordinate(void);
    void assign_radius(double r);
    double get_radius(void);
private:
    double x_centre, y_centre, z_centre, radius;
};

inline void sphere::assign_centre(double x, double y, double z)
{
    x_centre = x;
    y_centre = y;
    z_centre = z;
}

inline double sphere::get_x_coordinate(void)
{
    return x_centre;
}

inline double sphere::get_y_coordinate(void)
{
    return y_centre;
}
```

```
inline double sphere::get_z_coordinate(void)
{
    return z_centre;
}

inline void sphere::assign_radius(double r)
{
    radius = r;
}

inline double sphere::get_radius(void)
{
    return radius;
}

int main()
{
    sphere s;

    s.assign_centre(1.1, 2.2, 3.3);
    s.assign_radius(101.1);
    double x_centre = 0.0;
    double y_centre = 0.0;
    double z_centre = 0.0;
    double radius = 0.0;

    x_centre = s.get_x_coordinate();
    y_centre = s.get_y_coordinate();
    z_centre = s.get_z_coordinate();
    radius = s.get_radius();

    cout << "x_centre = " << x_centre <<
            "\ny_centre = " << y_centre <<
            "\nz_centre = " << z_centre <<
            "\nradius = " << radius << "\n";

    // The results show that the body of main() does
    // not have visibility of private class members.

    return(EXIT_SUCCESS);
}
```

This program demonstrates that although names, such as **x_centre**, are used as members of the **sphere** class, there is no visibility of these names within the body of the function **main()**, where they can safely be reused. In fact a class provides the third kind of scope, the other two being file scope and block scope, as described in Section 5.4.

It is worth pointing out that the concept of `private` class members is intended to encourage safer programming techniques. The `private` access specifier does not provide either secrecy or protection against malicious programming. Once we have the address of an object, a cast to a pointer of a different type would enable us to probe supposedly `private` regions of memory. Of course, there is little purpose in using a sophisticated programming language if the programmer intends to use such subversive techniques.

8.12 Static Members

Each class object has its own set of class data members. For example, every object belonging to our `sphere` class has its own radius together with coordinates for its centre. However, in some circumstances we may want a data member to be common to all objects of a class. For instance, we may want to keep track of the total number of spheres. Updating multiple copies of this number every time we defined a new `sphere` object would clearly be wasteful and likely to introduce errors. A *static data member* is useful in this situation since there is only one such member for a class, irrespective of the number of objects. The appropriate way of accessing such data is to use a *static member function*. We now consider these two concepts in more detail.

8.12.1 Static Data Members

As an example of a `static` data member, consider the following suitably modified `sphere` class:[6]

```
#include <iostream>
#include <cstdlib>
using namespace std;

class sphere {
public:
    void increment_spheres(int new_spheres);
    int total_spheres(void) { return total; }
    // Other public members go here.
private:
    static int total;
};

inline void sphere::increment_spheres(int new_spheres)
{
    total += new_spheres;
}

int sphere::total;
```

---

[6]It would be more satisfactory if `total` were incremented with the creation of each `sphere` object. The techniques for achieving this will be introduced in Section 8.13.

```
int main()
{
    sphere s1, s2;
    s1.increment_spheres(2);
    cout << "Total number of spheres known to s1 = " <<
        s1.total_spheres() <<
        "\nTotal number of spheres known to s2 = " <<
        s2.total_spheres() << "\n";
    return(EXIT_SUCCESS);
}
```

We can access a `static` data member in the same way as any other data member, by using a single dot (one of the class member access operators) with an object as the left operand. The significant difference is that the same `static` data is seen by all objects. When you run the above program, you will find that the `increment_spheres()` function gives the same result for any `sphere` object.

The `static` key word in the statement:

```
static int total;
```

within the class declaration, declares `total` to be `static`. However, it is important to realize that this statement does *not* allocate any memory for the variable `total`. Such memory has to be allocated somewhere and this is done by means of the statement:

```
int sphere::total;
```

This statement is *not* placed in the function `main()` since this would be a re-declaration of the variable `sphere::total`.

Notice that we did not initialize `total`. This is because `total` is `static` and is therefore initialized to zero by default. We could initialize `total` if we wanted to. For example, if we wanted the initial value to be ten (even though no spheres had yet been created!) this would be achieved by the following statement:

```
int sphere::total = 10;
```

This initialization is allowed even if (as here) `total` is declared `private`. It should be emphasized that this statement is *not* an assignment. We cannot make an assignment to a private `static` member (except within a class member function or `friend`).[7] Hence, the following attempt at an assignment is incorrect:

```
sphere::total = 10;                    // WRONG: total is private.
```

A common error is attempting to initialize `static` data within a class declaration. This is illegal, as illustrated below.

```
class sphere {
public:
    void increment_spheres(int new_spheres);
    int total_spheres(void) { return total; }
private:
    static int total = 0;        // WRONG!
};
```

---

[7] The idea of a `friend` of a class will be introduced in Section 8.15.

This is entirely consistent with the idea that a class declaration only defines a type and does not allocate memory for objects of that type.

The allocation of memory for a `static` data member, together with an initialization (if required) can only be performed once within a single program. Consequently, the statement allocating memory for a `static` data member should be placed in the `.cxx` file that implements the non-`inline` member functions, rather than in the class header file. An example illustrating the initialization of a `static` data member is given in Section 8.17.

We could, of course, declare `total` to be a global variable instead of making it a member of the `sphere` class. However, the total number of spheres is a property of the class and clearly belongs in the class declaration.

Public `static` data members can be accessed directly by means of the class name and the scope resolution operator (`::`). There is no need for objects of the class to exist since there is always one (and only one) copy of a `static` data member. For the same reason such members can be accessed by pointers to the appropriate type, and these pointers can be defined without using the class name and scope resolution syntax. An example in which we declare `total` to be a `public` member of the `sphere` class is given below.

```
#include <iostream>
#include <cstdlib>
using namespace std;

class sphere {
public:
    static int total;
// Other members go here.
};

int sphere::total;

int main()
{
    sphere::total = 10;
    cout << sphere::total << "\n";
    sphere::total = 20;
    int *pt = &sphere::total;
    cout << *pt << "\n";
    return(EXIT_SUCCESS);
}
```

Of course, this example is given simply as an illustration. Exposing the data member of a class in this way isn't a good idea, since it means that the data can be manipulated in ways that aren't clear from the class declaration.

8.12.2 Static Member Functions

In the program given on page 219, the `increment_spheres()` and `total_spheres()` functions need to be invoked for a `sphere` object. This situation is not very satisfactory, since the total number of spheres is a property of the collection of all such objects, rather than any particular instance of the class. In particular, it would be nice if an appropriate `total_spheres()` function could give the number of spheres (presumably zero) even if there were no instances of the `sphere` class. This can be achieved by declaring `increment_spheres()` and `total_spheres()` to be `static` member functions. Static member functions are special in that they can only access `static` data members and need not be invoked by a class object, as demonstrated by the program given below.

```
#include <iostream>
#include <cstdlib>        // For EXIT_SUCCESS
using namespace std;

class sphere {
public:
    static void increment_spheres(int new_spheres);
    static int total_spheres(void) { return total; }
private:
    static int total;
};

void sphere::increment_spheres(int new_spheres)
{
    total += new_spheres;
}

int sphere::total;

int main()
{
    cout << "Initial number of spheres = " <<
        sphere::total_spheres() << "\n";
    sphere s1, s2;
    sphere::increment_spheres(2);
    cout << "Total number of spheres = " <<
        sphere::total_spheres() << "\n";
    return(EXIT_SUCCESS);
}
```

In this version of our **sphere** class, the way in which we keep a tally of the total number of spheres does not depend on accessing a particular instance of the class.

It is also possible to invoke **static** member functions for particular objects, as in s1.`total_spheres()`. However, the `sphere::total_spheres()` notation is better because it emphasizes that the number of spheres is a feature of the class rather rather than a particular instance of the class.

8.13 Constructor Functions

As with variables of the fundamental types, it is safer to initialize objects rather than to define them with arbitrary values. A class that only has `public` data members can be initialized by a comma-separated list. This is illustrated by the program given below.

```
#include <iostream>
#include <cstdlib>        // For EXIT_SUCCESS
using namespace std;

class coordinate {
public:
    double x, y, z;
};

int main()
{
    coordinate w = {1.0, 2.0, 0.0};
    cout << "Coordinates are " << w.x << ", " << w.y << ", " <<
        w.z << "\n";
    return(EXIT_SUCCESS);
}
```

However, classes like this are unusual.

As a change from our `sphere` class, we now introduce a complex arithmetic class with data members, `re` and `im`, representing the real and imaginary parts of an instance of this `complex` class. Initializers cannot be included in a class declaration, so we cannot initialize the data for this class by something like the following declaration:

```
class complex {
private:
    double re = 0.0;        // WRONG!
    double im = 0.0;        // WRONG!
};
```

Once again, this is consistent with the idea that a class declaration only defines a type and does not allocate memory for objects of that type. However, there is a special kind of function, known as a *constructor function* (or, more simply, a *constructor*) that is specifically designed for initializing objects. A constructor is declared by giving the class name to a member function, as in the declaration of the `complex` class given below.

```
class complex {
public:
    complex(double x, double y) { re = x; im = y; }
private:
    double re, im;
};
```

Notice that the constructor function, `complex()`, does not have a **return** type; constructors are special functions in that they cannot return a type (not even **void**). By using this constructor, objects that are members of the `complex` class can be initialized, as demonstrated by the following statements:

```
complex u = complex(1.1, 2.2);   // Initializes u.
complex v(1.1, 2.2);             // Initializes v.
```

In both examples, the constructor function initializes the real part of the object with the value 1.1 and the imaginary part with the value 2.2. Notice that the syntax in the second statement does *not* signify that v is a function: rather v is an object, implicitly invoking the constructor function with arguments 1.1 and 2.2.

Since the only constructor defined for the `complex` class takes two arguments, we cannot declare an uninitialized `complex` object. Consequently, the following statement is incorrect:

```
complex z;     // WRONG: no suitable constructor defined.
```

However, the constructor function can be modified to provide default initializations, as in the class declaration given below.[8]

```
class complex {
public:
    complex(double x = 0.0, double y = 0.0) {
        re = x; im = y; }
private:
    double re, im;
};
```

If we now have the statement:

```
complex z;
```

then the real and imaginary parts of z are initialized to zero. In order to demonstrate this, we need to introduce some way of listing the values of the **private** data members of the `complex` class. This is done in the following program by introducing the `print()` function:[9]

```
#include <iostream>
#include <cstdlib>        // For EXIT_SUCCESS
using namespace std;

class complex {
public:
    complex(double x = 0.0, double y = 0.0) {
        re = x; im = y; }
    void print(void);
```

---

[8]The default initializations are supplied right to left just as for the default arguments in Section 5.1.6.

[9]It would be better if this could be accomplished with something like `cout << z`. We will learn how to achieve this in Section 18.9.

```
private:
    double re, im;
};

void complex::print(void)
{
    cout << "(" << re << ", " << im << ")";
}

int main()
{
    complex u;                      // Default initialization.
    u.print();
    cout << "\n";
    complex v(2.0, 3.0);            // Initialized by 2.0 and 3.0.
    v.print();
    cout << "\n";
    complex w =
        complex(3.0, 4.0);          // Initialized by 3.0 and 4.0.
    w.print();
    cout << "\n";
    complex z(2.0);                 // Initialized by 2.0 and 0.0.
    z.print();
    cout << "\n";
    complex q = complex(3.0);       // Initialized by 3.0 and 0.0.
    q.print();
    cout << "\n";
    return(EXIT_SUCCESS);
}
```

A common error is to attempt to define an object, initialized with default values, by:

```
complex u();     // WRONG: this is a function declaration.
```

In fact, this declares a function, which takes no arguments and returns the `complex` type. However, the following statement does define (rather verbosely) a `complex` object with the default initialization:

```
complex u = complex();
```

We can overload the constructor function by having more than one member function with the same name as the class name, provided that the function arguments differ. As an example, we may want to initialize a complex object with another object belonging to the same class. This is achieved by introducing a constructor with the declaration:

```
complex::complex(const complex &z);
```

Both types of constructors are used in the program given below.

```
#include <iostream>
#include <cstdlib>         // For EXIT_SUCCESS
using namespace std;

class complex {
public:
    complex(double x=0.0, double y=0.0) { re=x; im=y;}
    complex(const complex &z) { re=z.re; im=z.im; }
    void print(void);
private:
    double re, im;
};

void complex::print(void)
{
    cout << "(" << re << ", " << im << ")";
}

int main()
{
    complex u(24.5, 17.6);
    complex v(u);
    v.print();
    cout << "\n";
    return(EXIT_SUCCESS);
}
```

In the first statement in main(), the real and imaginary parts of u are initialized to
24.5 and 17.6 respectively. In the second statement, the real and imaginary parts of u
are themselves used to initialize v.

The second type of constructor function introduced for the complex class has a
single argument of a *reference* to the class type. This type of constructor is very
common and is known as a *copy constructor*. Notice the emphasis on reference; a
constructor with the declaration:

```
class_x::class_x(class_x);
```

is illegal, but:

```
class_x::class_x(const class_x &);
```

as in:

```
complex::complex(const complex &z);
```

is both legal and common.

Exercise

Implement a constructor for the sphere class described in Sections 8.12.1
and 8.12.2. The total number of spheres (that is the static variable, total)

should be incremented each time the constructor is invoked. Create various numbers of spheres and check that the `static total_spheres()` function gives the correct result.

(Note that you won't need the `increment_spheres()` member function any more. In fact, this function is very undesirable since it can be used to make the value of the `total` data member inconsistent with the actual number of `sphere` objects.)

Our `complex` class is now sufficiently developed for us to try a simple test program. Since this class might be of use in more than one program, we put both the class definition and the implementation of `inline` functions in a header file, which is given below.

```
// source:  complex.h
// use:     Defines complex arithmetic class.

#ifndef COMPLEX_H
#define COMPLEX_H

class complex {
public:
    complex(double x = 0.0, double y = 0.0);
    complex(const complex &z);
    double real(void);
    double imag(void);
    void print(void);
private:
    double re, im;
};

inline complex::complex(double x, double y)
{
    re = x;
    im = y;
}

inline complex::complex(const complex &z)
{
    re = z.re;
    im = z.im;
}

inline double complex::real(void)
{
    return re;
}

inline double complex::imag(void)
```

```
{
    return im;
}
```

```
#endif   // COMPLEX_H
```

It is worth making a brief diversion to explain why we have:

```
#ifndef COMPLEX_H
#define COMPLEX_H
      // Code in here.
#endif   // COMPLEX_H
```

in the `complex.h` file. This sequence of preprocessor commands is known as an *inclusion guard* and is an important method of preventing multiple inclusions of the same file. The way inclusion guards work is that if `COMPLEX_H` is not defined, then all of the file is processed and `COMPLEX_H` is consequently defined. Whenever the preprocessor meets another copy of the include file, all of the code between the `#ifndef` and `#endif` directives will be omitted. If inclusion guards are not used, then an example of what can go wrong is that after a program has been through the preprocessor it may contain multiple declarations of the same class. This would be a compile-time error. It is traditional to use the capitalized header file name (with an underscore replacing the dot) for an inclusion guard. Although this is not a language requirement, it helps to avoid having the same guard in different header files. You should avoid using identifiers for inclusion guards with leading and trailing double underscores since identifiers like these are used in library header files and by the linker. Inclusion guards can also be used to avoid multiple definitions in different files. For instance, if `file1.h` and `file2.h` both contain the statement:

```
const double ln_pi = 1.14472988584940017;
```

then a source file that included both header files would give rise to a compile-time error.

Returning now to our `complex` class, the following program tests our current version of the class by setting:
$$z = 24.5 + 17.6i$$
and printing the real and imaginary parts of `z`.

```
#include <iostream>
#include <cstdlib>        // For EXIT_SUCCESS
#include "complex.h"
using namespace std;

int main()
{
    complex z(24.5, 17.6);
    cout << "Real part = " << z.real() <<
        " Imaginary part = " << z.imag() << "\n";
    return(EXIT_SUCCESS);
}
```

Exercise

Compile and run the above program. Make modifications so that the copy constructor is also tested.

8.14 Accessing const Class Objects

In order to access the data of `complex` objects, we could introduce `public` member functions, `real()` and `imag()`, that return the real and imaginary parts of a `complex` object. This is demonstrated in the following class declaration:

```
class complex {
public:
    complex(double x = 0.0, double y = 0.0);
    complex(const complex &z);
    double real(void) { return re;}
    double imag(void) { return im; }
private:
    double re, im;
};
```

We have previously emphasized the desirability of using the `const` specifier to indicate to the compiler that an instance of a fundamental type should not change. Such remarks also apply to instances of classes. For example, we could define:

```
const complex i(0.0, 1.0);
```

A problem now arises if (for example) we try to access the real part of i, using the `real()` member function, as illustrated by the following statement:

```
cout << "Real part of i = " << i.real();
```

This statement is rejected by the compiler since there is no way of telling from the function declaration whether or not `real()` modifies i. Member functions that do not modify any data member can include the `const` specifier at the end of the function declaration, as illustrated by the class declaration given below.[10]

```
class complex {
public:
    complex(double x = 0.0, double y = 0.0);
    complex(const complex &z);
    double real(void) const;
    double imag(void) const;
private:
    double re, im;
};
```

---

[10]This `const` notation is only possible for member functions and cannot be used for **friend** functions, which are introduced in the next section. This is because only member functions have a hidden pointer, called `this`. The special `this` pointer is introduced in Section 9.1.2.

However, if this is done, then the const specifier must also be placed before the body
of the function implementation. So to be consistent with the function declaration given
above, the function definition must be:

```
inline double complex::real(void) const
{
    return re;
}
```

Any attempt by a const member function to modify member data would be flagged as
an error by the compiler. This technique of placing const at the end of the function
declaration solves the problem, described above, of invoking i.real(). The following
program illustrates this technique:[11]

```
#include <iostream>
#include <cstdlib>        // For EXIT_SUCCESS

using namespace std;

class complex {
public:
    complex(double x = 0.0, double y = 0.0);
    complex(const complex &z);
    double real(void) const;
    double imag(void) const;
private:
    double re, im;
};

inline complex::complex(double x, double y)
{
    re = x;
    im = y;
}

inline complex::complex(const complex &z)
{
    re = z.re;
    im = z.im;
}

inline double complex::real(void) const
{
    return re;
}
```

---

[11]Since this restricted complex class is only given to illustrate a particular technique, there is no
point in introducing a complex.h header file.

```
inline double complex::imag(void) const
{
    return im;
}

int main()
{
    complex z(24.5, 17.6);
    cout << "Real part = " << z.real() <<
        " Imaginary part = " << z.imag() << "\n";
    const complex i(0.0, 1.0);
    cout << "Real part of i = " << i.real() <<
        "\nImaginary part of i = " << i.imag() << "\n";
    return(EXIT_SUCCESS);
}
```

It may be necessary to declare a member function as both `const` and non-`const`. In this case the function is overloaded and which function is actually invoked depends on whether or not the object is defined with the `const` specifier.

8.15   Friend Functions

Our `complex` class is defined to have `private` data (`re` and `im`) and in general this data cannot be accessed by non-member functions. However, a *friend* of a class has access to the data members and member functions of that class, irrespective of any access specifiers. For example, an alternative method for obtaining the real and imaginary parts of a `complex` object is to define `friend` functions. This is achieved by including the `friend` keyword in the function declaration given within the class declaration:

```
#include <iostream>
#include <cstdlib>        // For EXIT_SUCCESS
using namespace std;

class complex {
    friend double real(const complex &z);
    friend double imag(const complex &z);
public:
    complex(double x = 0.0, double y = 0.0);
    // More class declarations go in here.
private:
    double re, im;
};

inline complex::complex(double x, double y)
{
    re = x;
    im = y;
}
```

```
inline double real(const complex &z)
{
    return z.re;
}

inline double imag(const complex &z)
{
    return z.im;
}

int main()
{
    complex z(24.5, 17.6);
    cout << "Real part = " << real(z) <<
        " Imaginary part = " << imag(z) << "\n";
    return(EXIT_SUCCESS);
}
```

The functions real() and imag() are now friends of the complex class, rather than members. For this reason, they are invoked by real(z) and imag(z), instead of z.real() and z.imag().

Both member and non-member functions provide equally satisfactory techniques for accessing the real and imaginary parts of a complex object, although it could be argued that real(z) is a more natural syntax than z.real() in this case. In fact, it is possible to have both definitions in the same class implementation.

Notice that since these functions are relatively simple, it is worth making them inline in order to avoid relatively high function call overheads. It is also worth using reference variables, to reduce the call overhead still further, and the const keyword, to provide some protection for the member data.

The friend keyword indicates that a function has access to the private members (data and functions) of a class, but is not a member of that class. A function can be a friend of more then one class, or even a member of one class and a friend of many classes. Notice that friendship is granted by the class whose private members are accessed. One of the special features of constructor functions is that they cannot be declared friends of a class.

So far, we have only considered friend functions. However, it is also possible to declare an entire class to be a friend of another class. The statement:

```
class node {
    friend class list;
    // Other declarations.
};
```

declares list to be a friend of the node class and gives list access to all members of any node object. Both friend classes and functions are not affected by access specifiers, such as public and private, and can be placed anywhere within a class declaration. However, a good convention is to place any friend declarations immediately after the class header and before any explicit access specifiers.

Friendship is not transferred to a `friend` of a `friend`; that is, friendship is *not* transitive. In the following code fragment, the `print()` function has access to the `private` members of a `complex_vector` object, but not to the `private` members of any `complex` object. If this restriction did not exist, then granting friendship to one class would open up the entire class implementation to a whole hierarchy of unknown classes and functions.

```
class complex {
    friend class complex_vector;
    // Other declarations go here.
private:
    double re, im;
};

class complex_vector {
    friend void print(const complex_vector &v);
    // Other declarations go here.
private:
    complex x, y, z;
};

void print(const complex_vector &v)
{
    cout << v.x.re;      // WRONG: re is private.
    // etc.
}
```

8.16 Program Structure and Style

Now that we have introduced many of the basic ideas of C++, it is worth making a brief diversion to consider how programs should be structured. The simplest C++ program consists of one file, containing a single function, called `main()`. More realistic programs often contain many thousand lines of code and must be split up into separate files in order to control the complexity. Unlike some languages, C++ does little to enforce any structure on these files, but there are well-established conventions that it is sensible to follow.

8.16.1 Separate Compilation

A large program should be split up into a number of source files (with a `.cxx` or equivalent extension), each of which can be compiled independently. Each source file usually consists of a number of related function definitions and may use functions defined in other source files. For example, a source file may implement all of the non-`inline` functions for a particular class, or collection of related classes.

Before a source file can be compiled, it is passed through the preprocessor to produce what is known as a *translation unit*. The compiler then works on the translation unit to produce *object code*. In order to produce a file that can be executed, the object code must be *linked*, both with object code derived from other source files provided by the programmer and with the compiled system-provided library functions. This is carried out by means of a special program, known as a *linker*. Linking, preprocessing and compilation are usually invoked by a single command, with options available to inhibit one or more of these processes.

In C++, the existence of function overloading makes it particularly important for the argument types of the compiled functions to be known. For this reason, *type-safe linking* is used. The compiler *mangles* (that is it modifies) function names in a well-defined way in order to encode information on the number of function arguments and their types. In the case of a member function, the class name is also encoded. Name mangling gives the same unique name to a function that may be declared in many different translation units.[12] This means that overloaded functions are resolved on linking rather than on compiling.

In general, names can either have *internal linkage* (the visibility of their names is restricted to one translation unit) or *external linkage* (their names are visible throughout the program). The following have internal linkage:

- `typedef` names (see Appendix C)

- enumerations

- `inline` functions

- objects declared `const`.

External linkage occurs for:

- `static` class members

- non-inline functions

- global (file scope) objects that are not declared `const`.

If the functions in different translation units have the same names and argument types, then the names will clash on linking. This can be overcome by using different namespaces. Namespaces are introduced in Chapter 14.

Changing linkage using `extern`

The keyword, `extern`, is a specifier that may be used to declare an object without defining it, as illustrated by the following code fragment:

```
extern double pi;
extern sphere s;
extern int &velocity;
```

In such cases the name and type of the variable is known to the translation unit, but the definition, which allocates memory, is elsewhere.

It is possible to use the `extern` specifier in a function declaration, as in the following statement:

---

[12]A function may only be defined once in a program, but it may be used (and therefore declared) in many translation units.

```
extern double sqrt(double z);
```

However, this is redundant since the function is necessarily defined "elsewhere".

An initialized object that is declared `extern` is in fact a definition. The motivation for definitions like this is that although a `const` object has internal linkage by default, the `extern` specifier makes the object visible to other translation units. This following statement provides an example:

```
extern const double h = 6.6262e-27;    // Planck's constant.
```

8.16.2 Header Files

The proper use of header files can do a lot to ensure the consistency of classes and functions across translation units. Once again C++ does not enforce a particular style, but there are well-established conventions and these are worth following. C++ header files usually have a `.h` (or equivalent) extension. A header file is included (by means of the `#include` directive) in the source files for which it is relevant. A large program may have many header files, some of which `#include` other header files and controlling this hierarchy can be a significant task. The `complex` class in Section 8.13 provides an example of how to use the `#ifndef` directive to avoid the possibility of multiple copies of included files.

In general, header files can and should contain the following:[13]

- Class declarations, such as:

```
class sphere {
public:
    void assign_radius(double r) { radius = r; }
    void get_radius(double &r) { r = radius; }
    double radius;
};
```

- Function declarations, as in:

```
double sqrt(double);
```

- Inline function definitions, such as:

```
inline void sphere::increment_spheres(int new_spheres)
{
    total += new_spheres;
}
```

- Variable declarations, as in:

```
extern double pi;
```

---

[13]Note the important distinction between "define" and "declare" for both variables and functions.

- Constant definitions, as in:

    ```
    const double ln_pi = 1.14472988584940017;
    ```

- Enumerations, such as:

    ```
    enum colour {RED, GREEN, BLUE};
    ```

- Other header files, such as:

    ```
    #include <cstdlib>
    ```

In addition, header files can contain the following (which we haven't covered yet):

- Named namespaces. (See Section 14.2.)

- Template declarations. (See Sections 16.1 and 16.2.)

- Template definitions. (See Sections 16.1 and 16.2.)

By contrast, header files should *not* include the following:

- Variable definitions, such as:

    ```
    double relative_velocity;
    ```

- Non-inline function definitions, as in:

    ```
    int max(int a, int b)
    {
        return a >= b ? a : b;
    }
    ```

- Constant array definitions, such as:

    ```
    const float table[] = {0.0, 1.0, 2.0};
    ```

8.17   Using Classes

Many of the objects that occur in mathematics are obvious candidates for making into classes; the possibilities are endless: vectors, matrices, complex numbers, geometrical objects, quaternions, arbitrary length integer arithmetic and many more. Most of these classes would benefit considerably from more advanced techniques that are introduced in Chapters 9 and 10. However, in this section we consider a self-checking, self-describing one-dimensional array class, which illustrates most of the ideas introduced in this chapter.[14] Let us suppose that this array class has the following design requirements:

---

[14]The exceptions are the two pointer-to-member operators since these are rarely used. (See Section 8.10.)

- The array must store n values of type `double`, where n is set at run-time.

- The elements of the array are labelled 1 to n, rather than 0 to $n-1$.

- An array object has a record of its own size and any attempt to access an element outside the array bounds is flagged as a run-time error.

- A tally is kept of the number of initialized array objects.

A suitable class declaration is:

```
// source:   array.h
// use:      Defines array class.

#ifndef ARRAY_H
#define ARRAY_H

#include <iostream>
#include <cstdlib>        // For EXIT_SUCCESS
using namespace std;

class array {
public:
    array(int size);
    int get_size(void);
    double &element(int i);
    static int number_of_arrays(void);
private:
    int elements;
    static int total;
    double *pt;
};

inline int array::get_size(void)
{
    return elements;
}

inline double &array::element(int i)
{
    if (i < 1 || i > elements) {
        cout << "Array index " << i << " out of bounds\n";
        exit(EXIT_FAILURE);
    }
    return pt[i - 1];
}

#endif  // ARRAY_H
```

Most of this class declaration is self-documenting. The pointer, pt, is the base address of a dynamically allocated array, of sufficient size to store n values of type double. The static variable, total, is the total number of arrays that have been initialized. Notice that the entire member data are private and only accessible through the member functions. (This is an example of data hiding.) Also of interest is that the function, element(), returns a reference and can therefore appear on the left-hand side of assignment statements. A class definition is usually placed in a separate header file and in this case the file is called array.h.

Those member functions of the array class that are declared inline are implemented in the array.h header file. Suitable implementations of the other member functions are in a file called array.cxx, which is given below.

```
// source:   array.cxx
// use:      Implements array class.

#include "array.h"

int array::total = 0;    // Included to demonstrate technique.
                         // (Default is 0)

array::array(int size)
{
    elements = size;
    pt = new double[elements];
    ++total;
}

int array::number_of_arrays(void)
{
    return total;
}
```

Observe that the only way of accessing an array object is through the member function, element(), which checks the validity of the array index.

A program that tries out this class is given below.[15]

```
// source:   my_test.cxx
// use:      Tests array class.

#include <iostream>
#include <cstdlib>        // For EXIT_SUCCESS
#include "array.h"
using namespace std;

int main()
{
```

---

[15] As on page 193, avoid calling the program test.cxx because it is likely there is already a command with the name **test** on your system.

```
    int array_size = 20;

    // Define an object:
    array x(array_size);

    // Access the array size:
    cout << "The array size is " << x.get_size() << "\n\n";

    // Store some data:
    for (int i = 1; i <= array_size; ++i)
        x.element(i) = i * 25.0;

    // Retrieve some data:
    for (int i = 1; i <= array_size; ++i)
        cout << x.element(i) << "\n";

    // Define another object:
    array y(array_size);

    // Now repeat the above, using a pointer to an object:
    array *p = &y;

    // Use static member function to get number of objects:
    cout << "\nNumber of arrays initialized is " <<
        array::number_of_arrays() << "\n";

    // Access the array size:
    cout << "The array size is " << p->get_size() <<
        "\n\n";

    // Store some data:
    for (int i = 1; i <= array_size; ++i)
        p->element(i) = i * 250.0;

    // Retrieve some data:
    for (int i = 1; i <= array_size; ++i)
        cout << p->element(i) << "\n";

    // Try to go outside of the array bounds:
    p->element(array_size + 1) = 3.142;

    return(EXIT_SUCCESS);
}
```

To test the array class, the file, `array.cxx`, is first compiled to give an object file, such as `array.o`, and then `my_test.c` should be compiled and linked with `array.o`. You will have to consult your C++ compiler documentation in order to discover the exact commands required. They will probably be something like:

```
g++ -c array.cxx
g++ my_test.cxx array.o
```

Most systems will have something resembling the UNIX *make* utility and this can be used to simplify program maintenance. For the present example, a file (typically called *makefile*) is created with the following contents:[16]

```
my_test: my_test.cxx array.o array.h
        g++ my_test.cxx array.o -o my_test
array.o: array.cxx array.h
        g++ -c array.cxx
```

The first line implies that the file `my_test` (which is the executable file that we wish to create) depends on the three files `my_test.cxx`, `array.o` (the object file) and `array.h` (the header file). If any of these three files is more recent than `my_test` (or `array.o` doesn't exist), then `my_test` is created again. The second line is a rule stating how to create `my_test` and is just what we would type at the system prompt. It is the instruction to compile the file `my_test.cxx` and link it with the file `array.o` to produce the output file `my_test`. The third line states that `array.o` depends on `array.cxx` and `array.h`, so if any of these is more recent than `array.o` (or `array.o` doesn't exist), then `array.o` should be created again. The rule for creating `array.o` is given in the fourth line. Again, this line is what we would type at the system prompt and states that we should compile `array.cxx`. The `-c` flag indicates that the object code (which by default is a file called `array.o` in this case) should be produced but not linked. It is important to note that in lines 2 and 4 there is a tab character before the `g++` command. This is an essential part of the syntax and replacing the tab by the equivalent number of spaces will *not* work. Typing *make* is all that is necessary to keep `my_test` up to date. It is worth emphasizing that using such a utility is almost essential for anything but the simplest program, especially as C++ together with modern programming practice strongly encourages the use of separate compilation units for unrelated functions or classes. The make utility has a rich language of its own and this can be used to control the compilation and linking of large numbers of source and header files. Have a look at your system documentation in order to get a feel for what is available.

Our `array` class still has some weaknesses, all of which can be solved by more advanced techniques:

- The default assignment operator, as in:

    ```
    array x, y;
    x.set_size(10);
    y = x;
    ```

    performs a simple copy operation, which is probably not very useful. In this example, `y.pt` points to the same memory as `x.pt`, so that making assignments to the elements of `y` will also change the `x` array.

- The notation:

---

[16]It is worth pointing out that some programmers consistently use the filename `Makefile` instead of `makefile` so that it appears near the start of a directory listing. (This only works on certain operating systems.)

```
x.elements(i) = 10.0;
```

is rather clumsy. Something like:

```
x[i] = 10.0;
```

would be much better.

- If an array object goes out of scope, the memory is not reclaimed.

8.18 Summary

- A class is a user-defined data type, complete with its own data, functions and operators:[17]

```
class circle {
    double x_centre, y_centre;
    double area(void);
};
```

- An object is an instance of a class:

```
circle one_circle, many_circles[100];
```

- The default assignment operator is a simple copy operator, which may inappropriately copy addresses (or other data):

```
circle new_circle = one_circle;
```

- Access to members of a class can be declared to be **private** (the default) or **public**:

```
class circle {
public:
    double area(void);
    void set_centre(double x, double y);
    void set_radius(double radius) { r = radius; }
    double give_radius(void);
private:
    double x_centre, y_centre;
    double r;
};
```

The data members, x_centre, y_centre and r are examples of data hiding.

---

[17]User-defined or, more correctly, overloaded operators are introduced in Chapter 9.

- A function defined within a class, such as `set_radius()`, is implicitly declared `inline`. The scope resolution operator, `::`, is used to define a member function outside the class definition:

  ```
  double circle::give_radius(void) { return r; }
  ```

  Functions defined in this way are not implicitly `inline`.

- Reference arguments eliminate unnecessary temporary copies:

  ```
  void circle::translate_circle(circle &c, double d_x,
      double d_y)
  {
      c.x_centre = x_centre + d_x;
      c.y_centre = y_centre + d_y;
      c.radius = radius;
  }
  ```

- There are two fundamental ways of specifying an object:

  ```
  circle c;                 // An object.
  circle *pt_object;        // A pointer to an object.
  ```

  and two ways of specifying a class member:

  ```
  radius = 10.0;            // A class member.
  double circle::*pt_member; // A pointer to a member.
  ```

  Hence, four different operators can be used to access class members:[18]

  1. A class member access operator:
     ```
     c.radius = 10.0;
     ```

  2. A second class member access operator:
     ```
     pt_object->radius = 10;
     ```

  3. A pointer-to-member operator:
     ```
     c.*pt_member = 10.0;
     ```

  4. A second pointer-to-member operator:
     ```
     pt_object->*pt_member = 10.0;
     ```

- There is only one copy of a `static` data member, which is declared as in:

  ```
  class circle {
      // ...
      static int total;
  };
  ```

---

[18]These examples assume that `radius` is a `public` member.

The declaration of a static data member of a class does *not* allocate memory. This is done by a statement such as:

```
int circle::total;
```

All instances of a class have access to this one copy:

```
circle c1, c2;
++c1.total;
++c2.total;
```

- A `static` data member is initialized to zero by default, but can be initialized explicitly:

```
int circle::total = 24;
```

- A `static` member function can only access `static` data and need not be invoked by an instance of a class:

```
class circle {
    // ...
    static int total_circles(void) { return total; }
    // ...
    static int total;
};

int number_of_circles = circle::total_circles();
```

- A constructor is a member function with the same name as its class. A constructor cannot return a type but can be overloaded:

```
class complex {
    // ...
    complex (double x, double y);
    complex (const complex &z);
};
```

For a class, X, a constructor with the declaration, `X::X(const X&)`, is known as a copy constructor.

- If we want to use `const` class objects, then we must define appropriate `const` member functions:

```
inline double complex::real(void) const
{
    return re;
}
```

- A function or class specified to be a **friend** of a class has access to all of the member functions and data members, irrespective of any access specifiers:

```
class complex {
// ...
    friend double real(const complex &z);
    friend class complex_matrix;
private:
    double re, im;
};
```

8.19   Exercises

1. Implement and test a **cylinder** class. The class should have **private** data members for storing the radius, height, position of the centre of the cylinder, and number of cylinders created. The public members should include a contructor, copy constructor and volume functions, together with functions to return values for the **private** data members. You should also implement a **static** member function that gives the number of cylinders created.

2. Improve the **array** class given in Section 8.17 by adding a copy constructor. Memory should be dynamically allocated for the new array and the original data elements copied into this memory.

3. Extend the **array** class, which we discussed in Section 8.17, to classes for two-dimensional and three-dimensional arrays. In each case write a program to test the new class.

Chapter 9

Operator Overloading

We already know how to overload functions. As was pointed out in Section 5.6, a function is overloaded if there is more than one function with the same name, but the functions have different numbers of arguments or different argument types. Using similar techniques, we can overload the built-in operators (such as assignment, addition, multiplication etc.) so that they perform user-defined operations. This chapter is concerned with the details of how to implement overloaded operators.

9.1 Introducing Overloaded Operators

Operators are overloaded in most languages; that is the meaning of an operator depends on its context. For instance, the operations performed by the binary plus operator, +, on the bits representing two floating point numbers are very different from those for two integers. However, the crucial feature of C++ is that most operators can be given a user-defined meaning. The only operators that cannot be overloaded are:

$$. \qquad .* \qquad :: \qquad ?: \qquad \texttt{sizeof}$$

and the preprocessor operators, `#` and `##`.

In order to prevent operator overloading getting out of control, there are a number of restrictions:

- An operator can only be overloaded if its operands include at least one class type, or it is a member function. This restriction is necessary so that a non-overloaded operator can be distinguished from its overloaded variant. As a consequence we cannot change the way operators work for the fundamental types; that is we cannot make something like:

    ```
    i = 10 + 20;
    ```

    really mean:

    ```
    i = 20 - 10;
    ```

245

Such redefinitions of built-in operators would make programs almost impossible to understand.

- The associativity and precedence of an operator, together with the number of operands, cannot be changed by overloading. Again, if these restrictions were removed, then programs could easily become incomprehensible.

- We cannot introduce new operators (such as ** for exponentiation).

- If we want to overload any of the operators:

$$= \qquad ( ) \qquad [ ] \qquad ->$$

then the overloaded operator must be implemented as a non-static member function.

9.1.1 Overloading the Assignment Operator

The assignment operator is rather special in that if we don't define an overloaded assignment operator for a class, then a default definition is provided by the compiler. This default operator simply does a copy of the data members of one object to another. Consequently, we are able to make assignments using the simple `complex` class introduced in the previous chapter, even though no assignment operator was defined. This is illustrated below.

```
complex u(24.5, 17.6), v;
v = u;
```

Exercise

Write a program to demonstrate that the values 24.5 and 17.6 are indeed assigned to the real and imaginary parts of v.

We can explicitly overload the assignment operator by making the following modification to the declaration of the `complex` class:

```
class complex {
public:
    void operator=(const complex &z);
    // More class declarations go in here.
};

inline void complex::operator=(const complex &z)
{
    re = z.re;
    im = z.im;
}

// More class implementations go in here.
```

This example illustrates the general syntax for overloading operators. In the declaration:

```
void operator=(const complex &z);
```

the `operator` keyword is followed by the operator itself and then parentheses containing the operand (or operands). The assignment operator is a binary operator, but only one operand (the second) is given explicitly, since the first operand is implicit. If we have the statement:

```
v = u;
```

for `complex` objects `v` and `u`, then the second operand is `u`. The expressions `z.re` and `z.im`, in the operator implementation, correspond to `u.re` and `u.im` respectively. The expressions `re` and `im` (without the class member operator) are the data members of `v`. In other words, because the operator is a *member* function, it has direct access to the member data. Don't forget that, since white space is ignored, the following overloaded operator declarations are all equivalent:

```
void operator = (const complex &z);
void operator=(const complex &z);
void operator= (const complex &z);
```

However, the second version is more usual.

The type that appears as the left-most part of the operator declaration (in this case `void`) is *not* the type of the left operand, but rather the type returned by the operator. Even expressions such as:

```
v = u
```

can return a type. In the above implementation we have chosen the return type to be `void`, but this does not mimic the normal situation for assignment of the fundamental types, as we now demonstrate.

9.1.2 The `this` Pointer

Consider the following statement for `i` and `j` of type `int`:

```
i = j = 1;
```

The assignment operator associates right to left, so that the first expression to be evaluated is:

```
j = 1
```

the result of which is of type `int` and value 1. It is not just `j` that has the value 1, but also the expression itself. The value of this expression is then assigned to `i`. An analogous situation does not exist for our current definition of the `complex` assignment operator, as can be seen by trying:

```
complex u, v, z(24.5, 17.6);
u = v = z;  // WRONG for current assignment operator.
```

However, we can mimic the assignment operator for the fundamental types by using the definition of the member function given below.

```
class complex {
public:
    complex &operator=(const complex &z);
    // More class declarations go in here.
};

inline complex &complex::operator=(const complex &z)
{
    re = z.re;
    im = z.im;
    return *this;
}
```

```
// More class implementations go in here.
```

In fact, this is equivalent to the default operator. In the last statement of the operator function body, we return a dereferenced pointer called `this`. The pointer, `this`, is a keyword, which is used by the C++ compiler to point to the object that the function was invoked on.[1] In the expression:

```
u = v
```

`this` points to u and:

```
return *this;
```

in fact returns u. Notice that we return a reference, which is permissible here since the reference is to something defined outside the operator function body.

The `this` pointer can only appear inside the body of a class member function and it is rare to use the pointer explicitly. Since the `this` pointer is a constant, it cannot be assigned to. The pointer is necessary because, although every instance of a class has its own data, there is only one copy of each member function. However, each non-`static` member function has its own `this` pointer, which holds the address of the object that the function was invoked on. Static member functions do not have a `this` pointer since they can only access `static` data members and there is only one instance of each such data member.

Exercise

Test the new assignment operator for the `complex` class. In particular, check that the operator behaves correctly for statements such as:

```
u = v = w = z;
```

and demonstrate that it is the explicitly defined assignment operator that is being used.

---

[1]Since overloaded operators are actually implemented by function calls, remarks concerning functions also apply to operators. (See Section 9.3 for more details.)

9.1.3 Overloading the Addition Operator

We would like to be able to write expressions involving the addition of `complex` objects in a similar way to integer and floating point expressions. For example, it would be useful to able to write the following expression for `complex` objects z, u and v:

```
z = u + v;
```

This is easily achieved by overloading the binary + operator, as shown below.

```
class complex {
    friend complex operator+(const complex &z,
        const complex &w);
    // More class declarations go in here.
};

inline complex operator+(const complex &z,
    const complex &w)
{
    return complex(z.re + w.re, z.im + w.im);
}
```

Notice that the operator is implemented as a **friend** function, rather than a member function. It therefore takes two arguments with the left and right operands corresponding to z and w respectively. We can test this overloaded operator by examining the results of the following program:

```
#include <iostream>
#include <cstdlib>        // For EXIT_SUCCESS
using namespace std;

class complex {
    friend double real(const complex &z);
    friend double imag(const complex &z);
    friend complex operator+(const complex &z,
        const complex &w);
public:
    complex &operator=(const complex &z);
    complex(double x = 0.0, double y = 0.0);
private:
    double re, im;
};

inline complex &complex::operator=(const complex &z)
{
    re = z.re;
    im = z.im;
    return *this;
}
```

```
inline complex::complex(double x, double y)
{
    re = x;
    im = y;
}

inline double real(const complex &z)
{
    return z.re;
}

inline double imag(const complex &z)
{
    return z.im;
}

inline complex operator+(const complex &z,
    const complex &w)
{
    return complex(z.re + w.re, z.im + w.im);
}

int main()
{
    complex u(1.1, 2.2), v(10.0, 20.0), w;
    w = u + v;
    cout << "Real part of w = " << real(w) <<
        "  Imaginary part of w = " << imag(w) << "\n";
    return(EXIT_SUCCESS);
}
```

Other binary operators, such as -, *, / etc. can be implemented in a similar manner.

Exercise

Implement and test overloaded binary operators for multiplication, *, and equality, ==, for the complex class.

The addition operator for the fundamental types has two significant features:

- The operator is symmetric with respect to its operands; that is (u + v) is equivalent to (v + u).

- The operands are not required to be lvalues. This is because the expression (u + v) does not assign a value to either u or v.

In order to avoid confusion, an overloaded addition operator should also possess these features.

The member function implementation of the complex addition operator:

```
class complex {
public:
    complex operator+(const complex &v);
    // ...
};

inline complex complex::operator+(const complex &v)
{
    return complex(re + v.re, im + v.im);
}
```

has the disadvantage that it is not symmetric with respect to the operands. For instance, the expression:

```
(u + v)
```

would be equivalent to:[2]

```
u.operator+(v)
```

This means that although an expression such as:

```
u + 10
```

would be valid, the apparently equivalent:

```
10 + u
```

would fail to compile since it is actually:

```
10.operator+(v)          // WRONG!
```

Consequently, the overloaded `complex` addition operator should be implemented as a `friend` function rather than a member function. Similar considerations apply to other binary operators, such as `-`, `*`, `/`, `&&` etc.

Overloaded operators that are defined as non-member functions are usually declared `friends`. This `friendship` is not directly connected with operator overloading, but is required whenever the non-member function implementing an overloaded operator needs to access `private` class data.

9.1.4   Overloading the Unary Minus Operator

The minus operator, `-`, in the context:

```
complex u, v(10.0, 20.0);
u = -v;
```

is a (prefix) unary, rather than binary, operator. For the `complex` class this operator can be implemented by the following member function:

---

[2]More detail on this operator function call notation is given in Section 9.3.

```
class complex {
public:
    complex operator-( ) const;
    // More class declarations go in here.
};

inline complex complex::operator-( ) const
{
    return complex(-re, -im);
}
```

Here, the unary minus operator is defined as a class member operator, rather than a `friend`. For this reason the operator takes no explicit arguments (the parentheses are empty) and the only "argument" is the hidden `this` pointer. The `const` at the end of the function declaration for the overloaded unary minus operator has a similar significance to `const` in the context of an ordinary member function; an object that invokes the operator cannot be changed by it.[3] In the above case an expression such as `-v` cannot change the values stored by `v`. A complete program demonstrating this implementation of the unary minus operator is given below.

```
#include <iostream>
#include <cstdlib>          // For EXIT_SUCCESS
using namespace std;

class complex {
    friend double real(const complex &z);
    friend double imag(const complex &z);
public:
    complex &operator=(const complex &z);
    complex(double x = 0.0, double y = 0.0);
    complex operator-( ) const;
private:
    double re, im;
};

inline complex &complex::operator=(const complex &z)
{
    re = z.re;
    im = z.im;
    return *this;
}

inline complex::complex(double x, double y)
{
    re = x;
    im = y;
}
```

---

[3]See Section 8.14.

```
inline complex complex::operator-( ) const
{
    return complex(-re, -im);
}

inline double real(const complex &z)
{
    return z.re;
}

inline double imag(const complex &z)
{
    return z.im;
}

int main()
{
    complex u, v(10.0, 20.0);
    u = -v;
    cout << "Real part of u = " << real(u) <<
        "  Imaginary part of u = " << imag(u) << "\n";
    return(EXIT_SUCCESS);
}
```

An alternative way to implement the unary minus operator is as a `friend` function, as shown below.

```
class complex {
    friend complex operator-(const complex &z);
    // More class declarations go in here.
};

inline complex operator-(const complex &z)
{
    return complex(-z.re, -z.im);
}
```

In contrast to binary operators, both member and non-member functions are acceptable ways of overloading the unary minus operator.

Exercise

Test the above definition of the unary minus operator implemented as a `friend` of the `complex` class.

9.2   User-defined Conversions

The fundamental data types have built-in conversions, which may be either implicit or explicit. As an example of an implicit cast, consider the following statement:

```
double x = 10;
```

In this case, the compiler supplies a cast that converts 10, of type int, to a double. A cast is explicit when it is supplied by the programmer, as demonstrated below.

```
int i, j;
// Assignments to i and j go in here.
double x = static_cast<double>(i) / static_cast<double>(j);
```

Such conversions are not just a cosmetic change of type; the bit pattern representing 10 is very different from that representing 10.0. Moreover, the built-in conversions are an essential simplification, without which the compiler would have to provide a different implementation of every binary operator for each legal combination of the fundamental types.

It is also possible to perform conversions from a class or fundamental data type to a class. Again, such conversions may be either implicit or explicit, but we must supply functions that specify how the conversions are to be made. There are two ways in which conversions involving classes can be performed: by constructors accepting a single argument and by conversion functions. Before we continue our discussion of operator overloading, we examine these conversions in more detail.

9.2.1   Conversion by Constructors

A constructor accepting a single argument, not of the class type, converts that argument to the class type. Such conversions are relatively common. For instance, an example of a constructor that accepts a single argument is given below.

```
class complex {
public:
    // Additional constructor:
    complex(double x) { re = x; im = 0.0; }
    // More class declarations go in here.
};
```

This constructor would apply implicit and explicit conversions from the type double to complex. However, the various versions of our complex class considered in this chapter have a constructor (originally defined in Section 8.13) with the declaration:

```
complex::complex(double x = 0.0, double y = 0.0);
```

Because default arguments are supplied right to left, this implies that we have already implemented a constructor taking a single argument of type double. Moreover, it would be an error to supply another such constructor since this would introduce ambiguous function overloading. Consequently, the complex class as already defined will perform the conversions given below.

```
complex u, v, z;
v = complex(3.6);       // Explicit conversion by constructor.
u = 1.2;                // Implicit conversion by constructor.
z = u + 3 + v;          // Implicit conversion by constructor.
```

The assignment to v clearly involves an explicit conversion by the `complex` constructor. In the statement assigning 1.2 to u, the constructor is implicitly used to perform a conversion. Moreover, z is evaluated correctly, even if we have only declared:

```
friend complex operator+(const complex &z, const complex &w);
```

and have not given versions for operands of type `double`. This is because 3 is implicitly converted to `double`, which is in turn implicitly converted to `complex` by the constructor function. These conversions are demonstrated in the program given below.

```cpp
#include <iostream>
#include <cstdlib>        // For EXIT_SUCCESS
using namespace std;

class complex {
    friend double real(const complex &z);
    friend double imag(const complex &z);
    friend complex operator+(const complex &z, const complex &w);
    friend complex operator-(const complex &z);
public:
    complex &operator=(const complex &z);
    complex(double x = 0.0, double y = 0.0);
private:
    double re, im;
};

inline complex &complex::operator=(const complex &z)
{
    re = z.re;
    im = z.im;
    return *this;
}

inline complex::complex(double x, double y)
{
    re = x;
    im = y;
}

inline double real(const complex &z)
{
    return z.re;
}

inline double imag(const complex &z)
{
    return z.im;
}
```

```
inline complex operator+(const complex &z, const complex &w)
{
    return complex(z.re + w.re, z.im + w.im);
}

inline complex operator-(const complex &z)
{
    return complex(-z.re, -z.im);
}

int main()
{
    complex u, v, z;
    v = complex(3.6);   // Explicit conversion by constructor.
    cout << "Real part of v = " << real(v) <<
        "  Imaginary part of v = " << imag(v) << "\n";
    u = 1.2;            // Implicit conversion by constructor.
    cout << "Real part of u = " << real(u) <<
        "  Imaginary part of u = " << imag(u) << "\n";
    z = u + 3;          // Implicit conversion by constructor.
    cout << "Real part of z = " << real(z) <<
        "  Imaginary part of z = " << imag(z) << "\n";
    return(EXIT_SUCCESS);
}
```

9.2.2   Conversion Functions[††]

A constructor accepting a single argument can only convert to its own class and this has two consequences:

- A constructor cannot convert to one of the fundamental types.

- Suppose class Y has been implemented and we define a new class, X. A class X constructor can convert from Y to X, but we must modify class Y in order that a class Y constructor can convert from X to Y. However, changing the source of class Y may not be an option.

The solution to both of these problems is to use a *conversion function*.

A conversion function is a class member declared as in:

```
class my_class {
public:
    operator type ();
    // More class declarations go in here.
};
```

where type is the type returned, which could be either a fundamental type (or a simple derivation, such as a pointer to a fundamental type) or else a class. For example, if we need a conversion from my_class to int, then the appropriate conversion function declaration is given in the following class declaration:

```
class my_class {
public:
    operator int ();
    // More class declarations go in here.
};
```

As demonstrated below, it is illegal to specify an argument for a conversion function.

```
class my_class {
public:
    operator int(int);      // WRONG: cannot specify argument.
    // More class declarations go in here.
};
```

Moreover, since the return type is already part of the function name, you cannot repeat the return type in an attempt to follow the usual syntax for a function. Consequently, the following code fragment is not valid:

```
class my_class {
public:
    int operator int();     // WRONG: cannot specify return type.
    // More class declarations go in here.
};
```

Another restriction is that since conversion functions do not take arguments they cannot be overloaded.

A conversion function is a user-defined cast operator and can be used in two distinct ways:

1. **Conversion From a Class to a Fundamental Type**

   Suppose we define a `time` class, which stores the time in hours, minutes and seconds. We can define a conversion function that converts a `time` object into the `int` type, corresponding to seconds. This is demonstrated below.

   ```
   #include <iostream>
   #include <cstdlib>      // For EXIT_SUCCESS
   using namespace std;

   class time {
   public:
       // Define a constructor:
       time(int h, int m, int s) { hours = h; minutes = m;
           seconds = s; }
       // Define a conversion function:
       operator int() { return(seconds + 60 * (minutes +
           60 * hours)); }
   private:
       int hours, minutes, seconds;
   };
   ```

```
int main()
{
    time t(16, 21, 35); // 16 hours 21 mins. 35 secs.
    int s;
    s = int(t);           // Time in seconds.
    cout << "Time in seconds is " << s << "\n";
    return(EXIT_SUCCESS);
}
```

Exercise

Try running the above program and check the result that you obtain.
What happens if you comment out the conversion function definition?

2. **Conversion Between Classes**

   It is also possible to provide conversion functions that convert between classes.
   As an example, suppose we define an array class for objects with three elements
   as shown below.

```
class array3 {
public:
    array3(){ data[0] = data[1] = data[2] = 0.0; }
    array3(double x, double y, double z);
    array3 &operator=(const array3 &a);
    void list_data(void);
private:
    double data[3];
};

array3::array3(double x, double y, double z)
{
    data[0] = x;
    data[1] = y;
    data[2] = z;
}

array3 &array3::operator=(const array3 &a)
{
    data[0] = a.data[0];
    data[1] = a.data[1];
    data[2] = a.data[2];
    return *this;
}

void array3::list_data(void)
{
    cout << "(" << data[0] << ", " << data[1] << ", " <<
```

```
        data[2] << ")";
    }
```

An array class, `array2`, for objects with two elements can provide a conversion function from `array2` to `array3`, as shown in the following code:

```
class array2 {
public:
    array2(double x, double y);
    array2 &operator=(const array2 &a);
    operator array3();
    void list_data(void);
private:
    double data[2];
};

array2::array2(double x, double y)
{
    data[0] = x;
    data[1] = y;
}

array2 &array2::operator=(const array2 &a)
{
    data[0] = a.data[0];
    data[1] = a.data[1];
    return *this;
}

// Define a conversion function:
array2::operator array3()
{
    return array3(data[0], data[1], 1.0);
}

void array2::list_data(void)
{
    cout << "(" << data[0] << ", " << data[1] << ")";
}
```

If we define an object of the `array2` class, as in:

```
array2 a2(10.0, 20.0);
```

then we can make an explicit conversion to the `array3` class, as demonstrated below.

```
array3 a3 = array3(a2);
```

Exercise

Verify that this conversion gives the expected result and then change the initialization statement for `a3` so that an implicit conversion is performed.

9.2.3   Implicit Conversions[††]

User-defined conversions should be kept simple since they are usually applied implicitly and it may not be obvious which conversions are used. Conversions are typically applied implicitly in the following circumstances:[4]

1. **Function Arguments**

   A function declared as:

   ```
   complex f(complex z);
   ```

   may be invoked by:

   ```
   complex w = f(2.4);
   ```

   The function argument, `2.4`, is implicitly converted to a `complex` object and the statement is equivalent to:

   ```
   complex w = f(complex(2.4));
   ```

2. **Function Return Values**

   The body of the `f()` function, declared above, could correctly include the statement:

   ```
   return(x);
   ```

   where x has type `double`. This `return` statement would then be equivalent to:

   ```
   return(complex(x));
   ```

3. **Operands**

   If we declare:

   ```
   friend complex operator*(const complex &u, const complex &v);
   ```

   then we do not need to declare versions of the * operator with a `double` operand, such as:

   ```
   friend complex operator*(const double &x, const complex &v);
   ```

---

[4]Notice that in all three examples the required conversion is from `double` to `complex` and therefore must be performed by a constructor rather than a conversion function. However, in other circumstances a conversion function may be appropriate.

For example, the expression:

```
complex z = 2.0 * v;
```

has an implicit conversion from `double` to `complex`.

In principle, it is possible to define a constructor and a conversion function, both of which perform a conversion between the same two types. However, any attempt to perform an implicit conversion between the two types would be ambiguous and would therefore fail.

Not more than one implicit user-defined conversion is applied in any one instance. For example, we could extend the example of the previous section to include a class for objects with four elements:

```
class array4 {
public:
    array4(){data[0] = data[1] = data[2] = data[3] = 0.0;}
    array4(double w, double x, double y, double z);
    array4 &operator=(const array4 &a);
    void list_data(void);
private:
    double data[4];
};
```

and introduce a new conversion function into the `array3` class with the implementation:

```
array3::operator array4()
{
    return array4(data[0], data[1], data[2], 1.0);
}
```

We now have the following user-defined conversions:

$$\text{array2} \longrightarrow \text{array3}$$
$$\text{array3} \longrightarrow \text{array4}$$

Given objects, a2, a3 and a4, for the classes, `array2`, `array3` and `array4`, implicit conversions occur for the following statements:

```
a3 = a2;
a4 = a3;
```

However, the statement:

```
a4 = a2;         // WRONG: needs two implicit conversions.
```

is invalid since only one user-defined conversion can be performed implicitly. If this were not so then conversions could get completely out of control.

Exercise

Verify by means of a short program that although single implicit conversions are applied from a2 to a3, and from a3 to a4, a double implicit conversion is not applied from a2 to a4.

9.3   Operator Function Calls

So far, we have introduced operator overloading in the context of the `complex` class and described user-defined conversion in some detail. We are now ready to return to a more thorough treatment of the subject of operator overloading.

Operator overloading is actually implemented by an *operator function* call. A non-member operator, such as:

```
friend complex operator+(const complex &u, const complex &v);
```

is a function with the name "`operator+`", taking two `complex` arguments and returning the `complex` type. An expression of the form:

```
z = u + v;
```

where `z`, `u` and `v` are all instances of the `complex` class, is directly equivalent to:

```
z.operator=(operator+(u, v));
```

Exercise

Verify the equivalence between the operator and function call notations by listing the real and imaginary parts of `z` for explicit (complex) `u` and `v` in the above expression. Are the outermost parentheses significant in the operator function call? Can white space be inserted around the = or +?

It is very rare for operator function calls to be invoked explicitly, since the resulting expressions are more cumbersome and less intuitive than their operator counterparts. However, the ability to rewrite expressions in terms of operator function calls (as above) leads to an understanding of how overloaded operators really work and this comes in useful when actually defining such operators.

Most overloaded operators can be implemented by both non-`static` member functions and non-member functions (which are usually `friend`s). One set of exceptions consists of the `new`, `new[]`, `delete` and `delete[]` operators. However, since the overloading of these operators is only likely to be useful in advanced applications, we do not consider them any further in this chapter. The other exceptions consists of the assignment, function call, subscripting and class member access operators:

$$= \quad (\ ) \quad [\ ] \quad ->$$

which can only be overloaded by non-`static` member functions. The overloaded class member access operator, `->`, has the additional unique feature in that it is considered to be a unary operator; that is:

```
pt->x
```

is interpreted as:

```
(pt.operator->())->x
```

Consequently, `pt.operator->()` must return something that can be used as a pointer.

It is worth emphasizing that the overloading of composite operators, such as `+=`, is independent of other overloaded operators. For instance, if we have defined the assignment and binary plus operators for the `complex` class, then if we want to use the `+=` operator we still need to define it. For the fundamental types, such as `int`, the statement:

```
k += 10;
```

has exactly the same meaning as:

```
k = k + 10;
```

We don't actually need to overload an operator like the `+=` operator in a way that mimics its meaning for the fundamental types. However, it would certainly be sensible to do so.

9.3.1 Binary Operators

A binary operator is invoked by the expression:

```
A @ B
```

where `A` and `B` are the two operands, and the `@` symbol is used to denote a generic operator. The expression `A @ B` is equivalent to:

```
A.operator@(B)
```

for a non-`static` member function and:

```
operator@(A, B)
```

for a non-member function implementation. Whereas the member function has one explicit argument and the hidden `this` pointer, the non-member function has two arguments and no `this` pointer.

As an example, suppose `z` and `w` are instances of the `complex` class, then:

```
z += w;
```

is equivalent to:

```
z.operator+=(w);
```

for the member function, and:

```
operator+=(z, w);
```

for the non-member function. The member function would be declared as:

```
class complex {
public:
    complex &operator+=(const complex &z);
    // More class declarations go in here.
};
```

Assuming the data members of the `complex` class are `private`, then the non-member function must be declared a `friend` of the class, as shown below.

```
class complex {
public:
    friend operator+=(const complex &u, const complex &v);
    // More class declarations go in here.
};
```

A significant difference between non-member and member binary operators is that whereas user-defined conversions may be applied to both operands for non-member implementations, user-defined conversions are not applied to the first operand for member implementations. For example, suppose we declare an (incomplete) `complex` class as:

```
class complex {
    friend double real(const complex &z);
    friend double imag(const complex &z);
public:
    complex(double x = 0.0, double y = 0.0);
    complex operator+(const complex &z);
    complex &operator=(const complex &z);
private:
    double re, im;
};
```

and initialize variables, u and x by:

```
complex u(1.0, 2.0);
double x = 5.0;
```

Then, if z is declared `complex`, the statement:

```
z = u + x;
```

is valid because the constructor with the default imaginary argument converts the explicit argument of the `operator+` function from `double` to `complex`. However, because the user-defined conversion is not applied to the dereferenced hidden `this` pointer, the following statement does not compile:

```
z = x + u;
```

This explains why Section 9.1.3 emphasized that the `complex` addition operator should be implemented as a `friend` function rather than as a member function.

Exercise

(a) Implement the `complex` class defined above and verify the claims made.

(b) Modify the `matrix` class introduced in Section 7.8.1 so that two matrices can be added by using the overloaded binary plus operator.

9.3.2   Prefix Unary Operators

A prefix unary member operator @ is invoked by the expression:[5]

```
@ A
```

where A is the single operand. This expression is equivalent to:

---
[5]Don't forget that we use the @ symbol to stand for a generic operator.

```
A.operator@()
```

for a non-static member function and:

```
operator@(A)
```

for a non-member function implementation. Whereas the member function has the hidden `this` pointer and no explicit argument, the non-member implementation has no `this` pointer and a single argument.

As an example, if `z` is an instance of the complex class, then:[6]

```
++z;
```

is actually invoked by:

```
z.operator++();
```

for the member implementation, and:

```
operator++(z);
```

for the non-member implementation. The member function would be declared as:

```
complex class {
 public:
    complex operator++();
    // More class declarations go in here.
};
```

and the non-member function as:

```
complex class {
 public:
    friend complex operator++(complex &z);
    // More class declarations go in here.
};
```

A program demonstrating the member function implementation is given below.

```
#include <iostream>
#include <cstdlib>        // For EXIT_SUCCESS
using namespace std;

class complex {
    friend double real(const complex &z);
    friend double imag(const complex &z);
 public:
    complex(double x = 0.0, double y = 0.0);
    complex operator++();
 private:
```

---

[6]It is not clear that we would want to overload the increment operator for the `complex` class. For the purpose of illustration, we could suppose that `++` increments the real part of `complex` objects.

```
        double re, im;
};

inline complex::complex(double x, double y)
{
    re = x;
    im = y;
}

// Member function implementation of prefix unary operator:
inline complex complex::operator++()
{
    return complex(++re, im);
}

inline double real(const complex &z)
{
    return z.re;
}

inline double imag(const complex &z)
{
    return z.im;
}

int main()
{
    complex z(3.0, 2.0);
    ++z;
    cout << "Real part of z = " << real(z) <<
        "  Imaginary part of z = " << imag(z) << "\n";
    z.operator++();
    cout << "Real part of z = " << real(z) <<
        "  Imaginary part of z = " << imag(z) << "\n";
    return(EXIT_SUCCESS);
}
```

Of course, in real code we would invoke the member function by using ++z rather than
z.operator++().

Exercise

(a) Check that the output from the above program is what you would
    expect.

(b) Modify the program so that it uses the non-member implementation
    of the prefix ++ operator.

9.3.3 Postfix Unary Operators

The increment and decrement operators have postfix, as well as prefix, variants. These postfix operators can be either non-member functions or non-`static` class members, and in both cases are distinguished from their prefix versions by having an extra argument of type `int`.

In the case of a member function implementation, an expression of the form:

```
A@
```

where `@` is either `++` or `--`, is equivalent to:

```
A.operator@(0)
```

Notice that the function is invoked with an argument of value zero and it is this that distinguishes the postfix from the prefix operators. If `z` is an instance of the `complex` class, then:

```
z++;
```

is invoked by:

```
z.operator++(0);        // Postfix operator.
```

However, the statement:

```
++z;
```

corresponds to:

```
z.operator++();        // Prefix operator.
```

For the complex class, the postfix `++` member operator would be declared as:

```
class complex {
 public:
    complex operator++(int);
    // More class declarations go in here.
};
```

In the case of a non-member implementation, the expression:

```
A@
```

is equivalent to:

```
operator@(A, 0)
```

The function is invoked with two arguments, the second of which has the value zero. If `z` is a `complex` object, then `z++` is invoked by:

```
operator++(z, 0);        // Postfix operator.
```

However, `++z` corresponds to:[7]

---

[7]Once again, in real code we wouldn't invoke the operators by using `operator++(z, 0)` or `operator++(z)`; we would simply use `z++`.

```
    operator++(z);           // Prefix operator.
```

The postfix ++ non-member operator would be declared as:

```
class complex {
 public:
    complex complex::operator++(complex &z, int);
    // More class declarations go in here.
};
```

A program demonstrating the non-member version of the postfix ++ operator is given below.

```
#include <iostream>
#include <cstdlib>        // For EXIT_SUCCESS
using namespace std;

class complex {
    friend double real(const complex &z);
    friend double imag(const complex &z);
    friend complex operator++(complex &z, int);
public:
    complex(double x = 0.0, double y = 0.0);
private:
    double re, im;
};

inline complex::complex(double x, double y)
{
    re = x;
    im = y;
}

inline double real(const complex &z)
{
    return z.re;
}

inline double imag(const complex &z)
{
    return z.im;
}

// Non-member function implementation of postfix unary operator:
inline complex operator++(complex &z, int)
{
    return complex(++z.re, z.im);
}
```

```
int main()
{
    complex z(3.0, 2.0);
    z++;
    cout << "Real part of z = " << real(z) <<
        " Imaginary part of z = " << imag(z) << "\n";
    operator++(z, 0);
    cout << "Real part of z = " << real(z) <<
        " Imaginary part of z = " << imag(z) << "\n";
    return(EXIT_SUCCESS);
}
```

Exercise

(a) Check that the above program gives the output that you would expect.

(b) Modify the program so that it uses the member function implementation of the postfix -- operator.

9.4 Some Special Binary Operators

There are three special binary operators that can only be overloaded by non-static *member* functions.[8] These are the assignment, subscripting and function call operators. Since we have already considered the assignment operator in some detail, we now turn our attention to the subscripting and function call operators.

9.4.1 Overloading the Subscripting Operator

As we have already mentioned, the built-in concept of an array in C++ is very primitive. An array element, a[i], means nothing more than *(a+i) and multi-dimensional arrays, such as b[i][j][k], are merely successive left to right applications of the subscripting operator, []. Moreover, indexing goes from zero up to one less than the number of elements and no checking is done to ensure that a program keeps to this range.

The overloaded subscripting operator has the form A[B], where A must be a class object and B can have any type. In the function call notation, the expression, A[B], is equivalent to:[9]

A.operator[](B)

There is no necessity for the overloaded operator to have any connection with the concept of an array; we simply have a function, with the name "operator[]". The function has two arguments; the first is the hidden this pointer and the second is B, which is not limited to the integral or even fundamental types. In spite of this freedom of definition, it is advisable to make the overloaded subscripting operator have some connection with its built-in counterpart. As a simple example, suppose we need objects

---

[8]Since they are overloaded by non-static member functions, their first operands are lvalues.

[9]Any white space inserted within the square brackets of operator[] is ignored.

that are one-dimensional arrays with the index going from 1 to 3. Also assume that
we want any array access to be checked, in case the index goes out of bounds. The
following class implements these features:

```
class array {
public:
    double &operator[](int i);
private:
    double data[3];
};

double &array::operator[](int i)
{
    if (i < 1 || i > 3) {
        cout << "Index = " << i << " out of range\n";
        exit(EXIT_FAILURE);
    }
    return data[i-1];
}
```

Notice that we return a reference for `operator[]()`. This enables us to have `array`
objects on the left-hand side as well as the right-hand side of assignment statements,
as the program given below demonstrates.

```
#include <iostream>
#include <cstdlib>        // For exit()
using namespace std;

class array {
public:
    double &operator[](int i);
private:
    double data[3];
};

double &array::operator[](int i)
{
    if (i < 1 || i > 3) {
        cout << "Index = " << i << " is out of range.\n";
        exit(EXIT_FAILURE);
    }
    return data[i-1];
}

int main()
{
    array x;
    // Assign values to elements of the array:
    for (int i = 1; i <= 3; ++i)
```

```
        x[i] = 10.0 * i;
    // Test assignment:
    for (int i = 1; i <= 3; ++i)
        cout << "x[" << i << "] = " << x[i] << "\n";
    // Try setting the index out of range:
    x[0] = 3.142;
    return(EXIT_SUCCESS);
}
```

Exercise

The `array` class introduced in Section 8.17 used the `element()` function to access elements of an `array` object. Replace this function by the subscripting operator so that statements such as

```
    x[i] = 10.0;
```

are valid (where x is an `array` object).

9.4.2 Overloading the Function Call Operator

The function call operator can also be overloaded. An expression such as `A(x, y, z)` is equivalent to:

```
    A.operator()(x, y, z)
```

where A must be a class object, but x, y and z can be of any type. There is no restriction on the number of arguments. For a single argument, the overloaded function call operator can play the same role as the overloaded subscripting operator. The function call operator could also be used to access the elements of a multi-dimensional `array` class. Suppose we want to extend our example in the previous section to two-dimensional arrays, with the data stored by columns, rather than rows. A suitable class implementing this idea is as follows.

```
class array {
public:
    double &operator()(int i, int j);
private:
    double data[3][3];
};

double &array::operator()(int i, int j)
{
    if (i < 1 || j < 1 || i > 3 || j > 3) {
        cout << "Index out of range:  i = " << i << "  j = " <<
            j <<"\n";
        exit(EXIT_FAILURE);
    }
    return data[j-1][i-1];
}
```

Notice that the indices are reversed in the return statement for `operator()` in order to implement storage by columns.

A simple test program for this class is given below.

```
#include <iostream>
#include <cstdlib>        // For exit()
using namespace std;

class array {
public:
    double &operator()(int i, int j);
private:
    double data[3][3];
};

double &array::operator()(int i, int j)
{
    if (i < 1 || j < 1 || i > 3 || j > 3) {
        cout << "Index out of range:   i = " << i << "   j = " <<
            j << "\n";
        exit(EXIT_FAILURE);
    }
    return data[j-1][i-1];
}

int main()
{
    array x;
    // Assign values to elements of the array:
    for (int i = 1; i <= 3; ++i)
        for (int j = 1; j <= 3; ++j)
            x(i, j) = 10.0 * i * j;
    // Test assignment:
    for (int i = 1; i <= 3; ++i)
        for (int j = 1; j <= 3; ++j)
            cout << "x(" << i << ", " << j << ") = " <<
                x(i, j) << "\n";
    // Try setting the index out of range:
    x(0, 2) = 3.142;
    return(EXIT_SUCCESS);
}
```

Exercise

(a) Compile and run the above program. Modify both the class and test program to handle three-dimensional arrays.

(b) In the `matrix` class introduced in Section 7.8.1, an element of a matrix was accessed by means of the `element()` function. Use the overloaded function call operator to modify this class so that an element a_{ij} of a matrix is accessed with the notation `A(i, j)` instead of the `element()` function.

9.5 Defining Overloaded Operators

There are some general guidelines that are worth following when defining overloaded operators:

- Overloaded operators should mimic their built-in counterparts; for example:

  ```
  z = u + v;
  ```

  should in some sense correspond to addition for whatever class (or classes) `z`, `u` and `v` belong.

- Operators that have related meanings for fundamental types should continue to do so for their overloaded counterparts. For instance:

  ```
  z += u;
  ```

  should be completely equivalent to:

  ```
  z = z + u;
  ```

- In general, overloaded operators can be defined as either member functions or non-member functions (usually `friends`). The exceptions are that:

  $$= \qquad (\ ) \qquad [\ ] \qquad ->$$

  must be non-`static` member functions. (The operators `new`, `new[]`, `delete` and `delete[]` are also exceptions, but the overloading of these operators is not considered in this chapter.)

- If a binary operator for a fundamental type requires an lvalue for the left operand, then the overloaded operator should be declared as a member function. For example, the `+=` operator for the `complex` class should be declared as:

  ```
  class complex {
  public:
      complex &operator+=(const complex &u);
      // More class declarations go in here.
  };
  ```

  rather than:

  ```
  class complex {
  public:
      friend void operator+=(complex &u, const complex &v);
      // More class declarations go in here.
  };
  ```

In the second case, counter-intuitive statements of the form:

```
1.4 += z;        // WRONG
```

would be accepted by the compiler.

- When the operator does not require an lvalue for a fundamental type, the operator should be a non-member function, rather than a class member. For instance, as discussed in Section 9.1.3, the + operator for the complex class should be declared as:

```
class complex {
public:
    friend complex operator+(const complex &u,
        const complex &v);
    // More class declarations go in here.
};
```

rather than:

```
class complex {
 public:
    complex operator+(const complex &v);
};
```

- In contrast to binary operators, member and non-member implementations of unary operators are often equally acceptable. However, as noted previously, one feature distinguishing member and non-member functions is that user-defined conversions are *not* applied to an argument which is effectively the dereferenced hidden this pointer.

9.6   Using Overloaded Operators

9.6.1   Complex Arithmetic

In this section we outline the implementation of a complex arithmetic class; it is left as an exercise to complete the project. The header file for the class is given below.

```
// source:   complex.h
// use:      Defines complex arithmetic class.
//           Implements inline functions.

#ifndef COMPLEX_H
#define COMPLEX_H

#include <cmath>           // For sqrt(), cos(), sin(), exp()
using namespace std;
```

```
class complex {
    friend complex operator+(double x, const complex &v);
    friend complex operator+(const complex &u, double x);
    friend complex operator+(const complex &u, const complex &v);
    friend double real(const complex &z);
    friend double imag(const complex &z);
    friend double mod(const complex &z);
    friend complex conj(const complex &z);
    friend complex exp(const complex &z);
public:
    double &real(void);
    double &imag(void);
    complex(void) { }
    complex(double r, double i);
    complex(const complex &z);
    complex &operator=(double x);
    complex &operator=(const complex &z);
    complex &operator+=(double x);
    complex &operator+=(const complex &z);
    complex operator-() const;
private:
    double re, im;
};

// friend functions:
inline complex operator+(double x, const complex &v)
{
    return complex(x + v.re, v.im);
}

inline complex operator+(const complex &u, double x)
{
    return complex(u.re + x, u.im);
}

inline complex operator+(const complex &u, const complex &v)
{
    return complex(u.re + v.re, u.im + v.im);
}

inline double real(const complex &z)
{
    return z.re;
}

inline double imag(const complex &z)
{
```

```
        return z.im;
}

inline double mod(const complex &z)
// Modulus of z.
{
        return(sqrt(z.re * z.re + z.im * z.im));
}

inline complex conj(const complex &z)
// Complex conjugate of z.
{
        return complex(z.re, -z.im);
}

inline complex exp(const complex &z)
// Exponential function for complex argument.
{
        double temp = exp(z.re);

        return complex(temp * cos(z.im), temp * sin(z.im));
}

// Member functions and operators:
inline double &complex::real(void)
{
        return re;
}

inline double &complex::imag(void)
{
        return im;
}

inline complex::complex(double r, double i)
{
        re = r;
        im = i;
}

inline complex::complex(const complex &z)
{
        re = z.re;
        im = z.im;
}

inline complex &complex::operator=(double x)
```

```
{
    re = x;
    im = 0.0;
    return *this;
}

inline complex &complex::operator=(const complex &z)
{
    re = z.re;
    im = z.im;
    return *this;
}

inline complex &complex::operator+=(double x)
{
    re += x;
    return *this;
}

inline complex &complex::operator+=(const complex &z)
{
    re += z.re;
    im += z.im;
    return *this;
}

inline complex complex::operator-() const
{
    return complex(-re, -im);
}
```

```
#endif  // COMPLEX_H
```

Notice how `complex` arguments are passed as `const complex &z` wherever possible. This is in order to improve efficiency. However, the `double` arguments are passed by value since using reference arguments would not be any quicker. Also notice that the library header, `<cmath>`, provides declarations for many common mathematical functions, such as `sqrt()`, `cos()`, `sin()` and `exp()`. Finally, it is worth pointing out that there are two different `exp()` functions. One is declared in `<cmath>` as:

```
double exp(double x);
```

The second `exp()` function is a `friend` of the `complex` class and is declared as:

```
complex exp(const complex &z);
```

These overloaded functions are distinguished by their arguments.[10]

---

[10]Don't forget that the potential ambiguity in overloaded functions is not resolved by different return types, but by their arguments. (See Section 5.6.1.)

Since all functions are `inline`, there is no `complex.cxx` file, although it could reasonably be argued that `mod()` and `exp()` should not be `inline`, as they involve non-trivial computation. A short program testing this class is given below.

```
// source:  my_test.cxx
// use:     Tests complex arithmetic class.

#include <iostream>
#include <cstdlib>        // For EXIT_SUCCESS
#include "complex.h"
using namespace std;

void print(const complex &z)
{
    cout << real(z) << " + i * (" << imag(z) << ")\n";
}

int main()
{
    complex z1(1, 2), z2(3, 3), z3;
    const complex i(0, 1);

    cout << "z1 = ";
    print(z1);
    cout << "z2 = ";
    print(z2);
    cout << "i = ";
    print(i);

    z3 = z1 + z2;
    cout << "z1 + z2 = ";
    print(z3);

    cout << "1.0 + z1 = ";
    print(1.0 + z1);
    cout << "z1 + 1.0 = ";
    print(z1 + 1.0);

    z3.real() = 3.4;
    z3.imag() = 4.5;
    cout << "Real and imaginary parts assigned: ";
    print(z3);

    cout << "mod(z1) = " << mod(z1) << "\n";

    cout << "conj(z1) = ";
    print(conj(z1));
```

```
        cout << "exp(z1) = ";
        print(exp(z1));

        return(EXIT_SUCCESS);
}
```

There is one feature of C++ that would improve this class. We have used:

```
print(z);
```

to print a complex number, but it would be better if this statement could be replaced by:

```
cout << z;
```

Techniques for achieving this are described in Chapter 18.

Exercise

Extend the `complex` class by implementing and testing overloaded `==` and `!=` operators. The meaning of these operators should mimic the equal and not equal operators for the fundamental types.

9.6.2 Strings

The idea of a string that C++ inherits from C simply consists of a `char` array with a string terminator, `'\0'`. Manipulation of such strings is prone to error. For example, a common mistake is to forget that the library function, `strlen()`, does not count the terminator when returning the length of a string. However, by implementing a `string` class we can provide both a safer and a more natural way of manipulating strings.[11] A suitable class declaration, together with the `inline` function definitions, is given below.[12]

```
// source:      string.h
// use:         Defines self-describing string class.

#ifndef SELF_DESCRIBING_STRING_H
#define SELF_DESCRIBING_STRING_H

#include <cstring>      // For memcmp()
using namespace std;

class string {
friend string operator+(const string &s1,const string &s2);
friend bool operator==(const string &s1, const string &s2);
friend bool operator!=(const string &s1, const string &s2);
public:
```

---

[11] The class described here is *not* the same as the class of the same name that is part of the Standard Library, described in Chapter 17.

[12] Do not confuse `string.h` with `<cstring>` (or `<string.h>` in pre-ANSI versions of C++).

```
        string(void) { characters = 0; pt = NULL; }
        string(const string &s);
        string(char *s);
        string &operator=(const string &s);
        char *char_array (void) const;
        int length(void) const { return characters; }
        void print(void);
private:
        string(int set_length);
        int characters;
        char *pt;
};

inline bool operator==(const string &s1, const string &s2)
{
        if (s1.characters == s2.characters &&
            !memcmp(s1.pt, s2.pt, s1.characters))
            return true;
        else
            return false;
}

inline bool operator!=(const string &s1, const string &s2)
{
        if (s1.characters == s2.characters &&
            !memcmp(s1.pt, s2.pt, s1.characters))
            return false;
        else
            return true;
}

#endif   // SELF_DESCRIBING_STRING_H
```

A string object consists of a char array (*without* a string terminator) together with the number of characters in the string, stored by the variable called characters. Functions are provided to concatenate string objects, test for equality and inequality, and to perform assignment. The char_array() function returns a pointer to an array that stores a *copy* of the string object, s, as a standard C++ null terminated string. This function avoids exposing details of the string implementation in circumstances where a ordinary null terminated string is required. Notice the use of the memcmp() function, declared in <cstring> as:

```
        int memcmp(void *pt_1, void *pt_2, size_t count);
```

This function compares each successive byte pointed to by pt_1 with the corresponding byte pointed to by pt_2 until either they do not have the same value or the number of bytes specified by count have been compared. The function returns an integer less than, equal to, or greater than zero, depending of whether the last comparison done for the byte pointed to by pt_1 is less than, equal to, or greater than the byte pointed to by `pt_2`. Consequently, if the value returned by `memcmp()` in the above code is non-zero, we know that the strings are not equal.

The non-inline members of the `string` class can be implemented as shown below.

```
// source:     string.cxx
// use:        Implements string class.

#include <iostream>
#include "string.h"
using namespace std;

string::string(const string &s)
{
    characters = s. characters;
    pt = new char[characters];
    memcpy(pt, s.pt, characters);
}

string::string(char *s)
{
    characters = strlen(s);
    pt = new char[characters];
    memcpy(pt, s, characters);
}

string &string::operator=(const string &s)
{
    delete pt;
    characters = s. characters;
    pt = new char[characters];
    memcpy(pt, s.pt, characters);
    return *this;
}

char *string::char_array(void) const
{
    char *buffer = new char[1 + characters];
    memcpy(buffer, pt, characters);
    buffer[characters] = '\0';
    return buffer;
}

string::string(int set_length)
{
    characters = set_length;
    pt = new char[characters];
}
```

```
void string::print(void)
{
    for (int i = 0; i < characters; ++i)
        cout << pt[i];
}

// friend function implementation:
string operator+(const string &s1, const string &s2)
{
    int s1_length = s1.length();
    int s2_length = s2.length();
    string new_string(s1_length + s2_length);
    memcpy(new_string.pt, s1.pt, s1_length);
    memcpy(new_string.pt + s1_length, s2.pt, s2_length);
    return new_string;
}
```

Notice that this code uses the memcpy() and strlen() functions, both of which are declared in <cstring>. These functions are described in Section 7.8.2.

A simple program to try out this string class is given below.

```
// source:      my_test.cxx
// use:         Tests string class.

#include <iostream>
#include <cstdlib>        // For EXIT_SUCCESS
#include "string.h"
using namespace std;

string f(string s)
{
    string ss;
    ss = s;
    return ss;
}

int main()
{
    // Test string::string(char *s):
    string s1("My first string");

    // Test void string::print(void):
    s1.print();
    cout << "\n";

    // Test int string::length(void):
    cout << "String length: " << s1.length() << "\n";
```

```
// Test string::string(void):
string s2;

// Test string &string::operator=(const string &s):
string s3 = s1;
cout << "Copied string is: ";
s3.print();
cout << "\nWith string length: " << s3.length() << "\n";

// Test string::string(const string &s):
string s4 = f(s3);
cout << "String copied by function is: ";
s4.print();
cout << "\nWith string length: " << s4.length() << "\n";

// Test string operator+(const string &s1, const string &s2)
// and   string::string(int set_length):
string s5(" and this is my second");
string s6 = s1 + s5;
cout << "Adding two strings using +: ";
s6.print();
cout << "\nThis new string has length: " << s6.length() <<
    "\n";

// Test char *string::char_array(const string &s):
char *pt = s1.char_array();
cout << "Assigning a 'string' to a char* gives: " << pt <<
    "\n";

// Test bool operator==(const string &s1, const string &s2):
cout << "Does string: '";
s3.print();
cout << "' == '";
s4.print();
cout << "'?\n";
if (s3 == s4)
    cout << "Yes\n";
else
    cout << "No\n";
cout << "Does string: '";
s3.print();
cout << "' == '";
s6.print();
cout << "'?\n";
if (s3 == s6)
    cout << "Yes\n";
else
```

```
        cout << "No\n";

    // Test bool operator!=(const string &s1, const string &s2):
    cout << "Does string: '";
    s3.print();
    cout << "' != '";
    s4.print();
    cout << "'?\n";
    if (s3 != s4)
        cout << "Yes\n";
    else
        cout << "No\n";
    cout << "Does string: '";
    s3.print();
    cout << "' != '";
    s6.print();
    cout << "'?\n";
    if (s3 != s6)
        cout << "Yes\n";
    else
        cout << "No\n";

    return(EXIT_SUCCESS);
}
```

The current implementation of the **string** class has three unsatisfactory features:

- A **string** object is sent to the output stream by using the **print()** function, rather than the **<<** operator. The techniques of Chapter 18 enable the **print()** function to be replaced by an insertion operator.

- No function is provided for directly reading a **string** object from the input stream. This deficiency can also be overcome by using the techniques of Chapter 18.

- Memory for a string is not reclaimed when the object goes out a scope. It is straightforward to correct this feature by means of a destructor function. Destructor functions are introduced in Section 10.2.

9.7   Summary

- All of the built-in C++ operators can be overloaded, with the exception of:

  .   .*   ::   ?:   sizeof

  and the preprocessor operators, **#** and **##**.

- The associativity, precedence and number of operands of an overloaded operator cannot be changed by overloading:

```
        z = u ! v;              // WRONG!
```

- With few exceptions, an overloaded operator can be either a non-member function (usually a `friend`):[13]

```
class X {
    friend X operator+(const X &u, const X &v);
    // More class declarations go in here.
};
```

or a member function:

```
class complex {
    complex &operator+=(const complex &u);
    // More class declarations go in here.
};
```

- Overloaded operators, when implemented as member functions, can use the `const` specifier, as in:

```
class complex {
    complex operator-() const;
    // More class declarations go in here.
};
```

to indicate that an object invoking the operator cannot be changed by it.

- A constructor accepting a single argument, not of the class type, performs a conversion to the class:

```
class complex {
    complex(const double &x);
    // More class declarations go in here.
};

complex z = 1.414;          // Implicit conversion.
complex w = complex(0.707); // Explicit conversion.
```

- A conversion function takes the form:

```
class complex {
    operator double();
    // More class declarations go in here.
};
```

Neither a `return` type nor an argument can be specified.

- Operator overloading is implemented by an operator function. Member functions have an implicit argument, which is the hidden `this` pointer. The relationship between an operator and the corresponding function call notation is summarized in Table 9.1, where @ stands for a generic operator.

---

[13] The exceptions are described in Section 9.3.

Operator	Function call	
	Member	Non-member
A @ B	A.operator@(B)	operator@(A, B)
@ A	A.operator@( )	operator@(A)
A @	A.operator@(0)	operator@(A, 0)

Table 9.1: Operator function calls.

9.8   Exercises

1. Describe the difficulties you would encounter in trying to implement a class for two-dimensional arrays by overloading the subscripting, rather than function call, operator.

2. Further improve our **array** class (originally given in Section 8.17) by adding an overloaded assignment operator. Memory should be dynamically allocated for the new elements and the original data copied to this memory.

3. Extend the complex arithmetic class, described in Section 9.6.1, as far as you can. Some of the many omissions are subtraction, multiplication and division, together with the elementary mathematical functions for complex arguments.

4. In the **complex** arithmetic class, implemented in Section 9.6.1, there is neither a constructor accepting a single argument nor a conversion function. Consequently, explicit operators must be defined for each possible operation performing mixed arithmetic. Implement the alternative technique of only defining operators for **complex** arguments and relying on implicit conversion. Write a program that tests the modified class as thoroughly as you can. What are the relative merits of the two techniques?

Chapter 10

Constructors and Destructors

In this chapter we consolidate our knowledge of constructor functions and introduce destructor functions. Constructors and destructors are class member functions. A constructor is invoked whenever an object is created and a destructor is invoked whenever an object is destroyed. Constructors are often invoked by the compiler, rather than as an explicit part of a statement supplied by a user of the class. Destructors are also often invoked by the compiler and almost never as an explicit part of a statement supplied by a user of the class.

Constructors are necessary because objects can be created throughout a program, often implicitly and with their initial data unknown at compile-time. A constructor can initialize object data and this may include storing the address of dynamically allocated memory. Constructors are also often used to open files.

If there is no user-defined constructor, then a simple `public` default that takes no arguments is supplied by the compiler. If at least one constructor is defined for a class, but all the constructors must take arguments, then objects of this class cannot be defined without being initialized. Such a class prevents the otherwise common error of using objects that have not been initialized.

Constructors are usually `public`. However, if all constructors for a particular class are `private`, then only member functions and friends can create objects of the class. A class which has no `public` constructors is known as a *private class* and can be useful if objects of one class are only ever used as clients of another. In order to ensure there are no `public` constructors, a `private` copy constructor must be defined or else the compiler will provide a `public` default.

A destructor is called when an object goes out of scope, or a program terminates normally, and is used to tidy up before an object is destroyed. Often there is no need for an explicit destructor to be supplied for a class since the default is perfectly adequate. However, if a constructor dynamically allocates memory, then a destructor should be defined in order that this memory can be returned to the "free store" when the memory is no longer required. Another common use of a destructor is to close a file that has been opened by a constructor. Since destructors are invoked implicitly by the compiler, such a file is usually closed as its associated object goes out of scope, rather than as a result of a statement supplied by a user of the class.

10.1   More on Constructor Functions

10.1.1   Dynamic Memory Management

In many cases, the storage needed by an object is not known until run-time. For example, we may have a `matrix` class, which could include functions to perform the usual multiply, add and inversion operations. Instead of assigning sufficient memory for matrices up to a certain size when we write the code, the `matrix` class could have a constructor that uses the `new` operator to dynamically allocate precisely the required amount of memory. We have already met the idea of dynamically assigning memory for a matrix in Section 7.8.1, but this can now be improved by introducing classes and constructor functions. An outline `matrix` class putting together these ideas might have the following header file:

```
// source:    matrix.h

class matrix {
public:
    matrix(matrix &a);
    matrix(int rows, int columns);
    double &operator()(int i, int j);
 private:
    double *p;                    // Address of data.
    int m, n;                     // m x n matrix.
};
```

Function definitions for the member functions are given below.

```
// source:    matrix.cxx

#include <iostream>
#include <cstring>        // For memcpy()
#include "matrix.h"
using namespace std;

matrix::matrix(matrix &a)
{
    m = a.m;
    n = a.n;
    p = new double[m * n];
    memcpy(p, a.p, m * n * sizeof(double));
}

matrix::matrix(int rows, int columns)
{
    m = rows;
    n = columns;
    p = new double[m * n];
}
```

```
double &matrix::operator()(int i, int j)
{
    return p[i - 1 + m * (j - 1)];
}
```

Notice that the constructor assigns the address of dynamically allocated memory to the pointer, p. That is the constructor does *not* dynamically allocate the matrix object (whose data members are m, n and p) but rather initializes the object after memory for m, n and p has been allocated.

The following code provides a simple test program:

```
// source:  my_test.cxx

#include <iostream>
#include <cstdlib>       // For EXIT_SUCCESS
#include "matrix.h"
using namespace std;

int main()
{
    int rows, columns;
    cout << "Enter number of rows: ";
    cin >> rows;
    cout << "Enter number of columns: ";
    cin >> columns;

    // Create matrix:
    matrix m(rows, columns);

    // Assign values to elements of the matrix:
    for (int i = 1; i <= rows; ++i)
        for (int j = 1; j <= columns; ++j)
            m(i, j) = 10.0 * i * j;

    // Test assignment:
    for (int i = 1; i <= rows; ++i)
        for (int j = 1; j <= columns; ++j)
            cout << "m(" << i << ", " << j << ") = " <<
                m(i, j) << "\n";

    return(EXIT_SUCCESS);
}
```

A suitable makefile for this project is given below.

```
my_test: my_test.cxx matrix.o matrix.h
    g++ my_test.cxx matrix.o -o my_test
matrix.o: matrix.cxx matrix.h
    g++ -c matrix.cxx
```

Exercise

(a) Compile and test the above matrix class.

(b) Implement overloaded assignment and addition operators for the matrix class by using the ideas in Sections 7.8.1, 9.1.1 and 9.1.3.

(c) Check by supplying suitable data that for $m \times n$ matrices A, B, C the statement:

```
A = B + C;
```

gives the expected results. You should try different values of m and n, so that you test both square and non-square matrices.

10.1.2   Assignment and Initialization

In C++ it is important to distinguish between assignment and initialization since they are implemented by different functions. An assignment is implemented by an assignment operator. An initialization is performed by a constructor. In either case the compiler supplies defaults if necessary.

Assignment implies that an object already exists and its data members are changed through an assignment statement, as illustrated in the following code fragment:

```
int x;
x = 10;           // 10 is assigned to x.
complex u, v(1.1, 2.2);
u = v;            // Assignment to u.
```

An assignment is a copy operation which, with one exception, takes place whenever we use the = symbol. The exception is that use of the = symbol within a declaration is an initialization.

An initialization can occur in the following circumstances:

- when an object is declared, as in the statements:

  ```
  complex z(1.1, 2.2);
  double x = 2.7183;
  complex w = 1.1;
  int i = 10;
  ```

- when an object is passed by value as a function argument. For instance, if a function:

  ```
  void f(complex z)
  {
      // function body.
  }
  ```

is invoked by:

```
    f(u);
```

then z is initialized rather than assigned to.

- when an object is returned by value from a function:

```
    complex f(complex z)
    {
        complex u;
        // More code goes in here.
        return u;
        // Implicit temporary is initialized by return statement.
    }
```

If an object is passed by reference in a function argument or return value, then there is no assignment. For many objects, such as large matrices, this is important since large (and unnecessary) copying operations may be avoided by using pass by reference.[1]

Exercise

To illustrate the difference between assignment and initialization, implement functions for the following version of the complex class:

```
    class complex {
        friend complex operator+(const complex &u,
            const complex &v);
    public:
        complex(double x = 0.0, double y = 0.0);
        complex(const complex &z);
        complex &operator=(const complex &z);
        double re, im;
    };
```

As each function is invoked, it should insert an identifying message in the output stream. Use these implementations to write a program that enables you to distinguish initializations from assignments in the code fragment:

```
    complex u;
    u = complex(2.0, 3.0);
    complex v(1.0, 7.0);
    complex z;
    z = u + v;
```

---

[1] However, remember that it is *not* possible to return a reference to a local non-static object.

10.1.3   Member Objects with Constructors

The data members of a class are not restricted to the fundamental types; members can themselves be instances of another class. This is illustrated in the following code fragment:

```
class data {
public:
    data(const complex &z1, const complex &z2);
    // More class declarations go in here.
private:
    complex u, v;
};
```

In this example, the member objects have constructors that take no arguments. Since the complex() constructors are called before the data() constructor, this means that the member objects may be first initialized to zero and then be assigned the values z1 and z2. For instance, the constructors may take the form:

```
complex::complex(void)
{
    re = 0.0;
    im = 0.0;
}

data::data(const complex &z1, const complex &z2)
{
    u = z1;
    v = z2;
}
```

This sequence of initialization followed immediately by assignment is not very efficient, particularly for large or much used data structures. However, if we have a constructor taking arguments for the member objects, such as:

```
complex::complex(const complex &z)
{
    re = z.re;
    im = z.im;
}
```

then an alternative syntax for the data() constructor is:

```
data::data(const complex &z1, const complex &z2) : u(z1), v(z2)
{
    // The constructor body goes in here, but it may be empty.
}
```

This directly calls the copy constructor for u and v, and the complex(void) constructor is not invoked. In general, the colon is followed by a comma-separated list of constructors taking arguments. This list, which is followed by the constructor body, can only occur in a constructor definition and not a declaration:

```
data::data(const complex &z1, const complex &z2) : u(z1), v(z2);
// WRONG: a declaration can't have a constructor list.
```

The order in which the constructors are invoked is not influenced by this list since the rule is that the constructors for member objects are invoked first, in the order in which they are declared within the class, and then the constructor for the class is invoked.[2]

Exercise

Modify the `complex()` and `data()` constructors so that they insert identifying messages in the output stream when they are called. Write short programs to demonstrate that:

(a) If the (unmodified) `data()` constructor has the definition:

```
data::data(const complex &z1, const complex &z2)
{
    u = z1;
    v = z2;
}
```

then u and v are first initialized to zero and then assigned the values of z1 and z2.

(b) The (modified) `data()` constructor with the declaration:

```
data::data(const complex &z1, const complex &z2)
    : u(z1), v(z2) { }
```

does not first initialize u and v to zero, but rather invokes the copy constructor for the `complex` class.

10.2 Destructor Functions

A destructor is a class member function with the name `~X`, where X is the name of the class. An example of a destructor function is given in the following header file for a `matrix` class:

```
// source:    matrix.h

class matrix {
public:
    matrix(matrix &a);
    matrix(int rows, int columns);
    ~matrix();            // Declares destructor function.
    double &operator()(int i, int j);
  private:
    double *p;
    int m, n;
};
```

---

[2]Destructors, which are introduced in the next section, are called in the reverse order.

There is often no need to declare a destructor. The `complex` class is a typical example; memory is allocated when a `complex` object is defined and deallocated when the object goes out of scope. This is illustrated in the following function:

```
complex f(void)
{
    complex z;        // Memory allocated for z.
    // More code goes in here.
    return z;         // z goes out of scope; memory is deallocated.
}
```

However, our `matrix` class is very different. Each time either of the two constructors is invoked, the `new` operator allocates sufficient memory to store a `matrix` object. When the object goes out of scope this memory is not deallocated, resulting in a gradual loss of the available memory. This is known as *memory leakage* and can cause unnecessary exhaustion of the available free store. Eventually this could cause your computer to run more and more slowly, and may result in it stopping altogether! The solution to this problem is to define a destructor that uses the delete operator to release memory back to the free store. The definition of an appropriate destructor for the `matrix` class is the following simple function:

```
inline matrix::~matrix() { delete p; }
```

Notice that the destructor deletes the memory whose address is stored by the pointer, `p`; the constructor does *not* actually delete the `matrix` object, since this consists of the three data members, `m`, `n` and `p`. Memory used by these data members would in any case be available for reuse when the `matrix` object went out of scope.

Destructors are rather special functions; a destructor takes no arguments, does not have a return type (not even `void`) and cannot be declared `static`.[3] Since a destructor cannot have any arguments, it cannot be overloaded. Destructors are called implicitly, as in:

```
matrix f(void)
{
    matrix m(5, 6);
    // More code goes in here.
    return m;         // ~matrix() called implicitly
}                     // as m goes out of scope.
```

It is usually unnecessary and even a bad idea for a programmer to explicitly invoke a destructor; doing so typically results in an attempt to `delete` memory that has already been deleted and this has undefined consequences, as pointed out in Section 7.6.2. One of the few cases where it may be necessary to explicitly invoke a destructor is when an object has been placed at a specific address by means of the `new` operator. The implementation of such cases is not straightforward due to the fact that ~ is the unary bitwise complement operator, which is introduced in Chapter 11, and examples demonstrating how to explicitly invoke destructors are given in [3].

---

[3]The fact that a destructor takes no arguments *can* of course be indicated by using a `void` argument.

10.3 Using Constructors and Destructors

A common requirement in many applications is the manipulation of lists. The amount of data stored in an item of a list is frequently large and the number of data items is often not known until run-time. Moreover, the number of items may even change during program execution. An array is not an appropriate data structure for such problems, but a suitable alternative is a *linked list*. Linked lists are good examples of constructors and destructors in use.

10.3.1 Singly Linked Lists

For simplicity we first consider a singly linked list. This consists of a number of *nodes*, with each node storing a data item and the address of the next node in the list. The address of the first node is known as the *head* of the list. A typical list is shown in Figure 10.1.

Figure 10.1: Singly linked list.

It is convenient to have separate classes for the nodes and lists. These are declared in the following header file:

```
// source:  slist.h
// use:     Defines singly linked list and node classes.

#ifndef SLIST_H
#define SLIST_H

typedef int DATA_TYPE;

class node {
    friend class list;
public:
    DATA_TYPE data;
private:
    node *next;
};

class list {
public:
    list(void);
    ~list();
    void push(DATA_TYPE new_data);
    void pop(DATA_TYPE &old_data);
    bool is_not_empty(void);
private:
```

```
    node *head;
};

inline bool list::is_not_empty(void)
{
    if (head == 0)
        return false;
    else
        return true;
}


#endif  // SLIST_H
```

The following points are worth noting for the node class:

- The node class declares the entire list class (not just a single function) to be a friend. This means that the list functions have access to all member functions of the node class and to all data members of node objects.

- For simplicity, the data stored by a node object is a single int. In practice, this would be replaced by a large data structure, perhaps the data on some experimental observation or a standard component for an engineering project. Each time we change DATA_TYPE, the node and list classes must be recompiled, which is contrary to the spirit of code re-usability in object-oriented programming. However, using class templates would overcome this problem. Class templates are introduced in Chapter 16.

- The data member, called next, stores the address of the next node. Notice that it is legal to declare a pointer to node within the node class declaration, but it is not legal to declare another node object within the node class. For instance, the following code fragment would be invalid:

```
    class node {
        node another_node;      // WRONG!
        // More class declarations go in here.
    };
```

  This is because within a class definition the class is considered to be declared, but not defined.

- The node class has no user-defined constructors or destructor. This is because node objects do not dynamically allocate memory, initialize other objects or open files.

The following comments concern the list class:

- The list class uses the notation of a *stack*; that is we can add data to the head of the list (push()) and remove data from the head (pop()), very much like a stack of plates. There is no way of accessing data further down the stack, except by popping the entire stack. A stack is a "last in first out" (LIFO) list and is traditionally drawn vertically, as shown in Figure 10.2. (In fact this is the only difference between Figures 10.1 and 10.2.)

- The `list` class has a user-defined constructor and destructor. The constructor is needed because we want the head of an empty list to be set to `0`. A destructor is required so that the dynamically allocated memory is returned to the free store when a `list` object goes out of scope.

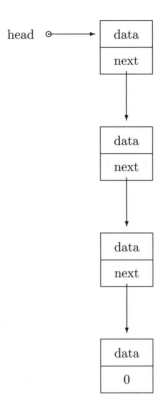

Figure 10.2: A stack.

Suitable implementations for the non-`inline` member functions are given below.

```
// source:  slist.cxx
// use:     Implements singly linked list class.

#include "slist.h"

list::list(void)
{
    head = 0;
}

list::~list()
{
```

```
    while (head != 0) {
        node *pt = head;
        head = head->next;
        delete pt;
    }
}

void list::push(DATA_TYPE new_data)
{
    node *pt = new node;
    pt->next = head;
    head = pt;
    pt->data = new_data;
}

void list::pop(DATA_TYPE &old_data)
{
    if (head != 0) {
        old_data = head->data;
        node *pt = head;
        head = head->next;
        delete pt;
    }
}
```

The following program provides a simple test of the singly linked list class by pushing values onto the stack and then printing the results obtained by popping the stack:

```
// source:  my_test.cxx
// use:     Tries out singly linked list class.

#include <iostream>
#include <cstdlib>        // For EXIT_SUCCESS
#include "slist.h"
using namespace std;

int main()
{
    list s;
    DATA_TYPE j;

    for (int i = 1; i <= 5; ++i)
        s.push(10 * i);
    while (s.is_not_empty()) {
        s.pop(j);
        cout << j << "\n";
    }
```

```
        return(EXIT_SUCCESS);
    }
```

A suitable makefile for this project is similar to previous makefiles and is given below.

```
my_test: my_test.cxx slist.o slist.h
    g++ my_test.cxx slist.o -o my_test
slist.o: slist.cxx slist.h
    g++ -c slist.cxx
```

Notice that we do not pop the stack by doing something like:

```
    for (int i = 1; i <= 5; ++i) {
        s.pop(j);
        cout << j << "\n";
    }
```

The list, s, has a state (either it is empty or it is not) so the best technique is to obtain the object's state from the object itself.

Looking at our test program, we might be tempted to replace the loop that prints the contents of the stack by the function call:

```
print(s);
```

where the print() function is defined by:

```
void print(list t)
{
    DATA_TYPE j;

    while (t.is_not_empty()) {
        t.pop(j);
        cout << j << "\n";
    }
}
```

This is a disaster. The problem is that, since the t in the function header:

```
void print(list t)
```

is a formal argument, a default copy constructor, which perform a simple copy of the original list, is invoked. Unfortunately both lists contain the address of the same dynamically allocated memory. As an item of the list is popped, the memory is deallocated. This is of no consequence as we return from print(), since t is empty but, when s goes out of scope, the destructor attempts to deallocate memory that is already deallocated. The result is an error message such as "Corrupted Heap" or "Segmentation Fault", depending on the operating system. Setting the pointers in list::pop() and list::~list() to zero after memory deletions is of no help here since the original list contains copies of the pointers.

A possible solution to the above problem is to define print() with a reference argument since this involves no copying:[4]

---

[4]Notice that the data is discarded as it is printed. This is a feature of any stack and can be overcome by introducing a function that traverses a list rather than popping the stack.

```
void print(list &t)
{
    // Same function body as before.
}
```

However, this does not prevent us from inadvertently implementing other functions that lead to similar disasters due to the lack of a suitable copy constructor. The lesson to be drawn is that when designing a class we should provide all functions for which inappropriate defaults could be supplied, *even* if we do not envisage using them. For an arbitrary class, X, there are three such functions:

X::X(const X &)	copy constructor
X::operator= (const &X)	assignment operator
X:: ˜X()	destructor

In some circumstances, invoking the copy constructor or assignment operator may actually constitute an error if, for example, only one instance of a class is supposed to exist. Such errors can be trapped by suitable implementations of these functions, as illustrated below for the assignment operator.

```
X &X::operator=(const X &x)
{
    cout << "No assignment implemented for class X\n";
    exit(EXIT_FAILURE);
    return *this;
}
```

This function should be a `private` member of the class, X, so that as many errors as possible are trapped at compile-time.

Exercise

Implement an appropriate copy constructor and assignment operator for the `list` class.

10.3.2   Doubly Linked Lists

Singly linked lists are a bit restricted since they can only be scanned efficiently in one direction. By contrast, doubly linked lists can be scanned in either direction with equal efficiency and only require the extra overhead of one additional pointer for each node. The following header file declares simple node and doubly linked lists classes:

```
// source:   dlist.h
// use:      Declares doubly linked list and node classes.

#ifndef DLIST_H
#define DLIST_H

#include <cstdlib>       // For exit()
using namespace std;
```

```
typedef int DATA_TYPE;

class node {
    friend class dlist;
public:
    DATA_TYPE data;
private:
    node *next, *last;
};

class dlist {
public:
    dlist(void);
    ~dlist();
    void add_head();
    void delete_head(void);
    node *forward(void);
    node *backward(void);
    void set_cursor_head(void);
    void set_cursor_tail(void);
    node *cursor_position(void);
private:
    node *head, *tail, *cursor;
};

inline node *dlist::forward(void)
{
    node *pt = cursor;
    if (cursor != 0)
        cursor = cursor->next;
    return pt;
}

inline node *dlist::backward(void)
{
    node *pt = cursor;
    if (cursor != 0)
        cursor = cursor->last;
    return pt;
}

inline void dlist::set_cursor_head(void)
{
    cursor = head;
}
```

```
inline void dlist::set_cursor_tail(void)
{
    cursor = tail;
}

inline node *dlist::cursor_position(void)
{
    return cursor;
}

#endif   // DLIST_H
```

Each node has a pointer to the next node (`next`) and a pointer to the previous node (`last`), as shown in Figure 10.3. The list has a pointer to `head` and `tail`, in addition to a cursor. The cursor is useful for list manipulations, such as sorting, and can be moved backwards (`backward()`) and forwards (`forward()`) along the list. These last two functions also return the position of the cursor (that is the address of the node) before the cursor was moved. There are two other functions that manipulate the cursor position (`set_cursor_head()`, `set_cursor_tail()`) and a function to give the current position (`cursor_position()`).

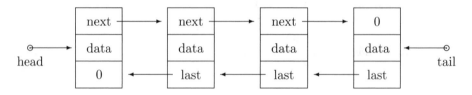

Figure 10.3: Doubly linked list.

The `dlist` class implementation, for those functions that are not `inline`, is given below.

```
// source:    dlist.cxx
// use:       Implements doubly linked list class.

#include "dlist.h"

dlist::dlist(void)
{
    head = 0;
    tail = 0;
    cursor = 0;
}

dlist::~dlist()
{
    while (head != 0) {
        node *pt = head;
```

```
        head = head->next;
        delete pt;
    }
}

void dlist::add_head(void)
{
    node *pt = new node;
    if (head == 0) {
        tail = pt;
    }
    else
        head->last = pt;
    pt->next = head;
    pt->last = 0;
    head = pt;
    cursor = pt;
}

void dlist::delete_head(void)
{
    if (head == 0)
        return;
    node *pt = head;
    head = head->next;
    if (head != 0)
        head->last = 0;
    cursor = head;
    delete pt;
}
```

These functions are straightforward generalizations of those for singly linked lists. Notice that we have had to make a choice as to where the cursor points after some of these operations.

A simple test program is given below.

```
// source:    test.cxx
// use:       Tests doubly linked list class.

#include <iostream>
#include <cstdlib>        // For EXIT_SUCCESS
#include "dlist.h"
using namespace std;

void print_forward(dlist &s)
{
    node *pt;
    s.set_cursor_head();
```

```
        cout << "Forward list: ";
        if ((pt = s.forward()) != 0)
            cout << pt->data;
        else
            cout << "empty";
        while ((pt = s.forward()) != 0)
            cout << " -> " << pt->data;
        cout << "\n";
}

void print_backward(dlist &s)
{
    node *pt;
    s.set_cursor_tail();
    cout << "Backward list: ";
    if ((pt = s.backward()) != 0)
        cout << pt->data;
    else
        cout << "empty";
    while ((pt = s.backward()) != 0) {
        cout << " -> " << pt->data;
    }
    cout << "\n";
}

int main()
{
    // Create an empty list:
    dlist s;

    // Print the list forwards and backwards:
    print_forward(s);
    print_backward(s);

    // Put some items in the list:
    for (int i = 1; i <= 5; ++i) {
        s.add_head();
        (s.cursor_position())->data = 10 * i;
    }

    // Print the list forwards and backwards:
    print_forward(s);
    print_backward(s);

    // Delete an item:
    s.delete_head();
```

```
    // Print the new first item:
    cout << "Head of list is now " <<
         (s.cursor_position())->data << "\n";

    // Print the list forwards and backwards:
    print_forward(s);
    print_backward(s);

    return(EXIT_SUCCESS);
}
```

A suitable makefile for this project is similar to the one given for the singly linked list. As with the singly linked list example, changing &s to s in the function headers for `print_forward()` and `print_backward()` invokes an inappropriate default copy constructor.

Exercise

Modify the copy constructor and assignment operator that you provided for the `list` class so that they are suitable for the `dlist` class.

10.4 Summary

- A constructor is invoked whenever an object is created. A class typically requires a user-defined constructor or destructor when there is initialization, dynamic memory allocation or file manipulation.

- A destructor is a class member function with the name of the class, prefixed by the tilde symbol:

```
    class matrix {
        ~matrix();
        // More class declarations go in here.
    };
```

- A destructor cannot have an argument, return a type or be declared `static`.

- It is very rare to explicitly invoke a destructor.

10.5 Exercises

1. Provide a destructor for the `string` class described in Section 9.6.2.

2. Replace the function header for `print_forward(dlist &s)` in Section 10.3.2 by the following (inappropriate) version:

```
    void print_forward(dlist s)
```

and add statements to the relevant `dlist` functions so that the addresses of `node` objects are printed as they are created and destroyed. Run the test program for `dlist` and work out exactly where `node` objects are created and destroyed. Use this information to work out how this `print_forward(dlist s)` function goes wrong.

3. Modify the `dlist` class (given in Section 10.3.2) to include the following functions:

   - Add or delete a node at the tail:

     ```
     void add_tail(void);
     void delete_tail(void);
     ```

   - Add a node before or after the current cursor position:

     ```
     void add_before(void);
     void add_after(void);
     ```

   - Delete the node at the current cursor position:

     ```
     void delete_node(void);
     ```

   (You will have to decide where the cursor should be placed after these operations.)

4. Implement a sorting algorithm (either a bubble sort or a more efficient technique) using your modified `dlist` class of the previous exercise.[5]

5. An instructive, but complicated, application of linked lists occurs in the implementation of a sparse matrix class. A much simpler, but related, problem is a sparse vector class. Use the `dlist` class to design and implement a sparse vector class. Each node should store the index of a component of a vector, together with the component itself, as a `double`. The list should be sorted on the basis of the index.

   Use overloaded operators to access individual vector components and to implement the addition of two vectors. You should also write a function that returns the dot product of two vectors.

---

[5]Almost any sorting algorithm is better than a bubble sort. For example, try a *merge sort* as described in [15].

Chapter 11

Bitwise Operations[††]

In this chapter we are concerned with the manipulation of single bits, each one of which can only store a value of zero or one. Explicit bitwise operations are not common in numerical applications, but such operations are sometimes used to reduce memory requirements. It is also important to understand how the fundamental data types are represented in terms of bits, and what restrictions these representations impose on our calculations.

Bitwise operations are part of the techniques that C++ inherited from C. We have delayed introducing bitwise operations until this stage since we can now demonstrate how the C++ concept of a class can be used to hide some of the details of such operations.

11.1 Bitwise Operators

There are six bitwise operators, as shown in Table 11.1, and these operators only operate on integral types. The precedence and associativity of the bitwise operators are given in Appendix B. The complement operator, ~, is unary, but all the other operators are binary.

Logical operators	
bitwise complement	~
bitwise AND	&
bitwise exclusive OR	^
bitwise OR	|

Shift operators	
left shift	<<
right shift	>>

Table 11.1: Bitwise operators.

It must be understood that some features of bit operations are compiler-dependent. Also the way in which the "sign bit" is handled for some bitwise operations can depend on the particular compiler, so it is safer to use the **unsigned** types for such operations where possible.

307

11.1.1   Bitwise Complement

The *bitwise complement* (or *one's complement*) operator, ~, changes all of the ones in the operand to zero and all of the zeros to ones. For example, if x stores the bit pattern:[1]

```
1 0 1 1 1 1 0 0 0 0 0 0 1 1 1 1
```

then ~x corresponds to:

```
0 1 0 0 0 0 1 1 1 1 1 1 0 0 0 0
```

Notice that ~x is an expression and consequently the value of x itself remains unchanged.

11.1.2   Bitwise AND

The bitwise AND operator, &, compares the two operands bit by bit. If both bits are set, the result is 1, otherwise the result is 0. For example, if x stores the bit pattern:

```
1 0 1 1 1 1 0 0 0 0 0 0 1 1 1 1
```

and y stores:

```
1 1 1 1 0 0 0 0 1 1 1 0 1 0 1 0
```

then x & y corresponds to:

```
1 0 1 1 0 0 0 0 0 0 0 0 1 0 1 0
```

Notice that it is important not to confuse & with &&. The expression, x && y, gives the integer one if at least one bit in each of x and y is set. So for the given bit patterns for x and y, x && y would give:

```
0 0 0 0 0 0 0 0 0 0 0 0 0 0 0 1
```

which is very different from the result for x & y obtained above.

11.1.3   Bitwise Exclusive OR

The bitwise exclusive OR operator, ^, compares the two operands bit by bit. If either, but not both, bits are set, then the result is 1, otherwise it is 0. If x stores the bit pattern:

```
1 0 1 1 1 1 0 0 0 0 0 0 1 1 1 1
```

and y stores:

```
1 1 1 1 0 0 0 0 1 1 1 0 1 0 1 0
```

then x ^ y corresponds to:

```
0 1 0 0 1 1 0 0 1 1 1 0 0 1 0 1
```

---

[1]In order to reduce the number of bits we need to consider, some of the examples given in this chapter are for 16-bit integers.

11.1.4 Bitwise Inclusive OR

The bitwise inclusive OR operator, |, compares the two operands bit by bit. If either bit is set, then the result is 1, otherwise it is 0. If x stores the bit pattern:

    1 0 1 1 1 1 0 0 0 0 0 0 1 1 1 1

and y stores:

    1 1 1 1 0 0 0 0 1 1 1 0 1 0 1 0

then x | y corresponds to:

    1 1 1 1 1 1 0 0 1 1 1 0 1 1 1 1

Be careful not to confuse | with ||. The expression, x || y, gives the integer one if any bit of x or y is set. So for the given bit patterns for x and y, x || y would give:

    0 0 0 0 0 0 0 0 0 0 0 0 0 0 0 1

which is very different from the result for x | y obtained above.

11.1.5 Shift Operators

The *left shift* operator, <<, shifts the bits in the left operand to the left by the number of bits specified in the right operand. If x has the bit pattern:

    1 0 1 1 1 1 0 0 0 0 0 0 1 1 1 1

then x << 3 is:

    1 1 1 0 0 0 0 0 0 1 1 1 1 0 0 0

The vacated bits are always filled with zeros.

The *right shift* operator, >>, shifts the bits in the left operand to the right by the number of bits specified in the right operand. If the left operand has an **unsigned** type or is not negative, then the vacated bits are filled with zeros, otherwise the way they are filled is dependent on the compiler. If x is an **unsigned** type, with the bit pattern:

    1 0 1 1 1 1 0 0 0 0 0 0 1 1 1 1

then x >> 3 is:

    0 0 0 1 0 1 1 1 1 0 0 0 0 0 0 1

Notice that both x >> 3 and x << 3 are expressions; the value of x (or equivalently, the bit pattern) does not change. To shift the bits in x, we must make an assignment, as in:

    x = x >> 3;

At this stage, you may be worried that these shift operators are the same as those used by the `iostream` library. But the fundamental operators, as defined by both the C and C++ languages, are actually the left and right shift operators. These operators are simply overloaded by the `iostream` library. In fact, we have been using overloaded operators from our very first C++ program! The only consequence of this overloading is that the `iostream` operators must have the same associativity and precedence as the left and right shift operators. This is why, for example, successive items can be sent to the output stream:

```
cout << "Output " << "More output " << "Yet more output";
```

and occasionally parentheses are needed to overcome the built-in precedence or associativity:

```
cout << (x && y);
```

11.1.6   Bitwise Assignment Operators

All of the bitwise operators, with the exception of the bitwise complement, have an associated assignment operator:

$$<<= \qquad >>= \qquad \&= \qquad \hat{} = \qquad |=$$

For instance, the statement:

```
i <<= 4;
```

is equivalent to:

```
i = i << 4;
```

These operators introduce nothing new and are simply shorthand for a bitwise operation on a variable, followed by an assignment to the same variable. However, don't forget that if we overload an operator, then the corresponding assignment operator is not necessarily defined. For example, we have made much use of the overloaded left shift operator to insert objects in the output stream, as in:

```
cout << "The answer to everything is " << 47;
```

but the `iostream` library does not overload the `<<=` operator.

11.2   Bit-fields

A *bit-field* is a class member with a width specified in bits, as in:

```
class packed {
public:
    int b1:4;
    int b2:8;
    int b3:20;
};
```

where each statement of the form:

```
int b1:4;
```

is a *field declaration*. In this example a bit-field of 4 bits is declared for `b1`. Bit-fields can be accessed by using the usual class member access operator (a single dot), as illustrated below.

```
packed p;
p.b2 = '\0';
```

For the class declared above, the bit-fields may be packed into a single 32-bit word, as shown in Figure 11.1.[2] However, such details are dependent on the compiler and the assumptions underlying any use of bit-fields should be carefully documented.

Figure 11.1: Example of bit-fields.

Bit-fields are sometimes used to access particular, system-dependent, memory locations, such as for memory mapped I/O drivers, or to save on storage. However, using bit-fields to reduce storage requirements is often misguided. Different compilers pack the bit-fields into words in different ways and this may even result in a waste of space. Moreover, there is usually a performance overhead in accessing bit-fields, so it is possible to end up with a slower program that uses more memory!

Bit-fields can be of any integral type. The main restrictions are:

- The address-of operator cannot be applied to bit-fields:

```
packed p;
&p.b1;          // WRONG:   cannot use &.
```

- Pointers to bit-fields are not legal.

- References to bit-fields are not allowed.

These restrictions are all very reasonable, since memory is addressable in bytes rather than bits.

Having introduced bit-fields, we won't meet them again in this book; they are best forgotten until you need to implement something like a memory mapped I/O driver.

---

[2]We use the convention that the *least significant bit* (lsb) is shown on the right and the *most significant bit* (msb) is on the left. The bits are labelled upwards from zero (corresponding to the least significant bit in a word).

11.3 Unions

A *union* is a special class in which the data members have overlapping memory locations. The following statement defines the identifier U to be a union:

```
union U {
    char c;
    int i;
    float f;
    double d;
};
```

A possible storage scheme for this union is given in Figure 11.2. (Assumptions have been made about the number of bits required to store the different data types.)

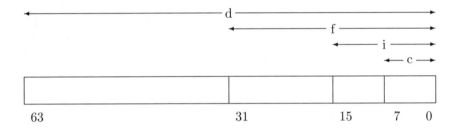

Figure 11.2: Example of a union.

One motivation for defining a union is to reduce memory requirements. For example, if you need to store an int and a float in the same application and are certain that you don't need to store them at the same time, then the union defined below will save memory.

```
union int_float {
    int i;
    float f;
};
```

The members of a union are accessed in the same way as other class members, by using the member access operator (a single dot). This is demonstrated by the following program:

```
#include <iostream>
#include <cstdlib>       // For EXIT_SUCCESS
using namespace std;

union int_float {
    int i;
    float f;
};
```

```
int main()
{
    int_float x;
    x.i = 10;
    cout << x.i << "\n";
    cout << x.f << "\n";
    x.f = 3.142;
    cout << x.f << "\n";
    cout << x.i << "\n";
    return(EXIT_SUCCESS);
}
```

Notice that there is no conversion between the two data types; the bits are interpreted as an `int` for `x.i` and as a `float` for `x.f`. Also, members of a `union` are `public` by default. Members of a `union` can be declared `private`, but this is not particularly useful.

Our simple `int_float` example doesn't save much memory, but if we have large objects, such as large arrays, then the saving can be significant. Unlike bit-fields, accessing the members of a `union` in this situation does not incur high overheads.

It is not necessary for a `union` to be named; an *anonymous union* is a `union` without a name, as illustrated by the following code fragment:

```
union {
    int i;
    float f;
};
```

An anonymous `union` simply makes the members occupy the same memory and member names must be used directly, without a member access operator. Anonymous unions cannot have member functions and, if global, must be declared `static`. A program demonstrating an anonymous union is given below.

```
#include <iostream>
#include <cstdlib>        // For EXIT_SUCCESS
using namespace std;

int main()
{
    union {
        int i;
        float f;
    };
    i = 10;
    cout << i << "\n";
    cout << f << "\n";
    f = 3.142;
    cout << i << "\n";
    cout << f << "\n";
    return(EXIT_SUCCESS);
}
```

Notice how the value of one member of the `union` is changed by changing the value of the other member. A more significant application of an anonymous `union` is given in Section 11.4.4.

A named `union` can have member functions, including constructors and destructors. The `union` in the following program has two constructors, one taking an `int` argument and the other taking a `float` argument:

```cpp
#include <iostream>
#include <cstdlib>        // For EXIT_SUCCESS
using namespace std;

union int_float {
    int i;
    float f;
    int_float(int j) { i = j; }
    int_float(float x) { f = x; }
};

int main()
{
    int_float x(10);         // x.i = 10
    cout << x.i << "\n";
    cout << x.f << "\n";
    int_float y(3.142f);     // y.f = 3.142
    cout << y.f << "\n";
    cout << y.i << "\n";
    return(EXIT_SUCCESS);
}
```

Once again, notice how the value of one member of the `union` is changed by changing the value of the other member.

There are many restrictions on unions, including the following:

- A `virtual` function is not allowed as a member of a `union`.

- A `union` cannot be a base class.

- A `union` cannot be a derived class.

- Data members cannot be `static`.

- If a data member is an instance of a class, then that class must not have constructors, destructors or a user-defined assignment operator.

These restrictions do not detract from the fundamental idea of a `union` as a low-level construct for saving memory.

11.4   Using Bitwise Operators

11.4.1   A Bit Array Class

A fairly common requirement is to store a large array of data, each element of which can only have the value zero or one. An obvious space-saving device is to store each data item as a single bit. However, accessing a data item by using bitwise operators is not very convenient and in fact is sufficiently cumbersome that care is needed to avoid programming errors. It would be much better if we could use an array-like notation, and this can be achieved by means of a *bit array* class, as described in this section. The header file for a bit array class is given below.

```
// source:    b_array.h
// use:       Declares bit array class.

#ifndef BIT_ARRAY_H
#define BIT_ARRAY_H

class bit_array {
public:
    bit_array(unsigned max_length);
    ~bit_array() { delete data; }
    void set(unsigned position);
    void clear(unsigned position);
    void set(void);
    void clear(void);
    void display(void);
    bool operator[](unsigned position);
    unsigned length(void) { return max_bits; }
private:
    unsigned *data;
    unsigned max_bits;
    static unsigned bits_per_word;
    unsigned max_words(void) const;
    unsigned max_bytes(void) const;
};

inline unsigned bit_array::max_bytes(void) const
{
    return(max_words() * sizeof(*data));
}

#endif  // BIT_ARRAY_H
```

Objects of the `bit_array` class store data items as single bits. Individual bits can be set (`set(unsigned)`) or cleared (`clear(unsigned)`) and the subscripting operator, `[]`, is overloaded to return `true` or `false`, depending on whether or not a bit is set. The number of bits that can be stored by an object is given by the `length()` function.

The following file gives an implementation of the `bit_array` class:

```
// source:     b_array.cxx
// use:        Implements bit array class.

#include <iostream>
#include <cstdlib>        // For exit(), system()
#include <climits>        // For CHAR_BIT
#include <cstring>        // For memset()
#include "b_array.h"
using namespace std;

const unsigned BIT_MASK = 1;

unsigned bit_array::bits_per_word = CHAR_BIT * sizeof(unsigned);

bit_array::bit_array(unsigned max_length)
{
    max_bits = max_length;
    data = new unsigned[max_words()];
}

void bit_array::set(unsigned position)
{
    if (position >= max_bits) {
        cerr << "bit_array set(" << position << ") out of range\n";
        exit(EXIT_FAILURE);
    }
    unsigned word = position / bits_per_word;
    unsigned bit = position % bits_per_word;
    data[word] |= (BIT_MASK << bit);
}

void bit_array::clear(unsigned position)
{
    if (position >= max_bits) {
        cerr << "bit_array clear(" << position <<
            ") out of range\n";
        exit(EXIT_FAILURE);
    }
    unsigned word = position / bits_per_word;
    unsigned bit = position % bits_per_word;
    data[word] &= ~(BIT_MASK << bit);
}

void bit_array::set(void)
{
    memset(data, ~0, max_bytes());
}
```

```
void bit_array::clear(void)
{
    memset(data, 0, max_bytes());
}

void bit_array::display(void)
{
    const unsigned display_width = 80;
    for (unsigned i = 0; i < max_bits; ++i) {
        if (!(i % display_width))
            cout <<"\n";
        if (operator[](i))
            cout << "x";
        else
            cout << ".";
    }
    cout << "\n";
}

bool bit_array::operator[](unsigned position)
{
    if (position >= max_bits) {
        cerr << "bit_array [" << position << "] out of range\n";
        exit(EXIT_FAILURE);
    }
    unsigned word = position / bits_per_word;
    unsigned bit = position % bits_per_word;
    if (data[word] & (BIT_MASK << bit))
        return true;
    else
        return false;
}

unsigned bit_array::max_words(void) const
{
    unsigned result = max_bits / bits_per_word;
    if (max_bits % bits_per_word)
        ++result;
    return result;
}
```

Notice that the static variable, bit_array::bits_per_word, is initialized (*not* assigned to) in this file. The constant, CHAR_BIT, which is defined in <climits>, is the number of bits for a char type. Since the sizeof operator gives the size of an object in units of char (that is, sizeof(char) is defined to be one), the expression for bits_per_word is entirely independent of the particular compiler being used. In the context of the bit_array class, we define a word to be the basic unit of storage for the `unsigned` type.

Looking at the constructor function, we can see that it dynamically allocates enough memory to store `max_length` number of bits. The number of words of memory required is calculated by the `max_words()` function. Notice that `max_bits` is set to `max_length`, and is not rounded up to fill an integral number of words.

Turning now to the `set(unsigned)` function, the variable called `word` gives the element of the array of type `unsigned` that stores a particular data item and `bit` gives the position within that word, as shown in Figure 11.3. The `BIT_MASK` constant has the single lowest bit set. A *mask* is an object that can be used to pick out particular bits from another object. In the `set(unsigned)` function we want to extract a single bit from an `unsigned` object. This can be achieved by left shifting `BIT_MASK` and taking the inclusive bitwise OR of the result with the `unsigned` object. An example is given in Figure 11.4. Notice that the left-shift operator doesn't actually shift the bits of `BIT_MASK` itself, which is in any case declared `const`. It is the expression resulting from `BIT_MASK << bit` that is left-shifted.

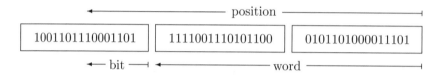

Figure 11.3: Accessing an element of a `bit_array` object.

Figure 11.4: Using BIT_MASK to set a bit.

Implementation of the `clear(unsigned)` function is similar to the implementation of `set(unsigned)`. The complement of the left-shifted `BIT_MASK` gives a mask that is all ones, except for a single zero in the required position. A bitwise AND then clears (sets to zero) the corresponding bit in an element of the `data` array, as shown in Figure 11.5.

Figure 11.5: Using BIT_MASK to clear a bit.

Two functions, `set(void)` and `clear(void)`, are provided to set and clear all the bits of a `bit_array` object. These functions make use of the `memset()` library function, which is declared in `<cstring>` as:

```
void *memset(void *pt, int val, size_t count);
```

The memset() function sets the first count characters pointed to by pt to the value specified by val. The function returns the pointer, pt.

The display() function shows the current state of a bit_array object. An x marks a set bit and a dot is used to denote a cleared bit. Since display() is a class member, the function notation can be used to invoke the bit_array operator as:

```
operator[](i)
```

Equally valid (but probably more obscure) alternative notations are:

```
this->operator[](i)
```

and:

```
(*this)[i]
```

A particular bit in a bit_array object can be accessed by using the overloaded sub-scripting operator, which is implemented by the operator[]() function. This function calculates the position of the bit within a word in a similar way to the set(unsigned) and clear(unsigned) functions. The bitwise AND operator, &, is then used to test whether or not the appropriate bit of the appropriate element of the data[] array is set.

The program given below demonstrates the bit_array class by storing an alternating series of bit_array elements and then displaying the data.

```
// source:    my_test.cxx
// use:       Tries out bit_array class.

#include <iostream>
#include <cstdlib>       // For EXIT_SUCCESS, atoi()
#include "b_array.h"
using namespace std;

void wait(void)
{
    cout << "\nHit <Enter> to continue." << "\n";
    cin.get();
}

int main(int argc, char *argv[])
{
    if (argc != 2) {
        cerr << "Usage:  my_test <length in bits>\n";
        exit(EXIT_FAILURE);
    }
    unsigned bits = atoi(argv[1]);
    bit_array b(bits);
    for (unsigned i = 0; i < bits; ++i)
```

```
            if (i % 2)
                b.set(i);
            else
                b.clear(i);
        cout << "bit array set\n" << "Array length available = " <<
            b.length() << "\n";
        wait();
        b.display();
        wait();
        b.clear();
        b.display();
        wait();
        b.set();
        b.display();
        wait();
        return(EXIT_SUCCESS);
    }
```

The function wait(), defined above, pauses the output until the user hits the Enter key. This function uses the get() library function, which is declared in <iostream> and extracts a single character from the input stream.

Exercise

Try modifying this program so that more interesting bit data is stored, such as data that appears as a disc when the display() function is invoked.

11.4.2   The Sieve of Eratosthenes

An interesting application of the bit_array class is to implement the Sieve of Eratosthenes, which finds all primes up to some specified maximum. Suppose we want to find the primes up to 20. The technique consists of writing down all the integers from 2 to 20 and striking out every second integer after 2. We then strike out every third integer after 3, every fourth integer after 4 and so on. This procedure is illustrated in Figure 11.6 where a dot indicates the start of a sweep through the integers and a cross marks integers that are struck out. Those integers that are not struck out are prime. Of course, in practice we would probably be interested in very large prime numbers and would need to use a more efficient technique.

The following program uses the bit_array class to implement the Sieve of Eratosthenes:

```
// source:    primes.cxx
// use:       Implements the Sieve of Eratosthenes using the
//            bit_array class.

#include <iostream>
#include <cstdlib>       // For EXIT_SUCCESS, atoi()
#include <cmath>         // For sqrt()
```

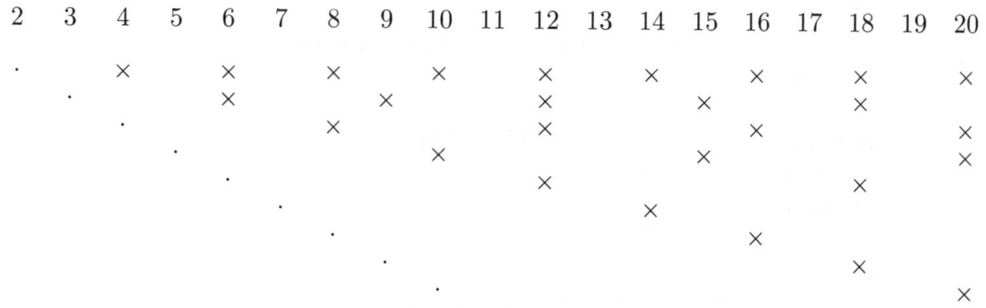

2	3	4	5	6	7	8	9	10	11	12	13	14	15	16	17	18	19	20
·		×		×		×		×		×		×		×		×		×
	·			×			×			×			×			×		
		·				×				×				×				×
			·					×					×					×
				·						×						×		
					·							×						
						·								×				
							·									×		
								·										×

Figure 11.6: Sieve of Eratosthenes.

```
#include "b_array.h"
using namespace std;

// function declarations:
void wait(void);
void list_primes(bit_array &b);

void wait(void)
{
    cout << "\nHit <Enter> to continue." << "\n";
    cin.get();
}

void list_primes(bit_array &b)
{
    const unsigned display_height = 24;
    unsigned primes_displayed = 1;
    unsigned i_max = b.length();
    for (unsigned i = 0; i < i_max; ++i) {
        if (b[i]) {
            cout << 2 * i + 3 << "\n";
            if (primes_displayed % display_height)
                ++primes_displayed;
            else {
                primes_displayed = 1;
                cout.flush();
                wait();
            }
        }
    }
}

int main(int argc, char *argv[])
{
```

```
    if (argc != 2) {
        cerr << "Usage:  primes <largest integer>\n";
        exit(EXIT_FAILURE);
    }
    unsigned max_int = atoi(argv[1]);
    // Only store odd numbers:
    unsigned bits = (max_int - 1) / 2;
    bit_array b(bits);
    b.set();
    unsigned max_m = static_cast<unsigned>(sqrt(max_int));

    // These 2 loops are the Sieve of Eratosthenes:
    for (unsigned m = 3; m <= max_m; m += 2) {
        unsigned max_n = max_int / m;
        for (unsigned n = m; n <= max_n; n += 2)
            b.clear(m * n / 2 - 1);
    }

    cout << "Bit array set for primes:\n";
    b.display();
    wait();
    cout << "Primes between 3 and " << max_int << ":\n";
    list_primes(b);
    wait();
    return(EXIT_SUCCESS);
}
```

The performance of the two loops in `main()` corresponding to the Sieve could be improved, but the version given here makes it easier to understand the algorithm. In any case, a number of techniques have already been used to speed up the program. These include the following:

- Only odd integers are stored since we know that no even integer (apart from two) is prime.

- If we are searching for all primes up to m, then we know that integers greater than $\sqrt{m}$ cannot divide m, and so there is no need to test for such integers.

- If, on a given sweep through the integers, we are striking out all those divisible by n, then we only need to start at n^2 because smaller multiples of n will already have been struck out.

In order to compile the program that tests the bit array class (on page 319) and the program to find primes, you may want to place all the relevant files in the same directory. You can then use the makefile given below. Notice that the first line states that both `my_test` and `primes` are targets for the `make` utility.

```
all: my_test primes
my_test: my_test.cxx b_array.o
        g++ my_test.cxx b_array.o -o my_test
```

```
primes: primes.cxx b_array.o
    g++ primes.cxx b_array.o -o primes
b_array.o: b_array.cxx b_array.h
    g++ -c b_array.cxx
```

Exercise

Compile and run the primes program. Use the program to list primes less than 100, and check that these are consistent with a standard table of primes.

11.4.3 Bit Representation of Integral Types

Bitwise operations enable us to investigate how the basic data types are stored on different systems. The following program gives the bit pattern for the unsigned type:

```
// source:    bits.cxx
// use:       Prints representation in bits for the unsigned type.

#include <iostream>
#include <cstdlib>      // For EXIT_SUCCESS
#include <climits>      // For CHAR_BIT
using namespace std;

void print_bits(unsigned i)
{
    unsigned bit_mask = 1U << (CHAR_BIT * sizeof(unsigned) - 1);
    for (int j = 0; j < sizeof(unsigned); ++j) {
        for (int k = 0; k < CHAR_BIT; ++k) {
            if (i & bit_mask)
                cout << "1";
            else
                cout << "0";
            bit_mask >>= 1;
        }
        cout << " ";
    }
    cout << "\n";
}

int main()
{
    print_bits(0);
    print_bits(1);
    print_bits(-1);
    print_bits(2000000000);
    print_bits(-2000000000);
    return(EXIT_SUCCESS);
}
```

The `print_bits()` function constructs a mask with only the most significant (left-most) bit initially non-zero. A bitwise AND of the mask and the function argument detects whether or not the left-most bit of `i` (the argument of `print_bits()`) is set. Progressively right-shifting the mask enables us to test the remaining bits of `i`.

The `print_bits(unsigned)` function also works for `signed int` if a two's complement representation is used, since conversion occurs with no change in the bit pattern. To understand this comment a brief diversion must be made to consider what is meant by a two's complement representation.

There are several ways of representing signed integers and two's complement is one of the most popular. For a positive integer, n, the most significant bit is 0 and the other bits take the expected form for a positive integer. To represent $-n$ we do the following:

- Find the binary representation of n.

- Take the bitwise complement.

- Add 1.

The consequence of these operations is that the most significant bit is always 1 for a negative integer. As an example consider -59 where, for the three stages listed above and assuming a 32-bit representation, we get the bit patterns given below.

Binary representation: 0000 0000 0000 0000 0000 0000 0011 1011

Bitwise complement: 1111 1111 1111 1111 1111 1111 1100 0100

Result of adding 1:   1111 1111 1111 1111 1111 1111 1100 0101

Typical bit patterns obtained on a particular system (using 32 bits to represent the `unsigned` type) are given in Table 11.2. Notice how all of the bits are set for -1; this is because two's complement arithmetic is being used.

Decimal number	Bit pattern
0	00000000 00000000 00000000 00000000
1	00000000 00000000 00000000 00000001
-1	11111111 11111111 11111111 11111111
2000000000	01110111 00110101 10010100 00000000
-2000000000	10001000 11001010 01101100 00000000

Table 11.2: Bit patterns given by `print_bits(unsigned)`.

Exercise

Write a program that prompts for an integer and uses the `print_bits()` function to give the bit pattern for the integer. Use your program to display the bit patterns for the largest and smallest integers that can be represented on your system by the `unsigned` and `int` types.

11.4.4 Bit Representation of Floating Point Types

The way in which floating point numbers are represented varies between compilers and processors. However, many systems use what is known as the "IEEE Standard for Binary Floating Point Arithmetic".[3] The following discussion is restricted to systems conforming to the IEEE Standard, and with the `float` type represented by 32 bits and `double` by 64 bits.

A floating point number can be written in binary form as:[4]

$$x = (-1)^s 2^E (b_0.b_1 b_2 \ldots b_{p-1})$$

where the significance of the symbols is explained below. To be definite, we start by restricting our discussion to the `float` type, where p has the value of 24. The way in which such numbers are stored in terms of bits is shown in Figure 11.7, where we adopt the convention of showing the least significant bits on the right.

Figure 11.7: Bit structure for a `float`.

There are three distinct parts to the binary representation of a floating point number:

The sign bit

The parameter, s, is a single bit, known as the *sign bit*. A floating point number is positive or negative, depending on whether s takes the value 0 or 1.

The exponent

Rather than storing the *exponent*, E, a *biased exponent* is used. That is, the exponent is shifted by a constant, known as the *bias*, so that the value actually stored is not negative. For the representation being considered here, the exponent is stored as 8 bits. Consequently, the biased exponent, e, is defined by:

$$e = E + 127.$$

and the bias is 127.

---

[3] ANSI/IEEE Std 745-1985, published by the Institute of Electrical and Electronics Engineers, Inc.

[4] It is a worth remembering that there is an infinity of floating point numbers that cannot be represented on any computer. This is because the sequence $b_1 b_2 \ldots b_{p-1}$ is finite and the range of E is limited.

The significand

The b_i are single bits, which can take the values 0 and 1. The sequence of b_i is known as the *significand* and consists of b_0 followed by a binary point and a *fractional* part (b_1 to b_{23}), which we will label as f. The value of b_0 is implicit, which means that only s, e and f are stored, as shown in Figure 11.7.

Combinations of the values of e and f are used to store five different types of information:

1. If $0 < e < 255$ then what is known as a *normalized* number is stored and this has the value given by:[5]

$$x = (-1)^s 2^{e-127}(1.f)$$

   Notice that b_0 is implicitly 1 and that f may or may not be zero.

2. If $e = 0$ and $f \neq 0$, then what is known as a *denormalized* number is stored and this has the value given by:

$$x = (-1)^s 2^{-126}(0.f)$$

   Notice that in this case b_0 is implicitly 0, rather than 1, and that 2 is raised to a fixed power. Denormalized numbers extend downwards the magnitude of a number that can be stored by a fixed length word. However, as the magnitude of the number gets smaller, f fills from the left with zeros and the precision decreases; in the limiting case only one bit is significant.

3. If $e = 0$ and $f = 0$, then zero is stored:

$$x = (-1)^s 0$$

4. If $e = 255$ and $f = 0$, then signed infinity is stored:

$$x = (-1)^s \infty$$

   A signed infinity is generated if a floating point number is either too big or too small to be represented by the particular type.

5. If $e = 255$ and $f \neq 0$, then what is known as a *NaN* (Not a Number) is stored. A NaN is the result of an invalid floating point operation, such as $0 \times \infty$ or $0 \div 0$. Once created, NaNs propagate through a computation and may appear in your (incorrect) final results.

It would be nice to be able to print bit patterns for the floating types. This is slightly complicated by the fact that bitwise operators only act on the integral types. However, an anonymous union enables us to store a `float` and then to treat it as an `unsigned int`.[6] The following program implements a function to print the bit pattern for an argument of type `float`, and then uses this function to list the bit patterns for several numbers:

---

[5]In this discussion, a number written as (1.f) or (0.f) is in binary notation.

[6]This assumes that an `unsigned int` has the same number of bits as a `float`. If this is not true on your system, then you should make the necessary simple modifications to `print_bits(float)`.

```cpp
#include <iostream>
#include <cstdlib>        // For EXIT_SUCCESS
#include <climits>        // For CHAR_BIT
using namespace std;

void print_bits(const float &x)
{
    union {
        unsigned int i;
        float y;
    };
    y = x;
    unsigned int bit_mask = 1U <<
        (CHAR_BIT * sizeof(unsigned int) - 1);
    // Find sign bit:
    if (i & bit_mask)
        cout << "1 ";
    else
        cout << "0 ";
    bit_mask >>= 1;
    // Find bits in exponent:
    for (int j = 1; j <= 8; ++j) {
        if (i & bit_mask)
            cout << "1";
        else
            cout << "0";
        bit_mask >>= 1;
    }
    cout << " ";
    // Find bits in significand:
    const int total_bits = CHAR_BIT * sizeof(unsigned int);
    for (int j = 9; j < total_bits; ++j) {
        if (i & bit_mask)
            cout << "1";
        else
            cout << "0";
        bit_mask >>= 1;
    }
    cout << "\n";
}

int main()
{
    print_bits(0.0f);
    print_bits(1.0f);
    print_bits(2.0f);
    print_bits(1e-7f);
```

```
        print_bits(-10.0f);
        return(EXIT_SUCCESS);
}
```

The results of this program are given in Table 11.3. Notice that spaces are put between
the bit patterns for the s, e and f terms.

Floating point number	Bit pattern
0.0	0 00000000 00000000000000000000000
1.0	0 01111111 00000000000000000000000
2.0	0 10000000 00000000000000000000000
1×10^{-7}	0 01100111 10101101011111110010101
-10.0	1 10000010 01000000000000000000000

Table 11.3: Bit patterns given by `print_bits(float)`.

Exercise

(a) Show mathematically that the bit patterns given in Table 11.3 are
    correct.

(b) Modify the program so that it prompts for a floating point number and
    then displays the corresponding bit pattern. Try out your program
    with various numbers.

The representation of the type `double` is very similar to `float`, except that e
and f are assigned additional bits, as shown in Figure 11.8. The biased exponent is
represented by 11 bits, with a bias of 1023. Consequently, the biased exponent, e, is
given in terms of the exponent, E, by:

$$e = E + 1023$$

The fractional part has 52 bits. Clearly a 64-bit `double` has a much greater range and
precision than a 32-bit `float`.

Figure 11.8: Bit structure for a `double`.

11.5 Summary

- The bitwise logical operators are ~, &, ^ and |.

- Don't confuse & with &&, or | with ||.

- The left shift operator is denoted by, <<, and the right shift operator by, >>.

- A bit-field is a class member with a specified width in bits. Bit-fields are very compiler-dependent and their use often carries a high performance penalty.

- A union is a class in which data members have overlapping memory locations:

```
union int_float {
    int i;
    float f;
};
```

11.6 Exercises

1. Implement a copy constructor, assignment operator and != operator for the bit-array class (Section 11.4.1).

2. Modify the Sieve of Eratosthenes program (Section 11.4.2) to make it perform as fast as you are able. (You should get some ideas from [12] and Figure 11.6.) Your changes could also include removing range checking from the bit_array class.

3. (a) Find out what bit pattern is used to represent the float type on your system, and then calculate the maximum and minimum normalized numbers that can be represented by this type. Verify your answers by using the print_bits(float) function.

   (b) How many decimal places are significant for a normalized float and what is the smallest denormalized float on your system? Again, calculate these numbers and then verify them by using print_bits(float). (You should also be able to verify your answers to this question by looking at the file <cfloat> on your system.)

4. Implement a function, print_bits(double), that prints the bit patterns for the three parts of the double type. Repeat Exercise 3 for this type.

Chapter 12

Single Inheritance

12.1 Derived Classes

One of the fundamental techniques in object-oriented programming is the use of *inheritance*. Inheritance is a way of creating new classes that extend the facilities of existing classes by including new member data and functions, as well as (in certain circumstances) changing existing functions. The class that is extended is known as the *base class* and the result of an extension is known as the *derived class*. The derived class *inherits* the data and function members of the base class.[1] Inheritance allows a class to be extended rather than modified. For a well-designed base class, only the base class interface rather than the implementation need be known. This avoids recompiling the base class and even means that a class can be extended without the source for its implementation being available. Moreover, both the base and derived classes can be used by a program. Without inheritance, the source for the original class would have to be modified. Consequently, either the original class would no longer be available, or else there would be code for two separate classes with the associated problem of maintaining consistency.

Inheritance also facilitates having consistent interfaces for a whole hierarchy of related classes. The same function interface may be used by objects of different classes, with the implementation of the function depending on the type of object by which it is invoked. In effect, this is a switch that depends on the object type. However, the switch is implemented automatically by the compiler, rather than by complicated control statements in the code.

Any class can be a base class. As an example, suppose we have a project involving general two-dimensional shapes, such as discs, squares, triangles etc. A shape class might take the form:

```
class shape {
public:
    int i_d;
    float x, y;
```

---

[1]Some authors use *superclass* and *subclass* (which are taken from the language, Simula) instead of base class and derived class. Other authors find the terms confusing and they are not used in this book.

```
        material_type material;
    };
```

where `i_d` is an identifying number for a particular shape and `x` and `y` are its position coordinates. The shape could be manufactured from various substances, which are defined by the following enumeration:[2]

```
    enum material_type {WOOD, STEEL, ALUMINIUM, PLASTIC};
```

The syntax for defining a class, `disc`, which is derived from `shape`, is illustrated by the following code:

```
    class disc : public shape {
    public:
        float radius(void) { return disc_radius; }
    private:
        float disc_radius;
    };
```

Other classes, such as `square` and `triangle` can be derived from `shape` in a similar manner to `disc`. The relationship between such classes is often clarified by a *directed acyclic graph* (or DAG),[3] as shown in Figure 12.1. A common convention, followed

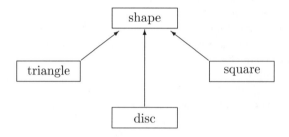

Figure 12.1: Classes derived from `shape`.

here, is to place the base class at the top of the diagram, with the arrows flowing from the derived classes towards the base class. Notice that a derived class possesses a "kind of" relationship with respect to its base class; that is a `disc` object is a "kind of" `shape`. Therefore it is both meaningful and useful to assign a derived class object to an instance of the base class, since this corresponds to a truncation of data. (See Figure 12.2.) An example of such an assignment is given below.

```
    shape s;
    disc d;
    s = d;
```

However, the assignment, without an explicit cast, of a base class object to an instance of a derived class is illegal since in general some data members will have no base class counterparts.

---

[2]Don't forget that the enumeration must come before the class declaration.

[3]A DAG is a graph in which the arcs have a direction, but there are no closed loops.

shape   disc

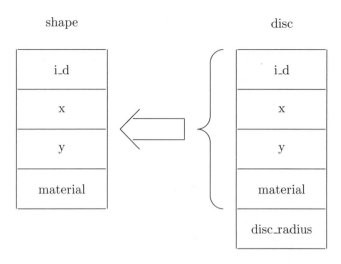

Figure 12.2: Assigning data for a disc object to a shape object.

The single colon in the first line of the disc class declaration is used to separate the derived class, disc, from the base class, shape. The derived class possesses all of the member functions and data specified in its own class definition, together with all those of the base class; inheritance therefore adds attributes to a base class. By default, all members of a base class are private in the derived class, but this can be overridden by putting an access specifier before the base class in the derived class declaration; in this example we have specified public. It is a useful piece of defensive programming to always include the access specifier, since a common error is to forget that the default is private. Member access privileges are considered in more detail in Section 12.6. However, it is important to realize that placing public before the base class in a class declaration does *not* permit a derived class to access the private members of a base class.

It is usually desirable that names given to members (particularly data members) of a derived class are different from those of a base class. However, in some cases there may be duplication of names and, in such situations, the concept of *dominance* often resolves any potential ambiguity.[4] If a class member name (either function or data) occurs in both a derived class and a base class, then the derived class name dominates. For instance, if we declare:

```
class disc : public shape {
public:
    int i_d;
};
```

with the shape class declaration given on page 331, then the following is not ambiguous:

```
disc d;
```

---

[4]Duplication of names is particularly likely in a complicated class hierarchy where the classes may be the work of different programmers. (See Section 12.4 and Chapter 13.)

```
    d.i_d = 100;              // Derived class member accessed.
```

In the above example, it is the derived class member of the disc object that is accessed. In order to access the base class member, we need to use the scope resolution operator, ::, together with the base class name, as in:

```
    d.shape::i_d = 10;        // Base class member accessed.
```

We can verify that dominance does indeed resolve the potential access ambiguity for the two variables, i_d, by running the following simple program:[5]

```
#include <iostream>
#include <cstdlib>          // For EXIT_SUCCESS
using namespace std;

enum material_type {WOOD, STEEL, ALUMINIUM, PLASTIC};

class shape {
public:
    int i_d;
    float x, y;
    material_type material;
};

class disc : public shape {
public:
    int i_d;
};

int main()
{
    disc d;

    d.i_d = 100;          // Derived class member accessed.
    d.shape::i_d = 10;    // Base class member accessed.
    cout << "Derived class i_d: " << d.i_d << "\n";
    cout << "Base class i_d: " << d.shape::i_d << "\n";

    return(EXIT_SUCCESS);
}
```

Exercise

Compile and run the above program. Why does your output demonstrate that two different data members are accessed?

It is quite common for a derived class to provide its own implementation of a base class function. Dominance then ensures that the derived class function is called for derived

---

[5]For the current shape and disc classes, the members are all public. Don't forget that these classes are given to illustrate ideas. Realistic classes would make use of data hiding.

class objects. The base class function is *hidden* rather than overloaded, even if the argument types differ. For example, consider the following program in which we we define a base class, shape, and a derived class, disc:

```
#include <iostream>
#include <cstdlib>        // For EXIT_SUCCESS
using namespace std;

class shape {
public:
    float give_area(void) { return area; }
    float area;
};

class disc : public shape {
public:
    float give_area(float radius) { return(3.142*radius*radius); }
};

int main()
{
    disc d;
    d.area = 1.0;
    cout << d.give_area(10.0)    // O.K.: uses function defined in
         << "\n";                // derived class.
    cout << d.give_area()        // WRONG: attempts to use function
         << "\n";                // defined in base class.
    return(EXIT_SUCCESS);
}
```

It might be supposed that give_area() is overloaded and that the d.give_area() expression (with no argument) invokes the base class function. What actually happens is that the derived class function hides the base class version. Invoking d.give_area() is therefore a compile-time error since the function has the wrong number of arguments. It is not the class definitions that are in error, but rather the attempt to invoke the base class function. This can be corrected by replacing the statement involving d.give_area() with the following statement:

```
cout << d.shape::give_area()    // Uses function defined in base
     << "\n";                    // class.
```

Now that we have introduced inheritance, we can distinguish the three ways in which a class is able to use other classes:

1. An object may contain objects that are instances of other classes. For example, the following data class has members u and v that are instances of the complex class:

```
class data {
```

```
public:
    data(const complex &z1, const complex &z2);
    // More class declarations go in here.
private:
    complex u, v;
};
```

A `data` object is said to be a *client* of the `complex` class; it is not a "kind of" `complex` object.

2. A class may use a pointer to an object of the same or another class. For example, a `node` class object has a pointer to a `node` object:

```
class node {
    friend class list;
public:
    DATA_TYPE data;
private:
    node *next;
};
```

A `node` object does not contain another `node`, which is in any case illegal, but rather accesses the data of another `node`.

3. A class may be derived from another class. For example, an instance of the `disc` class is a "kind of" `shape`:

```
class disc : public shape {
public:
    float radius(void) { return disc_radius; }
private:
    float disc_radius;
};
```

However, in no sense does a `disc` object use a `shape` object.

12.2   virtual Functions

It is often desirable for objects corresponding to different derived classes to respond differently to the same function call. This behaviour is known as *polymorphism* and we have already seen how it may be achieved through a derived class that provides its own implementation of a base class function. However, the exact kind of object on which a function acts may not be known at compile-time. For example, we may want to have an array of various derived `shape` objects, such as `squares`, `discs`, `triangles` etc., with the elements of the array being chosen at run-time. It would then be convenient if we could invoke different `give_area()` functions for the different kinds of elements in the array. All this can be achieved by using `virtual` functions.

A function is declared `virtual` by including the `virtual` keyword in the class declaration. This is illustrated by the `give_area()` function in the following declaration:

```
class shape {
public:
    virtual float give_area(void);
    float area;
};
```

A virtual function must be a non-static class member and the **virtual** specifier can *only* occur within a class body. A common error is to attempt to *define* a **virtual** function by including the **virtual** keyword, as shown in the following statement:[6]

```
virtual float shape::give_area(void) { return area; }    // WRONG!
```

This function is correctly defined as:

```
float shape::give_area(void) { return area; }            // O.K.
```

since it is sufficient to declare a function **virtual** in the base class for it to be **virtual** everywhere.

The **virtual** specifier in the **shape** class has no effect unless a derived class also defines a **give_area()** function, as in the following definition of a **disc** class:

```
class disc : public shape {
public:
    virtual float give_area(void) { return(3.142*radius*radius); }
    float radius;
};
```

Notice that both the return type and function arguments are identical in the two versions of **give_area()**. However, the situation here is subtly different from the duplication of names discussed in Section 12.1. To appreciate this distinction and to see the true significance of **virtual** functions, consider the following program:[7]

```
#include <iostream>
#include <cstdlib>        // For EXIT_SUCCESS
using namespace std;

class shape {
public:
    virtual float give_area(void);
    float area;
};

inline float shape::give_area(void) { return area; }

class disc : public shape {
public:
    virtual float give_area(void);
    float radius;
```

---

[6]Don't forget the distinction between define and declare.

[7]Notice that, as here, it is possible for a function to be both **inline** and **virtual**.

```
};

float disc::give_area(void)
{
    return(3.142 * radius * radius);
}

int main()
{
    shape *pt;
    shape s;
    s.area = 1.0;
    disc d;
    d.radius = 2.0;
    d.area = 3.0;
    pt = &s;
    cout << pt->give_area() << "\n";
    pt = &d;
    cout << pt->give_area() << "\n";
    return(EXIT_SUCCESS);
}
```

The first point to notice is that a base class pointer can store the address of a derived
class object. In this case, pt is first used to store the address of a shape object and then
a disc object. The second point is that the statements that output the areas of the
different kinds of object are identical. This is because if a virtual function is invoked
through a pointer, then which of the functions is actually called depends on the class
of the object pointed to. Consequently, the first statement involving give_area()
calls the shape function, and the second statement calls the disc version. This is
how virtual functions can be used to implement polymorphism and is an important
feature of object-oriented programming in C++.[8] The role of virtual functions in
polymorphism is crucial; for non-virtual functions, the base class pointer in the above
program would invoke the same shape version of the give_area() function in both
statements.

Exercise

Run the program given above and deduce which version of the give_area()
function is invoked by each of the:

```
        cout << pt->give_area()
```

statements. What happens if the virtual specifier is removed from either
one or both class declarations?

In the program given above, there was no need to include the virtual specifier in
the derived disc class declaration and we could use the following alternative declaration:

---

[8]Some authors, such as [3], describe the effect of virtual functions by saying that the derived class
function overrides the base class version. However, you should understand that overriding is not the
same as dominance; there is more to virtual functions than the mere hiding of base class names!

```
class disc : public shape {
public:
    float give_area(void);
    float radius;
};
```

This is because the `give_area()` function is implicitly `virtual` due to the declaration within the `shape` class. However, including the keyword in the derived class may be useful since it clearly indicates, without examining the base class, that the function is `virtual`.

One subtlety involving `virtual` functions is that a function declared in a derived class and a base class cannot differ only by the `return` type if the base class function is `virtual`. In order to understand this, have a look at the following classes:

```
class shape {
public:
    virtual float give_area(void);
    area;
};

class disc : public shape {
public:
    double give_area(void);      // WRONG!
    float radius;
};
```

We can see that the second declaration of `give_area()` cannot hide the first, since only the `return` types differ. However, neither can the second declaration be `virtual` since the `return` types *do* differ. As a result, the two declarations are inconsistent and there is a compile-time error.

Another subtlety is that if a `virtual` function is invoked through a pointer, then the access level is determined by the function declaration in the class corresponding to the pointer and not to the function actually invoked. This is illustrated by the following program. (Notice that, in contrast to previous classes in this chapter, the member data are now `private`.)

```
#include <iostream>
#include <cstdlib>      // For EXIT_SUCCESS
using namespace std;

class shape {
public:
    void set_area(float a) {area = a; }
    virtual float give_area(void) { return area; }
private:
    float area;
};

class disc : public shape {
```

```
public:
    void set_radius(float r) { radius = r; }
private:
    virtual float give_area(void) { return(3.142*radius*radius); }
    float radius;
};

int main()
{
    disc d;
    d.set_area(1.0);
    d.set_radius(2.0);
    shape *pt = &d;
    cout << pt->give_area() << "\n";
    return(EXIT_SUCCESS);
}
```

In the statement:

```
cout << pt->give_area() << "\n";
```

the pointer successfully invokes `disc::give_area()`, even though the latter is `private`. This is because `pt` is of type pointer to `shape` and the function, `shape::give_area()`, *is* `public`. This may appear to be a minor hole in the access protection mechanism, but it is consistent with the fact that which `virtual` function is actually invoked through a pointer is a run-time decision.

Exercise

(a) Run the above program and deduce which version of the `give_area()` function is invoked by the statement:

```
cout << pt->give_area() << "\n";
```

(b) What happens if the statement initializing `pt` in the above program is replaced by:

```
disc *pt = &d;
```

12.3   Abstract Classes

Suppose we have a project involving solids such as cubes, spheres, cones etc. Corresponding to each type of solid there is an appropriate equation for the volume and surface area. Moreover, suppose that we want to consider a collection of different kinds of solids and to perform calculations such as summing their volumes and surface areas. This project is an obvious candidate for inheritance; that is we could have a `solid` base class with derived `cone`, `sphere` and `cube` classes, as shown in Figure 12.3. Furthermore, if the `solid` class has `virtual` member functions for calculating the volume and surface area, then polymorphism can be used to find the total area and volume

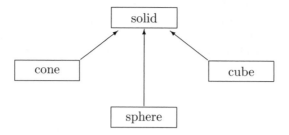

Figure 12.3: Classes derived from `solid`.

for a collection of `solid` objects. As an example, we can consider the class definitions given below.[9]

```
class solid {
public:
    virtual double volume(void);
    virtual double surface(void);
};

class sphere : public solid {
public:
    double volume(void) { return(4.0 * M_PI * r * r * r / 3.0); }
    double surface(void) { return(4.0 * M_PI * r * r); }
    void set_radius(double radius) { r = radius; }
private:
    double r;
};

class cube : public solid {
public:
    double volume(void) { return(side * side * side); }
    double surface(void) { return(6.0 * side * side); }
    void set_side(double length_of_side) { side = length_of_side; }
private:
    double side;
};
```

Notice that we have implemented member functions for the `sphere` and `cube` classes but not for the `solid` class. The problem is that the two member functions must return a value of type `double`, but without knowing the particular kind of solid it is not possible to calculate what this value should be. Returning an arbitrary volume and surface area is not very satisfactory; it would be much better if `solid` were an abstract class. In other words, we want the `solid` class to be used as a base class without

---

[9]The constant `M_PI`, which represents π, is defined in `<cmath>`.

implementing every member function. Moreover, we don't want it to be possible to have instances of the class since some functions would be undefined.

An *abstract class* is declared by including at least one member that is a *pure virtual function*. The way in which such a function is declared is illustrated by the following class declaration:

```
class X {
public:
      virtual void f(int i) = 0;        // A pure virtual function.
      // More class declarations go in here.
};
```

A pure `virtual` function is declared by using the strange "f() = 0" notation. This does not indicate that the function is numerically zero, but rather that it is not defined. A class that contains a pure virtual function is known as an *abstract class*. For example, the following declares an abstract `solid` class with two pure `virtual` functions:

```
class solid {
public:
      virtual double volume(void) = 0;
      virtual double surface(void) = 0;
};
```

Since the `volume()` and `surface()` functions are undefined, this class avoids the problem of how to implement these functions for an unknown solid.

The following program demonstrates the idea of an abstract class within the context of the `sphere` and `cube` classes:

```
#include <iostream>
#include <cstdlib>        // For EXIT_SUCCESS
#include <cmath>          // For M_PI
using namespace std;

class solid {
public:
      virtual double volume(void)=0;
      virtual double surface(void)=0;
};

class sphere : public solid {
public:
      sphere(double radius) { r = radius; }
      double volume(void) { return(4.0 * M_PI * r * r * r / 3.0); }
      double surface(void) { return(4.0 * M_PI * r * r); }
private:
      double r;
};

class cube : public solid {
```

```
public:
    cube(double side_length) {side = side_length; }
    double volume(void) { return(side * side * side); }
    double surface(void) { return(6.0 * side * side); }
private:
    double side;
};

int main()
{
    sphere s(2.0);
    cout << "Volume of sphere = " << s.volume() << "\n";
    cout << "Area of sphere = " << s.surface() << "\n";

    cube c(3.0);
    cout << "Volume of cube = " << c.volume() << "\n";
    cout << "Area of cube = " << c.surface() << "\n";

    return(EXIT_SUCCESS);
}
```

Notice how the `volume()` and `surface` functions are not defined for the `solid` class.

Exercise

Run the above program and check that the values given for the areas and volumes are correct. What happens if you modify the program to include the declaration of a `solid` object in the function `main()`?

There a few restrictions on abstract classes:

- It is not possible to have an instance of an abstract class:

```
    solid s;              // WRONG: solid is an abstract class.
```

- An abstract class cannot be an argument type:

```
    void f(solid s);    // WRONG!
```

- An abstract class cannot be a function return type:

```
    solid f(void);        // WRONG!
```

All of these restrictions are very reasonable since an abstract class has at least one undefined function. However, a pointer to an abstract class *is* legal, as in:

```
    solid *pt;              // O.K.
```

A reference to an abstract class is also valid, as illustrated by the following statement:

```
    double f(solid &s);    // O.K.
```

The reason why these statements are both legal and potentially useful is that a pointer to an abstract class can point to an instance of a derived class.

As an example of references to an abstract class, suppose we want to write a function that returns the total volume of two objects derived from the solid class. These objects may be instances of either the sphere or cube class, or even other classes (derived from solid) that we have not yet considered. A possible implementation using solid reference arguments is given below.

```
double volume_of_two_solids(solid &s1, solid &s2)
{
    return (s1.volume() + s2.volume());
}
```

This implementation should be contrasted with using non-reference arguments, which would necessitate four separate overloaded functions for just the sphere and cube classes. Also notice that we wouldn't have to modify (or even recompile) this function if we introduced new classes derived from the solid base class.

Another function of possible interest is one that returns the volume of a list of solid objects. Such a function is defined below.

```
double total_volume(solid *pt[], int solids)
{
    double vol = 0.0;
    for (int i = 0; i < solids; ++i)
        vol += pt[i]->volume();
    return vol;
}
```

The first argument in this function is an array of pointers to solid objects. This is perfectly legal, even though it is not possible to actually define a solid object. Notice that within the function body, the volume() function appropriate to the particular object is invoked. This is another example of polymorphism at work.

The following program demonstrates using the two functions defined above:

```
#include <iostream>
#include <cstdlib>        // For EXIT_SUCCESS
#include <cmath>          // For M_PI
using namespace std;

class solid {
public:
    virtual double volume(void)=0;
    virtual double surface(void)=0;
};

class sphere : public solid {
public:
    double volume(void) { return(4.0 * M_PI * r * r * r / 3.0); }
    double surface(void) { return(4.0 * M_PI * r * r); }
```

```
        void set_radius(double radius) { r = radius; }
private:
        double r;
};

class cube : public solid {
public:
        double volume(void) { return(side * side * side); }
        double surface(void) { return(6.0 * side * side); }
        void set_side(double length_of_side) { side = length_of_side; }
private:
        double side;
};

double volume_of_two_solids(solid &s1, solid &s2)
{
        return (s1.volume() + s2.volume());
}

double total_volume(solid *pt[], int solids)
{
        double vol = 0.0;
        for (int i = 0; i < solids; ++i)
            vol += pt[i]->volume();
        return vol;
}

int main()
{
        cube c;
        c.set_side(2.0);
        sphere s;
        s.set_radius(3.0);
        cout << "Volume of 2 solids = " <<
            volume_of_two_solids(c, s) << "\n";
        cube c1, c2;
        c1.set_side(2.0);
        c2.set_side(4.0);
        sphere s1, s2;
        s1.set_radius(3.0);
        s2.set_radius(5.0);
        solid *pt[4];
        pt[0] = &s1;
        pt[1] = &s2;
        pt[2]= &c1;
        pt[3] = &c2;
        cout << "Total volume = " << total_volume(pt, 4) << "\n";
```

```
        return(EXIT_SUCCESS);
    }
```

Notice how in the statement:

```
    cout << "Volume of 2 solids = " <<
        volume_of_two_solids(c, s) << "\n";
```

one reference argument is a `sphere` and the other is a `cube`. Also notice how the
`pt[]` array can store the addresses of different kinds of objects and how the `volume()`
function appropriate to the particular object is invoked automatically.

Exercise

Run the above program and check that the values given for the volumes
are correct.

A pure `virtual` function is inherited as a pure `virtual` function. As an example,
if we use the `solid` base class defined on page 342 to define a `cone` class:

```
    class cone : public solid {
    public:
        double volume(void) { return(M_PI * r * r * h / 3.0); }
        void set_radius(double radius) { r = radius; }
        void set_height(double height) { h = height; }
    private:
        double r, h;
    };
```

then `cone::surface()` is a pure `virtual` function. This is because `surface()` is a
pure `virtual` member function of the `solid` base class and is not defined by the `cone`
class. Consequently, this derived class is also an abstract class.

Exercise

Modify the `cone` class to include a definition of the `surface()` function and
then let the previously defined `total_volume()` function act on a list of
`sphere`, `cube` and `cone` objects. Likewise, implement and test an analogous
`total_surface()` function.

12.4   Class Hierarchies

A derived class can be the base class for another class. For instance, we could use the
`sphere` class, which was derived from the `solid` class, as a base class for a class of
coloured spheres (`c_sphere`). The following declares a coloured sphere class:

```
    enum colour {red, orange, yellow, green, blue, indigo, violet};

    class c_sphere : public sphere {
    public:
        void set_colour(const colour &col) { c = col; }
```

```
        void print_colour(void);
    private:
        colour c;
    };
```

The program given below defines the print_colour() function for the c_sphere class.

```
    #include <iostream>
    #include <cstdlib>        // For EXIT_SUCCESS
    #include <cmath>          // For M_PI
    using namespace std;

    class solid {
    public:
        virtual double volume(void)=0;
        virtual double surface(void)=0;
    };

    class sphere : public solid {
    public:
        double volume(void) { return(4.0 * M_PI * r * r * r / 3.0); }
        double surface(void) { return(4.0 * M_PI * r * r); }
        void set_radius(double radius) { r = radius; }
    private:
        double r;
    };

    enum colour {red, orange, yellow, green, blue, indigo, violet};

    class c_sphere : public sphere {
    public:
        void set_colour(const colour &col) { c = col; }
        void print_colour(void);
    private:
        colour c;
    };

    void c_sphere::print_colour(void)
    {
        switch (c) {
        case red:
            cout << "red\n";
            break;
        case orange:
            cout << "orange\n";
            break;
        case yellow:
```

```
                    cout << "yellow\n";
                    break;
                case green:
                    cout << "green\n";
                    break;
                case blue:
                    cout << "blue\n";
                    break;
                case indigo:
                    cout << "indigo\n";
                    break;
                case violet:
                    cout << "violet\n";
                    break;
                default:
                    cout << "Invalid colour\n";
                    exit(EXIT_FAILURE);
            }
    }

    int main()
    {
        c_sphere s[7];
        for (int i = 0; i < 7; ++i) {
            s[i].set_radius(i + 10.0);
        }
        s[0].set_colour(red);
        s[1].set_colour(orange);
        s[2].set_colour(yellow);
        s[3].set_colour(green);
        s[4].set_colour(blue);
        s[5].set_colour(indigo);
        s[6].set_colour(violet);
        for (int i = 0; i < 7; ++i) {
            cout << "Sphere " << i << " has colour ";
            s[i].print_colour();
            cout << "The volume is " << s[i].volume() <<
                "\n\n";
        }
        return(EXIT_SUCCESS);
    }
```

The `solid` class is known as an *indirect base class* of the `c_sphere` class and the resulting *class hierarchy* is shown in Figure 12.4. Notice that `volume()` is declared as a pure `virtual` function in the `solid` class, with the appropriate definition for the special case of a sphere being given in the `sphere` class. The `set_radius()` and `set_colour()` functions are defined in the `sphere` class and `c_sphere` classes respectively.

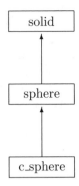

Figure 12.4: A simple class hierarchy.

The `enum colour` statement could alternatively be placed inside the class declaration, as shown below.

```
class c_sphere : public sphere {
public:
    enum colour {red, orange, yellow, green, blue, indigo, violet};
    void set_colour(const colour &col) { c = col; }
    void print_colour(void);
private:
    colour c;
};
```

Enumerations defined within a class have the scope of that class and are only accessible outside of this scope by using the explicitly qualified name, as in:

```
c_sphere s;
s.set_colour(c_sphere::red);
```

Such enumerations obey the same access rules as other class members; in this particular case `colour` is `public`. Placing an enumeration within a class declaration is consistent with the idea of data hiding and means that we can use the same enumeration names for different classes.

Exercise

Modify the above program so that the `enum colour` statement is defined within the `c_sphere` class. Verify that your program gives the same results as for the unmodified version.

The `print_colour()` function in the above program illustrates one of the main problems with conventional procedural languages. The function has a `switch` statement that depends crucially on the allowed colours. If we decide to add another colour to the enumeration, then we must change the `switch` statement. For the current example this change is simple, but in a complicated program there may be many functions containing similar `switch` statements, all of which need modifying when the permissible data changes. The distributed nature of these changes can make maintaining such code a huge problem for conventional languages. However, maintenance can be made much easier if we use object-oriented techniques. As a straightforward example, we could define a pure abstract `coloured_sphere` class and use this to derive `red_sphere` and `blue_sphere` classes. Defining a pointer to the `coloured_sphere` base class would enable us to store the address of the derived class objects. This pointer could then be used to select the `print_colour()` function appropriate to the type of object being pointed to. These ideas are put together in the program given below.

```cpp
#include <iostream>
#include <cstdlib>        // For EXIT_SUCCESS
#include <cmath>          // For M_PI
using namespace std;

class solid {
public:
    virtual double volume(void)=0;
    virtual double surface(void)=0;
};

class sphere : public solid {
public:
    double volume(void) { return(4.0 * M_PI * r * r * r / 3.0); }
    double surface(void) { return(4.0 * M_PI * r * r); }
    void set_radius(double radius) { r = radius; }
private:
    double r;
};

class coloured_sphere : public sphere {
public:
    virtual void print_colour(void) = 0;
};

class red_sphere : public coloured_sphere {
public:
    void print_colour(void) { cout << "red\n"; }
};

class blue_sphere : public coloured_sphere {
public:
    void print_colour(void) { cout << "blue\n"; }
};

int main()
{
    red_sphere r_s;
```

```
        blue_sphere b_s;
        coloured_sphere *pt = &r_s;
        pt->print_colour();
        pt = &b_s;
        pt->print_colour();
        return(EXIT_SUCCESS);
}
```

Notice that which function is actually called is determined by the type of object that the function is invoked on, rather than by a `switch` statement. In fact the type may not even be known until run-time. Type resolution of this form is called *dynamic* or *late binding* and is one the major contributions of object-oriented programming.[10] The significant feature is that extending an existing data type only involves defining an extra class, as in:

```
class green_sphere : public coloured_sphere {
public:
        void print_colour(void) { cout << "green\n"; }
};
```

rather than modifying `switch` statements in what may be a large number of functions. The changes are all in one place and manifest, rather than scattered and buried.

Exercise

Modify the program given on page 347 so that it uses the idea of a pure abstract `coloured_sphere` class in place of the `switch` statement. Verify that your program gives the same results as for the unmodified version.

Since any class may be the base class for any number of derived classes, class hierarchies can get very complicated. We can get a glimpse of the possible complication by adding a class for heavy spheres (an `h_sphere` class), as given below.

```
class h_sphere : public sphere {
public:
        void set_mass(double mass) { m = mass; }
        double mass(void) { return m; }
private:
        double m;
};
```

The resulting class hierarchy is shown in Figure 12.5.

Exercise

In Figure 12.5, replace `c_sphere` by our `coloured_sphere` class, which avoids using the `switch` statement, and construct the appropriate class hierarchy diagram.

---

[10]Which non-virtual function should be invoked is known at compile-time. This is called *static* or *early binding*. Dynamic binding should not be confused with polymorphism. Polymorphism means that the same function interface can perform different actions on different objects. Polymorphism may indeed be implemented by means of dynamic binding but in some circumstances the compiler may be able to decide which function to invoke (that is the binding may be static).

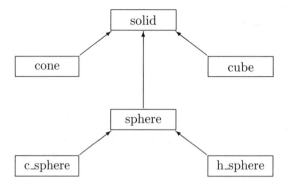

Figure 12.5: A more complicated class hierarchy.

12.5   Constructors and Destructors

Constructors are rather special member functions and, in particular, constructors are
not inherited and cannot be declared **virtual**. The compiler may generate a default
or a constructor may be defined by the programmer, in which case it must call the
direct base class constructor. This is achieved by a constructor definition containing a
colon followed by the base class constructor immediately before the constructor body.[11]
The syntax is demonstrated by the **coloured_sphere** and **red_sphere** classes in the
program given below.

```
#include <iostream>
#include <cstdlib>        // For EXIT_SUCCESS
using namespace std;

class sphere {
public:
    sphere(double radius) { r = radius; }
private:
    double r;
};

class coloured_sphere : public sphere {
public:
    virtual void print_colour(void) = 0;
    coloured_sphere(double radius) : sphere(radius){ }
};

class red_sphere : public coloured_sphere {
public:
    void print_colour(void) { cout << "red\n"; }
    red_sphere(double radius) : coloured_sphere(radius){ }
```

---

[11]This is the same syntax as described in Section 10.1.3 for member objects with constructors.

```
};

int main()
{
    red_sphere s(2.0);
    cout << "Sphere colour: ";
    s.print_colour();
    return(EXIT_SUCCESS);
}
```

Notice the empty constructor bodies and that a constructor is required for the abstract `coloured_sphere` class, even though there can be no instances of the class. In general, the base class constructor may well do something useful and the general rule is that the base class constructor is invoked before that of the derived class.

It is important to realize that it is the constructor function definition (not declaration) that invokes the base class constructor. For example, the following code defines the `red_sphere` constructor outside the class declaration:

```
class red_sphere : public coloured_sphere {
public:
    void print_colour(void) { cout << "red\n"; }
    red_sphere(double radius);
};

red_sphere::red_sphere(double radius) : coloured_sphere(radius){ }
```

Notice how the `red_sphere` constructor function declaration within the class differs from that given in the above program.

Destructors are also not inherited but, unlike constructors, they can be declared `virtual` and it is often useful to do so. The following program illustrates what can go wrong when a base class destructor is not declared `virtual`:

```
#include <iostream>
#include <cstdlib>          // For EXIT_SUCCESS
using namespace std;

class data {
public:
    data(int m);
    ~data();
private:
    double *p1;
};

class more_data : public data {
public:
    more_data(int m, int n);
    ~more_data();
private:
```

```
        double *p2;
};

data::data(int m)
{
    p1 = new double[m];
    cout << "data allocated: " << p1 << "\n";
}

data::~data()
{
    cout << "data deleting: " << p1 << "\n";
    delete p1;
}

more_data::more_data(int m, int n) : data(m)
{
    p2 = new double[n];
    cout << "more_data allocated: " << p2 << "\n";
}

more_data::~more_data()
{
    cout << "more_data deleting: " << p2 << "\n";
    delete p2;
}

int main()
{
    data *pt = new more_data(10, 20);
    delete pt;
    return(EXIT_SUCCESS);
}
```

The output on my computer is:

```
data allocated:       0x8049cf8
more_data allocated:  0x8049d50
data deleting:        0x8049cf8
```

As we can see, memory is allocated by the `data` and `more_data` constructor functions, but only deallocated by the base class destructor. Declaring `~data()` to be `virtual` ensures that the derived and base class destructors are both invoked. As a general rule it is usually worth declaring a base class destructor `virtual`.

Exercise

(a) Run the above program and check that you obtain similar output to that given.

(b) Modify the program so that the base class destructor is declared `virtual`. In what order are the `data` and `more_data` constructors and destructors invoked?

12.6 Member Access and Inheritance

All of the derived classes considered so far can access the `public` (but not the `private`) base class members. For many applications this "all or nothing" approach is very appropriate. However, C++ has techniques for fine tuning the access privilege and it is to these techniques that we now turn.

12.6.1 Access Specifiers[†]

The members of a class can have three levels of access: `public`, `protected` and `private`:

```
class A {
public:
    int a;
private:
    int b;
protected:
    int c;
};
```

We have already made use of the `public` and `private` specifiers. The reason we have not previously met the `protected` access specifier is that it is only relevant when we have a derived class.

The effect of the *protected* access specifier, used in the declaration of class A, is intermediate between the `public` and `private` specifiers. A `protected` member can be accessed by member functions and friends of the class, as well as by certain derived classes. In general, a derived class can be specified to have a `public`, `protected` or `private` base class so the effect of `access_specifier` in:

```
class X : access_specifier A {
// Declarations go in here.
};
```

is as follows:

- If `access_specifier` is public:

  > `public` members of A are `public` members of X.
  > `protected` members of A are `protected` members of X.
  > `private` members of A are `private` members of X.

- If `access_specifier` is protected:

public members of A are protected members of X.
protected members of A are protected members of X.
private members of A are private members of X.

- If access_specifier is private:

public members of A are private members of X.
protected members of A are private members of X.
private members of A are private members of X.

Notice that private members of a base class remain private and are not accessible to members of a derived class. This restriction is necessary since otherwise a derived class would be a trivial way of evading class access restrictions.

As an example, suppose we modify our shape class (as given on page 331) so that the members have the following access privileges:[12]

```
class shape {
public:
    int i_d;
protected:
    float x, y;
private:
    material_type material;
};
```

Then with the same disc class as on page 332, the following demonstrates the various ways in which we can attempt to access data members:

```
shape s;
s.i_d = 204;          // O.K.
s.material = WOOD;    // WRONG: material is private.
s.x = 54.37;         // WRONG: x is protected;

disc d;
d.i_d = 100;          // O.K.
d.material = STEEL;  // WRONG: material is private.
d.y = -71.3;          // WRONG: y is protected.
cout << d.x_coord();  // O.K. function has access to
                      // protected member, x.
```

Exercise

Make the code fragment given above into a short program and verify the stated access privileges.

As mentioned previously, constructors and destructors are usually declared public, but they can also be private. In fact they can also be declared protected. Instances of a class with all constructors protected can only be created by friends, members and

---

[12]This class declaration is *only* for illustration since there is no way of accessing the material data member.

derived classes; such classes are therefore useful when we don't want clients to construct objects. Analogous restrictions apply to classes with a `protected` destructor.

As a general rule, data in a base class should be `private` since `protected` data could be modified in unintended ways by derived classes. Moreover, it would be difficult to restructure protected data since the way in which the data may be used cannot be deduced from the base class; it would be necessary to examine all derived classes. In a large project this could lead to a big software maintenance headache.

12.6.2 Friendship and Derivation

In Section 8.15 we introduced the idea of an entire class being a `friend` of another class. A `friend` class and a derived class both have special privileges to access members of another class. However, friendship should not be confused with derivation. A `friend` class has access to *all* members of the class that has granted friendship. In the example:

```
class node {
    friend class list;
    // Other members go here.
private:
    node *next;
};

class list {
    // Members go here.
};
```

the `list` class has access to the `private` member, called `next`. By contrast, the `coloured_sphere` class has no access to the `private` member, `r`, of the `sphere` class in the following code fragment:

```
class sphere {
public:
    sphere(double radius) { r = radius; }
protected:
    static int spheres;
private:
    double r;
};

class coloured_sphere : public sphere {
public:
    virtual void print_colour(void) = 0;
    coloured_sphere(double radius) : sphere(radius){ }
};
```

However, the `coloured_sphere` class does have access to the `protected` and `public` members.

As pointed out in Section 8.15, friendship is not transitive. Moreover, neither is friendship inherited. For example, suppose we have the following class declarations:

```
class node {
    friend class list;
public:
    DATA_TYPE data;
private:
    node *next;
};

class list {
    // Members go here.
};

class superlist : public list {
    // Members go here.
};
```

Then superlist would have no access to any private or protected members of the node class. If this restriction did not exist, then granting friendship to one class would open up the entire class implementation to an arbitrary hierarchy of derived classes. As a general rule, if classes need access to the members of an indirect base class then protected access rather than friendship is the appropriate technique.

12.7   Using Single Inheritance

12.7.1   A Bounds Checked Array Class

In Section 8.17 we implemented a primitive self-describing, self-checking array class. This class was improved in subsequent exercises and further improvements can now be made by introducing destructors and virtual functions. The most fundamental change that we make is to remove the bounds checking from the array class and to introduce a derived class, called checked_array, that takes over this role. The modified header file is given below.

```
// source:   array.h
// use:      Defines array class.

#ifndef ARRAY_H
#define ARRAY_H

#include <iostream>
#include <cstdlib>        // For exit()
using namespace std;

class array {
public:
    array(int size);
    array(const array &x);
    virtual ~array() { delete pt; }
```

```
    array &operator=(const array &x);
    virtual double &operator[](int index);
    int get_size(void);
protected:
    int n;
    double *pt;
};


class checked_array : public array {
public:
    checked_array(int size) : array(size) { }
    double &operator[](int index);
private:
    void check_bounds(int index);
};


// inline array class implementations:

inline int array::get_size(void)
{
    return n;
}

inline double &array::operator[](int index)
{
    return pt[index - 1];
}


// inline checked_array class implementations:

inline double &checked_array::operator[](int index)
{
    check_bounds(index);
    return array::operator[](index);
}

inline void checked_array::check_bounds(int index)
{
    if (index < 1 || index > n) {
        cout << "Array index " << index << " out of bounds\n";
        exit(EXIT_FAILURE);
    }
}


#endif  // ARRAY_H
```

The improvements made to the original **array** class can be summarized as follows:

- An assignment operator and copy constructor are supplied since otherwise defaults would be generated by the compiler. Because an `array` object has a pointer to dynamically allocated memory, these defaults would not be appropriate.

- A destructor is introduced to delete the dynamically allocated memory.

- Instead of the `element()` function we overload the subscripting operator, which is declared `virtual` so that dynamic binding can be invoked. Overloading the function call operator would provide an equally natural interface.

The `checked_array` class, which is derived from the `array` class, has the following features:

- A constructor must be defined since constructors are not inherited. The new constructor simply calls the base class constructor and has an empty body.

- A new subscripting operator is defined. This operator invokes the base class function, which is safer than having two independent implementations.

- A function such as `check_bounds()` is sometimes called a *helper function* since it is only used internally by the `checked_array` class.

For efficiency, some of the smaller and much used functions are implemented `inline`. Implementations of the remaining functions are given in the following file:

```
// source:  array.cxx
// use:     Implements array class.

#include <cstring>      // For memcpy()
#include "array.h"
using namespace std;

array::array(int size)
{
    n = size;
    pt = new double[n];
}

array::array(const array &x)
{
    n = x.n;
    pt = new double[n];
    memcpy(pt, x.pt, n * sizeof(double));
}

array &array::operator=(const array &x)
{
    delete pt;
    n = x.n;
    pt = new double[n];
    memcpy(pt, x.pt, n * sizeof(double));
```

```
        return *this;
}
```

The program given below tries out the **array** and **checked_array** classes.

```
// source:  my_test.cxx
// use:     Tests array class.

#include <iostream>
#include <cstdlib>        // For EXIT_SUCCESS
#include "array.h"
using namespace std;

double sum(array &a, int first_index, int last_index)
{
    double result = a[first_index];
    for (int i = first_index + 1; i <= last_index; ++i)
        result += a[i];
    return result;
}

int main()
{
    const int n = 10;

    // Define an object:
    checked_array x(n);

    // Access the array size:
    cout << "The array x has " << x.get_size() << " elements.\n";

    // Store some data.
    // Also find the total directly
    // so it can be used for comparison.
    double total = 0.0;
    for (int i = 1; i <= n; ++i) {
        x[i] = i * 25.0;
        total += i * 25.0;
    }

    // Retrieve some data:
    cout << "The data stored in x are:\n";
    for (int i = 1; i <= n; ++i)
        cout << x[i] << "\n";
    cout << "\n";

    // Define another object using copy constructor:
    checked_array y = x;
```

```
// Check the array size:
cout << "The array y has " << y.get_size() << " elements.\n" <<
    "y was created by the copy constructor.\n\n";

// Check that the copy is identical:
int errors = 0;
for (int i = 1; i <= n; ++i)
    if (x[i] != y[i]) {
        cout << "x[" << i << "] != y[" << i << "]\n";
        ++errors;
    }
if (errors)
    cout << errors << " elements of the arrays x " <<
        "and y differ in their data.\n";
else
    cout << "Arrays x and y have identical data.\n\n";

// Define another object with half the size but no bounds
// checking:
array z(n/2);

// Check the array size:
cout << "The array z has " << z.get_size() << " elements.\n\n";

// Try out the assignment operator:

z = x;

// Check the array size:
cout << "After assignment of x to z, the array z has " <<
    z.get_size() << " elements.\n\n";

errors = 0;
// Check that the copy is identical:
for (int i = 1; i <= n; ++i)
    if (z[i] != x[i]) {
        cout << "z[" << i << "] != x[" << i << "]\n\n";
        ++errors;
    }
if (errors)
    cout << errors << " elements of the arrays x and z " <<
        "differ in their data.\n";
else
    cout << "Arrays x and z have identical data.\n\n";

// Find sum for z[i], going out of bounds. If you get the
```

```
    // correct result, try changing the third argument in sum()
    // to something else greater than n:
    cout << "The sum of the data in z is " << sum(z, 1, 2 * n) <<
        ".\nThe sum should be " << total << ".\n\n";

    // Find sum for x[i], going out of bounds:
    cout << "The sum of the data in x is " << sum(x, 1, 2 * n) <<
        ".\nThe sum should be " << total << ".\n\n";

    return(EXIT_SUCCESS);
}
```

Have a look at the sum() function in this program. Since this function accepts a *reference* to an array object, the function body can manipulate both derived and base class objects. The type of object that is passed as the function argument determines which overloaded subscripting operator is invoked. Putting the bounds check in a derived class has the advantage that we can choose to either have checked or unchecked arrays. We could also achieve this by having two distinct classes rather than using inheritance. However, this would complicate program maintenance, in addition to making it more difficult to mix checked and unchecked arrays.[13]

Exercise

Compile and run the above test program. Describe and explain what happens in the following circumstances:

(a) The virtual specifier is omitted from the base class declaration for the subscripting operator.

(b) The sum() function is changed so that the array argument is passed by value.

12.7.2 A Menu Class

A common method of "driving" a program is to use a menu. In this section we develop a general menu class. Typically a user initiates a program that displays a list of options, as shown below.

```
Options
    0   Exit menu
    1   Jacobi iteration
    2   Weighted Jacobi iteration
    3   Gauss-Seidel iteration
Select option:
```

A *menu* consists of a number of items, which we call *options*. An option may either be a menu, which itself has a list of options, or an *action*. An action does some specific task, such as setting the value of a parameter, running another program, opening a

---
[13]It could be argued that it is the client that should provide error checking rather than the class. A detailed discussion of this point is given in [7].

file, etc. When an action is completed, it may either invoke a menu or perform some other task, such as return to the operating system. This discussion leads naturally to the idea of having three classes, as shown in Figure 12.6.

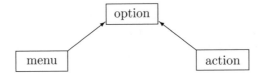

Figure 12.6: Classes for menus.

Suitable declarations for the three classes are given in the following header file:

```
// source:  menu.h
// use:     Defines menu classes.

#ifndef MENU_H
#define MENU_H

#include "string.h"

class option {
public:
    option(const string &option_label) { label = option_label; }
    virtual void activate(void) = 0;
    void print_label(void) { label.print(); }
private:
     string label;
};

class menu : public option {
public:
    menu(const string &menu_label, int max_number_of_options);
    virtual ~menu();
    void set_option(option *op_pt);
    void activate(void);
private:
    option **option_list;
    int options, max_options;
};

class action : public option {
public:
    action(const string &action_label, void (*function_pt)(void),
        option *option_pt = 0);
    void activate(void);
```

```
private:
    void (*f_pt)(void);
    option *op_pt;
};

#endif  // MENU_H
```

Notice how use is made of the **string** class, introduced in Section 9.6.2.[14] The **option** class is an abstract base class and, as such, we cannot define instances of the class. However, the two derived classes inherit the properties of this base class. Consequently, all three classes have an **activate()** function and a "label". The derived classes also add their own data and member functions in addition to those provided by the base class.

The **option_list** data member of the **menu** class is used to store the address of a dynamically allocated array of **option** pointers. The class also has data members to hold the size of this array (**max_options**) and the number of options set (**options**). The **set_option()** member function is used to assign an **option** address to an element of the array.

The **action** class has a pointer to the function that invokes the action (**f_pt**) and a pointer to an option (**op_pt**), which may be invoked after the action has completed.

Suitable class implementations are given below.

```
// source:  menu.cxx
// use:     Implements menu classes.

#include <iostream>
#include <cstdlib>       // For exit()
#include "menu.h"

using namespace std;

menu::menu(const string &menu_label, int max_number_of_options) :
    option(menu_label)
{
    max_options = max_number_of_options;
    if (max_options <= 0) {
        cout << "Error in menu(): options = " << max_options <<
            "\n";
        exit(EXIT_FAILURE);
    }
    else {
        option_list = new option *[max_options];
    }
    options = 0;
}
```

---

[14]The #include directive for string.h in menu.h assumes that both header files are in the same directory. If this is not true then you will have to give the complete path in the #include directive.

```cpp
menu::~menu()
{
    delete option_list;
}

void menu::set_option(option *op_pt)
{
    if (options < max_options) {
        option_list[options] = op_pt;
        ++options;
    }
    else {
        cerr << "Attempt to set " << options + 1 <<
            " menu options\nOnly " << max_options <<
            " options are allowed.\n";
        exit(EXIT_FAILURE);
    }
}

void menu::activate(void)
{
    int i, choice;

    do {
        cout << "\n";
        print_label();
        cout << "\n";
        for (i = 0; i < options; ++i) {
            cout << i << "\t";
            option_list[i]->print_label();
            cout << "\n";
        }
        cout << "Select option:   ";
        cin >> choice;
    } while (choice < 0 || choice >= options);
    option_list[choice]->activate();
}

action::action(const string &action_label,
    void (*function_pt)(void), option *option_pt) :
    option(action_label)
{
    f_pt = function_pt;
    op_pt = option_pt;
}
```

```
void action::activate(void)
{
    f_pt();
    op_pt->activate();
}
```

The `menu` constructor initializes the `option_list` pointer to sufficient dynamically allocated memory to store addresses for the maximum number of options and also initializes the current number of options to zero. Notice the way in which the base class, `option()`, constructor is called.

Initialization of the `action` data members, `f_pt` and `op_pt`, is carried out by the `action` constructor, which also invokes the base class constructor. The value stored by `op_pt` is the address of either an `action` or a `menu` object.

The `menu::activate()` function lists the menu items, accepts a menu choice and activates that particular option. Since the `option_list[]` array holds addresses of both `action` and `menu` objects, which `activate()` function is invoked depends on the object type. This is a run-time decision and another example of dynamic binding. It is:

```
option_list[choice]->activate();
```

in the `menu::activate()` function that replaces the `switch` statement.

The `action::activate()` function is very simple. Since no choice is involved, the appropriate function for the particular action is first invoked and then the option pointed to by `op_pt` is activated; it is this that allows us to do something after the action has terminated, such as going to a menu or exiting to the operating system.

The following program demonstrates how to use the `menu` class:

```
// source:  my_test.cxx
// use:     Tests menu.

#include <iostream>
#include <cstdlib>        // For exit()
#include "menu.h"

using namespace std;

void to_system(void);
void create_multi_grid(void);
void run_V_cycle(void);
void run_W_cycle(void);
void delete_multi_grid(void);

void to_system(void)
{
    cout << "\nReturning to system.\n";
    exit(EXIT_SUCCESS);
}
```

```
void create_multi_grid(void)
{
    cout << "\nMulti-grid created.\n";
}

void delete_multi_grid(void)
{
    cout << "\nMulti-grid has been deleted.\n";
}

void run_V_cycle(void)
{
    cout << "\nA single V-cycle has been run.\n";
}

void run_W_cycle(void)
{
    cout << "\nA single W-cycle has been run.\n";
}

int main()
{
    const int top_menu_options = 2;
    const int grid_menu_options = 3;
    // Define menus:
    menu top_menu("Top menu", top_menu_options);
    menu grid_menu("Multi-grid menu", grid_menu_options);
    // Set options for top menu:
    top_menu.set_option(new action("Return to system",
        to_system));
    top_menu.set_option(new action("Create multi-grid",
        &create_multi_grid, &grid_menu));
    // Set options for multi-grid menu:
    grid_menu.set_option(new action("Delete multi-grid",
        delete_multi_grid, &top_menu));
    grid_menu.set_option(new action(
        "Run V-cycle multi-grid",run_V_cycle, &grid_menu));
    grid_menu.set_option(new action(
        "Run W-cycle multi-grid",run_W_cycle, &grid_menu));
    // Start the menu system:
    top_menu.activate();
    return EXIT_SUCCESS;
}
```

For simplicity the functions do not actually do very much; they only send messages to the output stream declaring which action they should have performed. A typical session is shown below; bold face type represents entries typed in by the user and italics are used for the results of a particular action:

```
Top menu
0         Return to system
1         Create multi-grid
Select option:  1
```

Multi-grid created.

```
Multi-grid menu
0         Delete multi-grid
1         Run V-cycle multi-grid
2         Run W-cycle multi-grid
Select option:  1
```

A single V-cycle has been run.

```
Multi-grid menu
0         Delete multi-grid
1         Run V-cycle multi-grid
2         Run W-cycle multi-grid
Select option:  2
```

A single W-cycle has been run.

```
Multi-grid menu
0         Delete multi-grid
1         Run V-cycle multi-grid
2         Run W-cycle multi-grid
Select option:  0
```

The multi-grid has been deleted.

```
Top menu
0         Return to system
1         Create multi-grid
Select option:  0
```

Returning to system.

The action of this menu system is shown in Figure 12.7. The ovals contain descriptions of particular options that can be activated. The two boxes represent menus that are presented to the user. The arrows represent the flow of control. Execution of the program causes top_menu to be activated and a list of options to be displayed. In the above example the user chooses option 1, which in turn invokes the create_multi_grid() function and then causes the grid_menu options to be presented. The user next chooses option 1 again. This invokes the run_V_cycle() function and then returns to the same grid_menu. Eventually option 0 of top_menu is chosen and the program terminates.

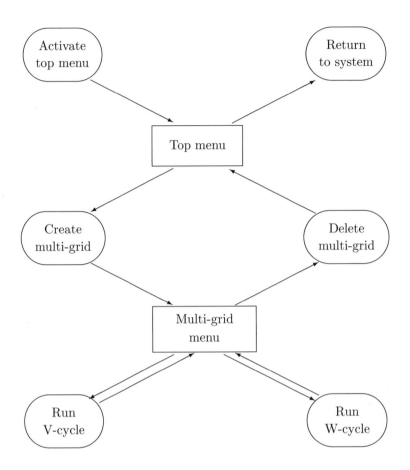

Figure 12.7: A typical menu session.

It is possible to build a complicated system of many sub-menus and options by simply adding more statements to the function, `main()`. It is dynamic binding that causes either a `menu` or an `action` object to be activated as a result of a run-time decision by the user; the member functions for the `option, action` and `menu` classes remain unchanged.

Exercise

Compile and run the above program using a menu. An appropriate makefile is given below.

```
my_test: my_test.cxx menu.o string.o
    g++ my_test.cxx menu.o string.o -o my_test
menu.o: menu.cxx menu.h
    g++ -c menu.cxx
string.o: string.cxx string.h
    g++ -c string.cxx
```

Try adding more sub-menus with different options.

12.8   Summary

- A class, X, is declared to be derived from a base class, A, by the statement:

```
class X : public A {
public:
     int x;
private:
     int y;
protected:
     int z;
};
```

The public access specifier before the base class, A, overrides the private default. Protected members of a class *may* be accessed by a derived class, whereas private members can never be accessed by a derived class.

- A derived class has the members given in its declaration, in addition to all of the members of the base class. (However, note that the latter members may not be accessible.)

- The scope resolution operator, ::, can be used to access a member declared in the base class:

```
class A {
public:
     int v;
};

class X : public A {
public:
     int v;
};

X x;
x.v = 1;          // Access v declared in class X.
x.A::v = 2;       // Access v declared in class A.
```

- The particular virtual function that is invoked through a pointer or reference is determined by the type of object:

```
class A {
public:
     virtual void f(void) { cout << "Class A\n"; }
};
```

```
class X : public A {
public:
    void f(void) { cout << "Class X\n"; }
};

A a;
X x;
A *pt = &a;
pt->f();          // Invokes A::f()
pt = &x;
pt->f();          // Invokes X::f()
```

- The declaration of an abstract class must contain a pure **virtual** function:

```
class solid {
public:
    virtual double volume(void) = 0;
};
```

A reference or pointer to an abstract class is allowed but an instance of an abstract class is illegal.

- A derived class can be the base class for another derived class:

```
class A {
    // Declarations go in here.
};

class B : public A {
    // Declarations go in here.
};

class C : public B {
    // Declarations go in here.
};
```

Class A is an indirect base class of C. Class B is a direct base class of C.

- An enumeration defined within a class has the scope of that class:

```
class coloured_cone : cone {
public:
    enum colour {red, yellow, blue};
};
```

- Constructors cannot be declared **virtual** and are not inherited. A derived class should define its own constructor and call the base class constructor if necessary:

```
class checked_array : public array {
public:
    checked_array(int size) : array(size){ }
};
```

- Destructors are not inherited, but it is usually a good idea to declare them virtual.

12.9 Exercises

1. Use the `menu` class to drive a test program for one of the previous classes that we have developed. Suitable examples are the complex arithmetic (Section 9.6.1) and string (Section 9.6.2) classes.

2. Implement an overloaded `!=` operator for the `array` class that we developed in Section 12.7.1. If `x` and `y` are arrays, then `x != y` must return `true` or `false`, as appropriate. You should provide both an unchecked `virtual` function for the `array` class and a bounds checked function for the `checked_array` class.

3. Use the doubly linked list of Section 10.3.2 as the base class for a sparse vector class providing the same member functions as Exercise 5 of Chapter 10.

Chapter 13

Multiple Inheritance

As has already been emphasized, inheritance is a natural technique to use when we have a "kind of" relationship; for instance, a cube is a "kind of" solid and a triangle is a "kind of" shape. Sometimes, we would like to construct a derived class with more than one base class. Simple examples abound in the natural world. For instance, if we classify animals into carnivores and herbivores, then we will probably also need an omnivores class, which is derived from both of these, as shown in Figure 13.1. In fact

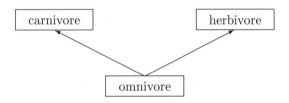

Figure 13.1: Multiple inheritance.

many hierarchical classification systems lead to this type of relationship. The ability to have a derived class with more than one base class is known as *multiple inheritance* and is the subject of this chapter.

13.1   Derived Classes

In Chapter 12 we introduced the idea of a derived class, such as `disc`, depending on a single base class:

```
class shape {
// Members go in here.
};

class disc : public shape {
// Members go in here.
};
```

375

Multiple inheritance consists, syntactically, of replacing the base class by a list of base classes, known as a *derivation list*.   As an example, suppose we are working on a communications project. We could reasonably decide that we needed a `transmitter` class and a `receiver` class, as shown below.

```
class transmitter {
// Members go in here.
};

class receiver {
// Members go in here.
};
```

However, our project might involve devices that both transmit and receive. Such devices could be relevant in a variety of situations, ranging from an RS232 interface to communication with an interplanetary probe. A device that can both transmit and receive (which we will call a *transceiver*) is a kind of transmitter and a kind of receiver. We can represent a transceiver as an instance of a `transceiver` class, which we declare as follows:

```
class transceiver : public transmitter, public receiver {
// Members go in here.
};
```

In this example, the derivation list is:

```
public transmitter, public receiver
```

The `transceiver` class illustrates the idea of multiple inheritance; the class inherits all data and member functions of the `transmitter` and `receiver` classes. The relationship between these classes is shown in Figure 13.2. As is the case for single inheritance, the arrows in the diagram point to the base classes.

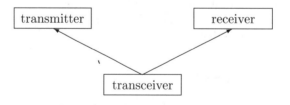

Figure 13.2: Derivation of a `transceiver` class.

The access restrictions on a derived class are the same for multiple inheritance as for single inheritance. Notice that in order for both `transmitter` and `receiver` to be `public` base classes, the `public` specifier *must* be repeated. A common mistake is to write the derivation list as:

```
public transmitter, receiver
```

with the assumption that public also applies to receiver. However, since no access specifier is given for receiver, the access for this class defaults to private. Even if a private base class is intended, it is standard practice to include the access specifier.

In order to give a demonstration of inheritance, we could suppose that transmitter and receiver objects each store their frequency, together with a flag to indicate whether or not they are currently transmitting or receiving data. A program to demonstrate the transceiver class is given below.

```cpp
#include <iostream>
#include <cstdlib>        // For EXIT_SUCCESS
using namespace std;

class transmitter {
public:
    void set_tx_frequency(double frequency);
    void set_tx_active(bool active);
    void list_tx_state(void);
private:
    double tx_f;
    bool tx_active;
};

inline void transmitter::set_tx_frequency(double frequency)
{
    tx_f = frequency;
}

inline void transmitter::set_tx_active(bool active)
{
    tx_active = active;
}

void transmitter::list_tx_state(void)
{
    cout << "Transmitter state:\n\tfrequency: " << tx_f <<
        "\n\tactive:     ";
    if (tx_active)
        cout << "true\n";
    else
        cout << "false\n";
}

class receiver {
public:
    void set_rx_frequency(double frequency);
    void set_rx_active(bool active);
    void list_rx_state(void);
private:
```

```cpp
        double rx_f;
        bool rx_active;
};

inline void receiver::set_rx_frequency(double frequency)
{
    rx_f = frequency;
}

inline void receiver::set_rx_active(bool active)
{
    rx_active = active;
}

void receiver::list_rx_state(void)
{
    cout << "Receiver state:\n\tfrequency: " << rx_f <<
        "\n\tactive:    ";
    if (rx_active)
        cout << "true\n";
    else
        cout << "false\n";
}

class transceiver : public transmitter, public receiver {
    // Other members go in here.
};

int main()
{
    transceiver trx;
    trx.set_tx_frequency(412.75);
    trx.set_tx_active(true);
    trx.set_rx_frequency(422.75);
    trx.set_rx_active(false);
    trx.list_tx_state();
    trx.list_rx_state();
    return(EXIT_SUCCESS);
}
```

Notice how the state of the transceiver can be set and accessed by using member functions from the transmitter and receiver classes.

Exercise

Implement the classes shown in Figure 13.1. The data for each base class should include a list of typical items that are eaten, with items chosen from two enumerations corresponding to the two base classes. Provide functions to list what is eaten by a particular animal. Is there any difficulty in having a data member to store the name of the particular species of carnivore, herbivore or omnivore?

Now suppose we have a two-channel receiver and that we want to develop a corresponding `two_channel_rx` class. We might be tempted to make the following declaration:

```
class two_channel_rx : public receiver, public receiver {
// Members go in here.
};       // WRONG: repeated base classes are not allowed.
```

However, this declaration is incorrect because we are not allowed to have a repeated base class in a derivation list. The reason for this restriction is that there would be no way of distinguishing members of the repeated class. The way to avoid this problem is by having multiple base classes, each of which has the same base class. This is demonstrated by the following code fragment and Figure 13.3.

```
class rx_1 : public receiver {
// More members may go in here.
};

class rx_2 : public receiver {
// More members may go in here.
};

class two_channel_rx : public rx_1, public rx_2 {
// More members may go in here.
};
```

As explained in Section 12.4, a base class of another base class is known as an *indirect base class*. In this example, `receiver` is a repeated indirect base class of `two_channel_rx`. The inheritance diagram (Figure 13.3) shows one class (`rx_1` or `rx_2`) between `two_channel_rx` and each repetition of the `receiver` class. As is the case for single inheritance, there is no limit on the possible number of classes between an

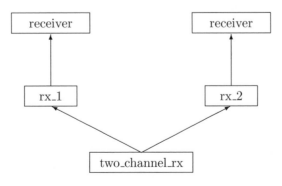

Figure 13.3: Repeated indirect base class.

indirect base class and a derived class. A program to demonstrate the `two_channel_rx` class is given below.

```
#include <iostream>
#include <cstdlib>        // For EXIT_SUCCESS
using namespace std;

class receiver {
public:
    void set_rx_frequency(double frequency);
    void set_rx_active(bool active);
    void list_rx_state(void);
private:
    double rx_f;
    bool rx_active;
};

inline void receiver::set_rx_frequency(double frequency)
{
    rx_f = frequency;
}

inline void receiver::set_rx_active(bool active)
{
    rx_active = active;
}

void receiver::list_rx_state(void)
{
    cout << "Receiver state:\n\tfrequency: " << rx_f <<
        "\n\tactive:    ";
    if (rx_active)
        cout << "true\n";
    else
        cout << "false\n";
}

class rx_1 : public receiver {
};

class rx_2 : public receiver {
};

class two_channel_rx : public rx_1, public rx_2 {
};

int main()
{
```

```
      two_channel_rx two_rx;
      two_rx.rx_1::set_rx_frequency(412.75);
      two_rx.rx_2::set_rx_frequency(422.75);
      two_rx.rx_1::set_rx_active(true);
      two_rx.rx_2::set_rx_active(false);
      cout << "Receiver channel 1 state:\n";
      two_rx.rx_1::list_rx_state();
      cout << "Receiver channel 2 state:\n";
      two_rx.rx_2::list_rx_state();
      return(EXIT_SUCCESS);
}
```

The member functions of the `two_channel_rx` class belong to the repeated (indirect) `receiver` base class. Consequently, invoking these functions is ambiguous, unless we specify the direct base class, as in the above program. Notice that the member access operator (a single dot) is used between a class and a derived class. The scope resolution operator (two semicolons) is used between a class and a member function. Also notice that when invoking member functions, it is only necessary to give sufficient classes to resolve any ambiguities. In this example there is no need to give the `receiver` indirect base class because specifying `rx_1` or `rx_2` is sufficient to resolve any ambiguity.

Exercise

Extend the above program to a `transceiver` class that has two receive and two transmit channels.

13.2 Virtual Base Classes

If we have a repeated indirect base class, then the derived class will have two or more copies of the base class. This may not always be appropriate. For example, suppose we are considering solids of various shapes and materials. We may want to have a general `solid` class with `sphere` and `plastic_solid` as derived classes. A `plastic_sphere` class could then be derived from the `sphere` and `plastic_solid` classes. However, it would clearly be inappropriate for a `plastic_sphere` object to have two copies of the `solid` data member. For instance, if a `solid` object had an associated identifier (perhaps called `id`) then we wouldn't want a `plastic_sphere` object to inherit two (possibly different) identifiers. Such multiple copies can be avoided by using the `virtual` specifier before the indirect base class, as illustrated below.

```
class solid {
public:
    int id;
};

class sphere : virtual public solid {
// More members may go in here.
};

class plastic_solid : virtual public solid {
```

```
// More members may go in here.
};

class plastic_sphere : public sphere, public plastic_solid{
// More members may go in here.
};
```

The class hierarchy is shown in Figure 13.4. There is only one copy of the `id` data

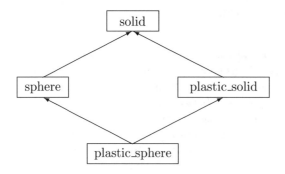

Figure 13.4: Virtual base class.

member of the `solid` class, which can therefore be accessed unambiguously, as in:

```
plastic_sphere s;
s.id = 1;
```

Notice that the indirect base class, `solid`, is an ordinary class, with the `virtual`
specifier only occurring at the next level.[1] Also, specifying `virtual` has no effect on
the `sphere` and `plastic_solid` classes.

    A simple program demonstrating how to access a `plastic_sphere` object is given
below.

```
#include <iostream>
#include <cstdlib>        // For EXIT_SUCCESS
using namespace std;

class solid {
public:
    void set_id(int identifier) {id = identifier;}
    int get_id(void) {return id;}
private:
    int id;
};
```

---

[1]The `virtual` keyword, used in the context of an indirect base class, is rather different from the
idea of a `virtual` function; the `solid` class could also contain `virtual` functions. The relative order
of `virtual` and `public` is of no consequence.

```
class sphere : virtual public solid {
public:
    void set_radius(double radius) { r = radius; }
    double get_radius(void) { return r; }
private:
    double r;
};

class plastic_solid : virtual public solid {
public:
    void set_density(double density) { rho = density; }
    double get_density(void) { return rho; }
private:
    double rho;      // This is the density.
};

class plastic_sphere : public sphere, public plastic_solid{
public:
    void list_properties(void);
};

void plastic_sphere::list_properties(void)
{
    cout << "\tIdentifier:\t" << get_id() << "\n\tDensity:\t" <<
        get_density() << "\n\tRadius:\t\t" << get_radius() <<
        "\n";
}

int main()
{
    plastic_sphere s;
    s.set_id(1);
    s.set_density(6.5);
    s.set_radius(10.7);
    cout << "Properties of sphere:\n";
    s.list_properties();
    return(EXIT_SUCCESS);
}
```

Notice how there is no ambiguity in accessing the id data member of the virtual base class.

An indirect base class can be both virtual and non-virtual, as in:

```
class A {
// Members go in here.
};

class W : public A {
```

```
// More members may go in here.
};

class X : virtual public A {
// More members may go in here.
};

class Y : virtual public A {
// More members may go in here.
};

class Z : public A {
// More members may go in here.
};

class P : public W, public X, public Y, public Z {
// More members may go in here.
};
```

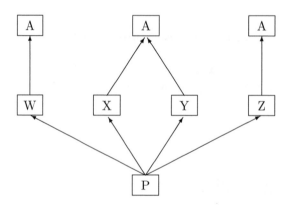

Figure 13.5: Indirect base class that is both `virtual` and non-`virtual`.

The class hierarchy for the declarations given above is shown in Figure 13.5. Notice that:

- If the derived class is to have `public` access to each base class, then the `public` access specifier must be repeated for each base class.

- The order of classes in a derivation list is not important. The order may effect such things as the storage layout, but this is compiler-dependent and as programmers we shouldn't need to worry about such things.

- There is no `virtual` keyword in the declaration of the class P; this keyword only occurs at the next level, in the declarations of the X and Y classes.

- There are three, rather than four, copies of the data members of class A.

Exercise

Implement the classes shown in Figure 13.5. The class A should have a single integer private data member, together with public functions to access this data. Write a program that sets the data members of a P object and then list the data stored by a P object. Demonstrate that there are only three items of data associated with a single P object.

13.3 Constructors and Destructors

As we remarked in Section 12.5, constructors and destructors are not inherited and the derived class constructor usually calls the base class constructor. The additional feature with multiple inheritance is that the derived class constructor can call a list of base class constructors. Only the *direct* base classes (as distinct from the indirect base classes) can normally be members of this list. The one exception is that virtual indirect base classes can be included. The classes forming the hierarchy shown in Figure 13.4 provide an illustration of this feature. The program given on page 382 can be modified to use constructor functions to initialize the data, as shown below.

```
#include <iostream>
#include <cstdlib>        // For EXIT_SUCCESS
using namespace std;

class solid {
public:
    solid(int identifier) { id = identifier; }
    int get_id(void) {return id;}
private:
    int id;
};

class sphere : virtual public solid {
public:
    sphere(int identifier, double radius) :
        solid(identifier) { r = radius; }
    double get_radius(void) { return r; }
private:
    double r;
};

class plastic_solid : virtual public solid {
public:
    plastic_solid(int identifier, double density) :
        solid(identifier) { rho = density; }
    double get_density(void) { return rho; }
private:
    double rho;       // This is the density.
```

```
};

class plastic_sphere : public sphere, public plastic_solid{
public:
    plastic_sphere(int identifier, double density, double radius) :
        solid(identifier), sphere(identifier, radius),
        plastic_solid(identifier, density) { }
    void list_properties(void);
};

void plastic_sphere::list_properties(void)
{
    cout << "\tIdentifier:\t" << get_id() << "\n\tDensity:\t" <<
        get_density() << "\n\tRadius:\t\t" << get_radius() <<
        "\n";
}

int main()
{
    plastic_sphere s(1, 6.5, 10.7);
    cout << "Properties of sphere:\n";
    s.list_properties();
    return(EXIT_SUCCESS);
}
```

Notice how the `solid` indirect base class is included in the initialization list for the `plastic_sphere` constructor. If we didn't do this, then the `sphere` and `plastic_solid` constructors would both attempt to initialize the `solid` base class. However, including `solid` in the initialization list overcomes this problem because a `virtual` base class is initialized by its most derived class (in this case, `plastic_sphere`).

13.4 Member Access Ambiguities

Part of the motivation for multiple inheritance is to reuse existing classes. However, the base classes may well have members with the same names. As in the case of single inheritance, potential ambiguities are often resolved by the concept of dominance; that is names in the derived class *dominate* those of the base classes. For example, the `transceiver`, `transmitter` and `receiver` classes could all use functions called `list_state()` to output the present state of the particular device. However, if we invoke a `list_state()` function for a `transceiver` object then it is the `transceiver` version that is invoked. We can illustrate this by modifying the program given on page 377 so that all three classes have a `list_state()` function. This is done in the program shown below, where we also implement constructor functions.

```
#include <iostream>
#include <cstdlib>        // For EXIT_SUCCESS
using namespace std;
```

```
class transmitter {
public:
    transmitter(double frequency);
    void set_tx_active(bool active){ tx_active = active; }
    void list_state(void);
private:
    double tx_f;
    bool tx_active;
};

inline transmitter::transmitter(double frequency)
{
    tx_f = frequency;
    tx_active = false;
}

void transmitter::list_state(void)
{
    cout << "Transmitter state:\n\tfrequency: " << tx_f <<
        "\n\tactive:    ";
    if (tx_active)
        cout << "true\n";
    else
        cout << "false\n";
}

class receiver {
public:
    receiver(double frequency);
    void set_rx_active(bool active){ rx_active = active; }
    void list_state(void);
private:
    double rx_f;
    bool rx_active;
};

receiver::receiver(double frequency){
    rx_f = frequency;
    rx_active = false;
}

void receiver::list_state(void)
{
    cout << "Receiver state:\n\tfrequency: " << rx_f <<
        "\n\tactive:    ";
    if (rx_active)
        cout << "true\n";
```

```
        else
            cout << "false\n";
    }

    class transceiver : public transmitter, public receiver {
    public:
        transceiver(double tx_frequency, double rx_frequency) :
            transmitter(tx_frequency), receiver(rx_frequency){ }
        void list_state(void);
    };

    void transceiver::list_state(void)
    {
        transmitter::list_state();
        receiver::list_state();
    }

    int main()
    {
        transceiver trx(412.75, 422.75);
        trx.set_tx_active(true);
        trx.list_state();
        return(EXIT_SUCCESS);
    }
```

If we actually want to access the members that are dominated by the derived class members then we need to specify the explicit base classes. We have already used this technique in the program demonstrating the `two_channel_rx` class on page 380. Examples for the `transceiver` class are given in the following code fragment:

```
    cout << "Access transmitter base function:\n";
    trx.transmitter::list_state();
    cout << "Access receiver base function:\n";
    trx.receiver::list_state();
```

Data members with duplicate names can be accessed in a similar manner to member functions.

Exercise

Modify the program demonstrating the `transceiver` class to include the code fragment given above. By compiling and running your modified program, demonstrate that the base class `list_state()` functions are accessed.

Use of a `virtual` base class is another way in which potential ambiguities are resolved since there is only one instance of that base class within a derived class object. The `plastic_sphere` class, which we discussed in Section 13.2, is an example of this. However, for `virtual` base classes there is also a potential access privilege ambiguity. As an example, we could modify the program on page 385 so that `solid` is a `private` base class of the `plastic_solid` class, as shown below.

```
class plastic_solid : virtual private solid {
public:
    plastic_solid(int identifier, double density) :
        solid(identifier) { rho = density; }
    double get_density(void) { return rho; }
private:
    double rho;       // This is the density.
};
```

There is now an apparent ambiguity since the `plastic_sphere` class can access the public members of `solid` via `sphere`, but not via `plastic_solid`. This ambiguity is resolved by the rule that the `public` access path always dominates.

13.5 Using Multiple Inheritance

One of the primary reasons for introducing multiple inheritance is to control the complexity that occurs in very large applications. This makes it difficult to find realistic examples that can be explained within a few pages; a description of carnivore, herbivore and omnivore classes is unlikely to convince an engineer of the true power of C++. What follows is a semi-realistic example, but you will only come to really appreciate multiple inheritance when writing large applications.

In this section we consider a set of *controller* classes. A `controller` base class is given below.

```
// source:  control.h
// use:     Defines controller class.

#ifndef CONTROLLER_H
#define CONTROLLER_H

#include "string.h"

class controller {
public:
    virtual void input(double set_data) { }
    virtual void output(double &give_data) { }
    virtual void display(void);
protected:
    controller(double set_max_data,const string &set_label,
        int set_scale_length);
    int cursor, scale_length;
    double max_data, data;
    string label;
};

#endif   // CONTROLLER_H
```

The basic idea is that an object of an appropriate class derived from this `controller` class accepts a controlling value via the `input()` function, displays this value using

the `display()` function and produces a controlled output by means of the `output()`
function. Such a controller could be used for anything from an audio system to a power
station. All three `controller` functions could take many different forms. For example,
the volume control on an audio system might accept inputs from 0 to 100 (in some
arbitrary units), but inputs from −50 to +50 would be more appropriate for the stereo
balance control. The output may be any function of the input; two obvious examples
are linear and logarithmic outputs.

For some applications there may be many different types of controller but, rather
than design each controller from scratch, we can use multiple inheritance to mix and
match the various controller classes, as shown in Figure 13.6. The `controller` base

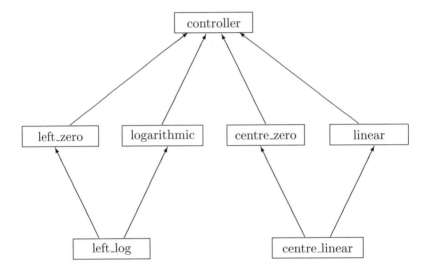

Figure 13.6: Controller classes.

class ensures a uniform interface, but it is not intended that we should be able to
create instances of this class. A number of derived classes, such as `left_zero` and
`centre_zero`, implement specific `input()` functions, corresponding to the various kinds
of input that may be appropriate for a particular project. A number of other classes,
such as `log` and `linear`, implement different versions of the `output()` function. Again,
it is not intended that objects of any of these classes can be created. However, one
class from each of the sets of input and output classes can be used as bases for the
required controller classes; `left_log` and `centre_linear` are the examples shown in
Figure 13.6. Instances of such classes *can* indeed be created.

The `controller` class header file has already been listed. The main points worth
noting are:

- The `input()` and `output()` functions are declared to be **virtual** rather than
  pure **virtual** since no derived class provides implementations for both of them.

- The one and only constructor has **protected** access, ensuring that a `controller`
  object cannot be created. This is a necessary restriction since the class does not
  provide satisfactory implementations for the `input()` and `output()` functions.

- Use is made of the **string** class, initially described in Section 9.6.2 and improved in Exercise 1 of Chapter 10.

The **controller** class implementation is straightforward and is given below.

```
// source:   control.cxx
// use:      Implements controller class.

#include <iostream>
#include <cstdlib>       // For exit()
#include "control.h"
using namespace std;

controller::controller(double set_max_data,
    const string &set_label, int set_scale_length)
{
    if (set_max_data <= 0.0) {
        cerr << "Maximum input to a controller " <<
            "must be positive.\n";
        exit(EXIT_FAILURE);
    }
    max_data = set_max_data;
    label = set_label;
    scale_length = set_scale_length;
    data = 0.0;     // Default initial value.
}

// The following function displays the value of data on an
// analogue scale. The cursor is set by the input()
// function:
//
//                     -------------
//                     |     *     |
//                     -------------
//                         label
//
void controller::display(void)
{
    cout << "\n";
    // Draw upper edge:
    int i;
    for (i =- 2; i <= scale_length; ++i)
        cout << "-";
    // Draw left edge:
    cout << "\n|";
    // Step to cursor position:
    for (i = 0; i < cursor; ++i)
        cout << " ";
```

```
    // Draw cursor:
    cout << "*";
    // Step to right edge:
    while (i++ < scale_length)
        cout << " ";
    // Draw right edge:
    cout << "|\n";
    // Draw lower edge:
    for (i =- 2; i <= scale_length; ++i)
        cout << "-";
    cout << "\n";
    // Shift label to centre of display:
    int offset = (scale_length + 3 - label.length()) / 2;
    if (offset < 0)
        offset = 0;
    for (i = 0; i < offset; ++i)
        cout << " ";
    label.print();
    cout << "\n";
}
```

One way in which the above code could be improved is by replacing the statement:

```
label.print();
```

by:

```
cout << label;
```

Since label is a **string** object, this would require the techniques of Chapter 18.

The class declaration for a controller accepting values from zero up to a positive maximum (called **max_data** in the **controller** base class) is given below.

```
// source:  left.h
// use:     Defines left_zero controller class.

#ifndef LEFT_ZERO_H
#define LEFT_ZERO_H

#include "control.h"

class left_zero : virtual public controller {
public:
    void input(double set_data);
protected:
    left_zero(double set_max_data, const string &set_label,
        int set_scale_length) :
        controller(set_max_data, set_label, set_scale_length) { }
};

#endif   // LEFT_ZERO_H
```

The following file gives the implementation of the `left_zero` controller class:

```
// source:  left.cxx
// use:     Implements left_zero controller class.

#include "left.h"

void left_zero::input(double set_data)
{
    if (set_data <= 0.0)
        data = 0.0;
    else if (set_data >= max_data)
        data = max_data;
    else
        data = set_data;
    // Calculate cursor position (0.5 ensures rounding):
    cursor = static_cast<int>((data * scale_length) / max_data +
        0.5);
}
```

Note the following:

- The `controller` base class of the `left_zero` class is `virtual`. This is because `left_log` objects should only have one copy of the data members of the `controller` class, as shown in Figure 13.6.

- The constructor is `protected` so that `left_zero` objects cannot be created. The base class constructor is specifically invoked since constructors are not inherited.

- The `input()` function ensures that only valid values are stored by the `data` variable, no matter what numbers are actually input.

The class declaration for a controller accepting input values between +max_data and −max_data is given in the following file and is similar to the `left_zero` class:

```
// source:  centre.h
// use:     Defines centre_zero controller class.

#ifndef CENTRE_ZERO_H
#define CENTRE_ZERO_H

#include "control.h"

class centre_zero : virtual public controller {
public:
    void input(double set_data);
protected:
    centre_zero(double set_max_data, const string &set_label,
        int set_scale_length) :
        controller(set_max_data, set_label, set_scale_length) { }
```

```
    };

    #endif  // CENTRE_ZERO_H
```

The `centre_zero` class can be implemented as shown below.

```
// source:   centre.cxx
// use:      Implements centre_zero controller class.

#include "centre.h"

void centre_zero::input(double set_data)
{
    if (set_data <= -max_data)
        data = -max_data;
    else if (set_data >= max_data)
        data = max_data;
    else
        data = set_data;
    // Calculate cursor position (0.5 ensures rounding):
    cursor = static_cast<int>(0.5 * (data + max_data) *
        scale_length / max_data + 0.5);
}
```

Note that:

- The `controller` base class is `virtual` so that each `centre_linear` object only has one copy of the data members of `controller`.

- The constructor is `protected` and this ensures that no `centre_zero` objects can be created. Since constructors are not inherited, the base class constructor is explicitly invoked.

- The `input()` function ensures that only values lying between $-$max_data and $+$max_data are stored by the `data` variable.

The class for a controller giving a linear output is completely defined by the following header file:

```
// source:   linear.h
// use:      Defines linear controller output class.

#ifndef LINEAR_H
#define LINEAR_H

#include "control.h"

class linear : virtual public controller {
public:
    void output(double &give_data) { give_data = data; }
```

```
    protected:
        linear(double set_max_data, const string &set_label,
            int set_scale_length) :
            controller(set_max_data, set_label, set_scale_length) { }
    };

    #endif  // LINEAR_H
```

Observe that:

- As for the `centre_zero` class, the `controller` base class is `virtual` and the `protected` constructor invokes the base class constructor.

- In order to avoid a comparatively large function call overhead, the `output()` function is implemented `inline`.

A controller class giving a logarithmic output is likewise very simple and is given below.

```
    // source:  log.h
    // use:     Defines log controller output class.

    #ifndef LOG_H
    #define LOG_H

    #include <cmath>        // For log()
    #include "control.h"
    using namespace std;

    class logarithmic : virtual public controller {
    public:
        void output(double &give_data) { give_data = log(1.0 + data); }
    protected:
        logarithmic(double set_max_data, const string &set_label,
            int set_scale_length) :
            controller(set_max_data, set_label, set_scale_length) { }
    };
    #endif  // LOG_H
```

From these four base classes we could construct four derived classes. Two of these classes are shown in the inheritance diagram given in Figure 13.6. The `left_log` class is defined in the following file:

```
    // source:  left_log.h
    // use:     Defines controller class with left-zero display and
    //          logarithmic output.

    #ifndef LEFT_LOG_H
    #define LEFT_LOG_H
```

```
#include "left.h"
#include "log.h"

class left_log : public left_zero, public logarithmic {
public:
    left_log(double set_max_data, const string &set_label,
        int set_scale_length) :
        left_zero(set_max_data, set_label, set_scale_length),
        logarithmic(set_max_data, set_label, set_scale_length),
        controller(set_max_data, set_label, set_scale_length) { }
};

#endif  // LEFT_LOG_H
```

The file given below defines the `centre_linear` class.

```
// source:  centre_linear.h
// use:     Defines controller class with centre-zero display and
//          linear output.

#ifndef CENTRE_LINEAR_H
#define CENTRE_LINEAR_H

#include "centre.h"
#include "linear.h"

class centre_linear : public centre_zero, public linear {
public:
    centre_linear(double set_max_data, const string &set_label,
        int set_scale_length) :
        centre_zero(set_max_data, set_label, set_scale_length),
        linear(set_max_data, set_label, set_scale_length),
        controller(set_max_data, set_label, set_scale_length) { }
};

#endif  // CENTRE_LINEAR_H
```

Notice that for both of these derived classes the `public` constructors invoke the direct
and indirect base classes, although in fact the `controller` data is only initialized once.
(See Section 13.3.) Since the constructors are `public`, we can create `left_log` and
`centre_linear` objects.

The following program demonstrates how to use the controller classes:

```
// source:    my_test.cxx
// use:       Tests controller classes.

#include <iostream>
#include <cstdlib>      // For EXIT_SUCCESS
#include "left_log.h"
```

```
#include "centre_linear.h"
using namespace std;

int main()
{
    double x, y, z;

    left_log controller_1(100, "volume", 10);
    cout << "\n";
    centre_linear controller_2(100, "balance", 10);
    cout << "\n";

    cout << "Input value to be displayed: ";
    cin >> x;
    controller_1.input(x);
    controller_2.input(x);

    controller_1.display();
    controller_2.display();

    controller_1.output(y);
    controller_2.output(z);
    cout << "controller_1 output = " << y <<
        "\ncontroller_2 output = " << z << "\n";

    return(EXIT_SUCCESS);
}
```

Since we have more files than in previous projects, it is worth introducing some new makefile features. The files can be compiled and linked by using the following makefile:[2]

```
objects=control.o left.o centre.o string.o
my_test: my_test.cxx $(objects)
    g++ my_test.cxx $(objects) -o my_test
control.o: control.cxx control.h
    g++ -c control.cxx
left.o: left.cxx left.h
    g++ -c left.cxx
centre.o: centre.cxx centre.h
    g++ -c centre.cxx
string.o: string.cxx string.h
    g++ -c string.cxx
clean:
    rm my_test $(objects)
```

New makefile features are:

---

[2]Don't forget that the `string.cxx` and `string.h` files are given in Section 9.6.2 and modified in Exercise 1 of Chapter 10.

- Since the list of object files (those files ending with .o) occurs several times, it is worth defining a "variable" to represent all of them. This is done on line 1 of the makefile. The traditional name for a variable used in this context is objects. (Another tradition is to use OBJECTS.) One advantage of defining objects is that if we want to add a new file name to the list, we only have to do so in one place. This reduces the risk of introducing errors into the makefile.

- Having defined the objects variable, we can use the $(objects) syntax whenever we would otherwise use the list of object files, as in line 2 of the makefile.

- Having tested our code, we may want to clean up the directory by removing executable and object files. The final two lines of the makefile define a rule for doing this. This rule has no effect if we simply type "make". However, typing "make clean" results in the required clean up.

Of course, this discussion of controller classes uses some very powerful language features to implement what are really very simple requirements. However, in a real application area, such as chemical engineering or power generation, object-oriented techniques would help to reduce software complexity and to increase maintainability.

13.6   Summary

- A class is declared to depend on multiple base classes by using a derivation list:

  ```
  class X : public A, public B, public C {
      // More members may go in here.
  };
  ```

- A direct base class cannot be repeated:

  ```
  class X : public A, public A { };        // WRONG!
  ```

- An indirect base class can be repeated:

  ```
  class A : public P {
      // More members may go in here.
  };
  class B : public P {
      // More members may go in here.
  };
  class X : public A, public B {
      // More members may go in here.
  };
  ```

- A derived class has only one copy of a virtual indirect base class:

  ```
  class A : virtual public P {
      // More members may go in here.
  ```

```
};
class B : virtual public P {
    // More members may go in here.
};
class X : public A, public B {
    X(int i, int j, int k) : P(i), A(j), B(k) { }
    // More members may go in here.
};
```

An indirect base class constructor can only appear in an initialization list if the indirect base class is `virtual`, as demonstrated by the `X` class constructor.

- If the same member name occurs in multiple base classes, or in a base class and a derived class, then the derived class dominates. In such cases, the base class names can be accessed by using the member access and scope resolution operators. For example, given the classes:

```
class T {
public:
    double f;
};

class R {
public:
    double f;
};

class TR : public T, public R {
public:
    double f;
};
```

then the following code fragment demonstrates member access:

```
TR x;
x.f = 412.75;       // Accesses TR member.
x.T::f = 422.75;    // Accesses T member.
x.R::f = 432.75;    // Accesses R member.
```

13.7 Exercises

1. Our `controller` classes do not define copy constructors or overloaded assignment operators. Define and test *suitable* copy constructors and overloaded assignment operators.

2. By developing the `controller` classes, described in Section 13.5, create treble, base, volume and balance controllers for an audio system. Should the system

have its own class? If so, would it be better for this `audio` class to be derived from the controller classes or to be a client of them?

(Your audio system should have an array of `controller` pointers and dynamically create objects for the specific derived classes.)

3. Suppose we now decide to upgrade the audio system, built in the previous exercise, so that it has a remote controller. This controller can only issue ± 1 to indicate whether a particular parameter should be increased or decreased. Make whatever changes are necessary to the `input()` functions, and use the + and − keys to issue "increase" or "decrease" commands.

Chapter 14

Namespaces

A namespace is a way of grouping together logically related functions and data. In this sense a namespace resembles a class, but it is a weaker concept than a class. In fact, a namespace is simply a named scope. Namespaces are a recent addition to the C++ language, so if you have an older compiler, you may find that it doesn't support this feature.

14.1 Name Clashes

Imagine the following situation. You have access to a header file, called `met_office.h`:

```
// source:  met_office.h

void temperature(void);
void pressure(void);
```

and another header file, called `weather_station.h`:

```
// source:  weather_station.h

void temperature(void);
void pressure(void);
```

You also have access to the corresponding object code (which we might refer to as `met_office.o` and `weather_station.o`). Although the function interfaces in the two files are the same, the functions might do different things, as in the implementations:[1]

```
// source:  met_office.cxx

#include <iostream>
#include "met_office.h"
using namespace std;
```

---

[1] In general, if functions do different things it is a good idea to give them different names. However, this is not always possible; for example the functions could be from different software libraries written by different programmers.

```
void temperature(void)
{
    cout << "Temperature (K) is:\n";
    // etc.
}

void pressure(void)
{
    cout << "Pressure (mb) is:\n";
    // etc.
}
```

and:

```
// source:  weather_station.cxx

#include <iostream>
#include "weather_station.h"
using namespace std;

void temperature(void)
{
    cout << "Temperature (C) is:\n";
    // etc.
}

void pressure(void)
{
    cout << "Pressure (mmHg) is:\n";
    // etc.
}
```

Moreover, suppose your program needs to use the `temperature()` function implemented in `met_office.cxx` together with the `pressure()` function implemented in `weather_station.cxx`. However, you are not allowed to edit the files `met_office.cxx` and `weather_station.cxx`. Of course, for the simple functions given here, you could easily type in your own versions, but for real code this may not be possible. This could be because the source code belongs to different software companies and they won't release it to you, or it could be because changing the source would break too much other code. You could attempt to do something similar to the program given below.

```
// source:  my_test.cxx

#include <iostream>
#include <cstdlib>        // For EXIT_SUCCESS
#include "met_office.h"
#include "weather_station.h"
using namespace std;
```

```
int main()
{
    temperature();
    pressure();
    return(EXIT_SUCCESS);
}
```

Unfortunately, this doesn't work as can be shown by using the following makefile:

```
my_test: my_test.cxx met_office.o weather_station.o
    g++ my_test.cxx met_office.o weather_station.o -o my_test
met_office.o: met_office.cxx met_office.h
    g++ -c met_office.cxx
weather_station.o: weather_station.cxx weather_station.h
    g++ -c weather_station.cxx
```

The two source files compile without any error, but the linker complains of multiple definitions of the functions `temperature()` and `pressure()`.

Software projects often involve very large amounts of code, written by many different programmers. Moreover, even if you are only writing a very modest program, you are likely to want to use libraries written by other programmers. Therefore name clashes, with the consequent multiple definitions of functions, classes etc., are fairly common. This potential problem can be resolved by the use of namespaces.

14.2 Creating a Namespace

The syntax for creating a *namespace* uses the `namespace` keyword and is illustrated below.

```
namespace weather_station
{
    void temperature(void);
    void pressure(void);
}
```

This declaration creates a namespace with the name `weather_station`. Notice that there is *no* semicolon following the closing brace. In the example given in the previous section, the duplicate names would have been avoided if the two different source files had used two different `namespaces`. For example, the header files:[2]

```
// source:  met_office.h

namespace met_office
{
    void temperature(void);
    void pressure(void);
}
```

---

[2]There is no need for the name appearing in the `namespace` to be the same as the header file, but often a good name for a header file is a good name for a namespace.

and:

```
// source:  weather_station.h

namespace weather_station
{
    void temperature(void);
    void pressure(void);
}
```

would avoid duplicate names.

14.3   Accessing Namespace Members

One way of accessing namespace members is by using the name of the appropriate namespace, together with the scope resolution operator. Using this technique, the two source files could have been written as:

```
// source:  met_office.cxx

#include <iostream>
#include "met_office.h"
using namespace std;

void met_office::temperature(void)
{
    cout << "Temperature (K) is:\n";
    // etc.
}

void met_office::pressure(void)
{
    cout << "Pressure (mb) is:\n";
    // etc.
}
```

and:

```
// source:  weather_station.cxx

#include <iostream>
#include "weather_station.h"
using namespace std;

void weather_station::temperature(void)
{
    cout << "Temperature (C) is:\n";
    // etc.
}
```

```
void weather_station::pressure(void)
{
    cout << "Pressure (mmHg) is:\n";
    // etc.
}
```

We can demonstrate the benefit of namespaces by means of the following program:

```
// source:  my_test.cxx

#include <iostream>
#include <cstdlib>        // For EXIT_SUCCESS
#include "met_office.h"
#include "weather_station.h"
using namespace std;

int main()
{
    met_office::temperature();
    weather_station::pressure();
    return(EXIT_SUCCESS);
}
```

Exercise

You should try compiling and linking this program. When you run `my_test` you will find that the messages demonstrate that the functions invoked are the `temperature()` function, defined in `met_office.cxx`, together with the `pressure()` function, defined in `weather_station.cxx`.

It is possible to put almost anything in a `namespace` (apart from the function `main()`). However, it is always a good idea to keep the interface distinct from the implementation. Consequently, the function implementations should be in a `.cxx` source file. The `namespace` should be declared in a header file and should contain things like class and function declarations (*not* definitions).

14.4 More on Creating Namespaces

Unlike a class, a namespace is "open"; that is we can declare new members and these need *not* be in the same translation unit as the original namespace. This is done by using the namespace syntax. For example, we could add a `humidity()` function to the previously defined `met_office` namespace by means of:

```
// source:  humidity.h

namespace met_office
{
    void humidity(void);
}
```

The humidity.h header file would have its own corresponding source code in a file such as humidity.cxx:

```
// source:  humidity.cxx

#include <iostream>
#include "humidity.h"
using namespace std;

void met_office::humidity(void)
{
    cout << "Humidity is: 98%\n";
}
```

It is also possible to nest namespaces, as in:

```
namespace environment
{
    namespace met_office
    {
        void temperature(void);
        void pressure(void);
    }
    namespace weather_station
    {
        void temperature(void);
        void pressure(void);
    }
}
```

In order to access members of the environment namespace, we need to use the scope resolution operator together with all of the namespace names in the order in which they appear in the namespace declaration, as in:

```
environment::met_office::temperature();
```

Exercise

Verify the above statement concerning the use of the scope resolution operator by using an appropriate environment.h header file and modifying the code in my_test.cxx.

14.5   Namespace Aliases

It is worth using a fairly long name when declaring a namespace since this reduces the risk of it clashing with the name used by another programmer. However, using long names gets very tedious and (even worse) can make the code very difficult to read. A useful feature is the *namespace alias* which has the syntax:

```
namespace alias_name = namespace_name;
```

For example, the namespace aliases:

```
namespace E = environment;
namespace M = met_office;
```

enable us to invoke the `environment::met_office::temperature()` function with:

```
E::M::temperature();
```

rather than:

```
environment::met_office::temperature();
```

Of course, overuse of this feature can make code very difficult to read and also increases the likelihood of name clashes. However, if a name clash does occur, it is relatively easy to declare a new namespace alias.

14.6   The using Directive

We can avoid using the scope resolution operator by employing the *using directive*. An example of the using directive is:

```
using namespace met_office;
```

This statement makes all of the names in `met_office` visible and accessible without qualification with the namespace name. The scope of a using directive starts at the directive and ends at the end of the current scope.

It is worth clarifying the application of the using directive by considering a detailed example. Suppose we have the two header files:

```
// source:  met_office.h

namespace met_office
{
    void temperature(void);
    void pressure(void);
}
```

and:

```
// source:  weather_station.h

namespace weather_station
{
    void temperature(void);
    void pressure(void);
}
```

You might imagine that we could have the using directive within each .cxx file, as in:

```
// source:  met_office.cxx

#include <iostream>
#include "met_office.h"
using namespace std;

using namespace met_office;

void temperature(void)
{
    cout << "Temperature (K) is:\n";
    // etc.
}

void pressure(void)
{
    cout << "Pressure (mb) is:\n";
    // etc.
}
```

and:

```
// source:  weather_station.cxx

#include <iostream>
#include "weather_station.h"
using namespace std;

using namespace weather_station;

void temperature(void)
{
    cout << "Temperature (C) is:\n";
    // etc.
}

void pressure(void)
{
    cout << "Pressure (mmHg) is:\n";
    // etc.
}
```

and then pick out the appropriate function in `my_test.cxx` by using the explicit namespace together with the scope resolution operator. However, what happens is that both
of the source files compile correctly, but the linker complains about a multiple definition of "temperature(void)" and a multiple definition of "pressure(void)". The lesson
to be learned from this is that the `using` directive doesn't just open up the namespace
to the file in which it appears; its effect is almost as if the namespace had never been

declared. Consequently, the `using` directive is a way of avoiding namespaces rather than a way of avoiding name clashes.

Exercise

Demonstrate that compiling and linking the following code with the version of `met_office.cxx` given above, does correctly invoke the `met_office` versions of the `temperature()` and `pressure()` functions.

```
//source: my_test.cxx

#include <iostream>
#include <cstdlib>        // For EXIT_SUCCESS
using namespace std;

void temperature(void);
void pressure(void);

int main()
{
    temperature();
    pressure();
    return(EXIT_SUCCESS);
}
```

What happens if you `#include "met_office.h"` instead of explicitly giving the function declarations in the `my_test.cxx` file?

14.7   The using Declaration

If we don't need to have access to all members of a namespace, then the `using` declaration is a better way of gaining access than the `using` directive. The statement:

```
using met_office::temperature;
```

is a *using declaration* and makes the `temperature()` function in the `met_office` namespace visible within the current scope. This means that we can access the `temperature()` function without using the namespace name. The scope of a `using` declaration starts at the declaration and ends at the end of the current scope. Notice that visibility of a function name is not the same as a function declaration and that we still need to include the appropriate header files.

Examples of the `using` directive are contained in the following file:

```
// source:  my_test.cxx

#include <iostream>
#include <cstdlib>        // For EXIT_SUCCESS
#include "met_office.h"
```

```
#include "weather_station.h"
using namespace std;

using met_office::temperature;
using weather_station::pressure;

int main()
{
    temperature();
    pressure();
    return(EXIT_SUCCESS);
}
```

The statement:

```
using met_office::temperature;
```

makes the `temperature()` function in the namespace `met_office` visible within the file. We can therefore access the `temperature()` function without using the namespace name. In a similar manner, the statement:

```
using weather_station::pressure;
```

makes the `pressure()` function in the namespace `weather_station` visible, and we can access the `pressure()` function without using the namespace name. However, as we learned in the previous section, the `met_office.cxx` and `weather_station.cxx` files should use the explicit namespace name together with the scope resolution operator.

Exercise

Compile and link the above code for `my_test.cxx`. Verify that the correct versions of the `temperature()` and `pressure()` functions are invoked.

It is worth emphasizing that the `using` directive does not provide a way out of using the namespace name and scope resolution operator in the file in which the namespace functions are defined. As an example, suppose we are only interested in the `met_office` version of the `temperature()` and `pressure()` functions. Then the following attempt at a source file for these functions fails to compile:

```
// source: met_office.cxx

#include <iostream>
#include "met_office.h"
using namespace std;

using met_office::temperature;
using met_office::pressure;

void temperature(void)
{
    cout << "Temperature (K) is:\n";
```

```
        // etc.
    }

    void pressure(void)
    {
        cout << "Pressure (mb) is:\n";
        // etc.
    }
```

14.8 The Standard Library

All C++ header files for the Standard Library add all of their function names etc. to a namespace called `std`.[3] For example, the `<cmath>` header file will contain something like:

```
    namespace std
    {
        double exp(double x);
        double pow(double x, double y);
        double cos(double x);
        double sqrt(double x);
        // etc.
    }
```

This explains why all of our programs so far have contained the statement:

```
    using namespace std;
```

In general, it isn't a good idea to do this since it makes the entire namespace of the Standard Library visible and it is highly unlikely that we really want to use everything in the Library. Of course, this is precisely what we have done so far in this book! Our excuse is that we wanted to keep our examples as short and uncomplicated as possible, and occasionally we will continue to use this technique. Moreover, making the entire `std` namespace visible is unlikely to cause any problems in small programs; namespaces really become important for large projects, probably involving many programmers.

As example of one way of accessing the `std` namespace, we can modify our program on page 105, so that we avoid the `using` directive with the help of the explicit namespace and the scope resolution operator:[4]

```
    #include <cmath>        // For sqrt()
    #include <cstdlib>      // For EXIT_SUCCESS
    #include <iostream>

    int main()
    {
```

---

[3]See Chapters 17 and 18.

[4]Note that since the preprocessor replaces `EXIT_SUCCESS` by 0, using `std::EXIT_SUCCESS` does not work.

```
    for (int i = 1; i <= 20; ++i) {
        std::cout << "square root of " << i << " is " <<
            std::sqrt(double(i)) << "\n";
    }
    return(EXIT_SUCCESS);
}
```

However, for large programs this can get tedious and the **using** declaration allows us
to have:

```
#include <iostream>
#include <cmath>
#include <cstdlib>
using std::cout;
using std::sqrt;

int main()
{
    for (int i = 1; i <= 20; ++i) {
        cout << "square root of " << i << " is " <<
            sqrt(double(i)) << "\n";
    }
    return(EXIT_SUCCESS);
}
```

14.9   Unnamed Namespaces

If the name for a namespace is omitted then we have what is known as an unnamed
namespace, as in:

```
// source:  file1.cxx

#include <iostream>
#include "file1.h"

namespace {
    int flag;
}

void set(void)
{
    flag = 10;
}

void list(void)
{
    std::cout << flag << "\n";
}
```

Within the translation unit, we can refer directly to members of the namespace. In this sense it is as if we had written:[5]

```
// source:  file1.cxx

#include <iostream>
#include "file1.h"

namespace XXX{
    int flag;
}
using namespace XXX;

void set(void)
{
    flag = 10;
}

void list(void)
{
    std::cout << flag << "\n";
}
```

For instance, the `list()` function accesses `flag` without any qualification within the translation unit. However, we don't have any knowledge of the name that goes in place of `XXX` and this name is unique to the translation unit. The important consequence of this is that members of an *unnamed* namespace are *not* accessible outside of the translation unit.[6] So in the above example, `flag` is not visible to the rest of the program.

14.10 Using Namespaces

Our aim in this section is to modify the singly linked list code, given in Section 10.3.1, so that the relevant declarations are given within a namespace. This can be achieved by modifying the header file, as shown below.

```
// source:  slist.h
// use:     Defines singly linked list class.

#ifndef SLIST_H
#define SLIST_H

namespace slist
```

---

[5]Notice that in this discussion the **using** directive appears in the *same* translation unit as the **using** declaration.

[6]Using an unnamed namespace replaces one of the uses of the C keyword `static`. Although this use of `static` is currently available in C++, its use is not encouraged by the ANSI Standard committee and may be removed from future versions of the Standard.

```
    {
        typedef int DATA_TYPE;

        class node {
            friend class list;
        public:
            DATA_TYPE data;
        private:
            node *next;
        };

        class list {
        public:
            list(void);
            virtual ~list();
            void push(DATA_TYPE new_data);
            void pop(DATA_TYPE &old_data);
            bool is_not_empty(void);
        private:
            node *head;
        };

        inline bool list::is_not_empty(void)
        {
            if (head == 0)
                return false;
            else
                return true;
        }
    }

#endif  // SLIST_H
```

The non-inline functions are defined in the modified source file, `slist.cxx`. Within this file, the functions are accessed by using the `slist` namespace name, together with the scope resolution operator. Because these functions are also members of the `list` class, the function names are a bit cumbersome, as can be seen by examining the following file:

```
// source:  slist.cxx
// use:     Implements singly linked list class.

#include "slist.h"

slist::list::list(void)
{
    head = 0;
}
```

```
slist::list::~list()
{
    while (head != 0) {
        node *pt = head;
        head = head->next;
        delete pt;
    }
}

void slist::list::push(DATA_TYPE new_data)
{
    node *pt = new node;
    pt->next = head;
    head = pt;
    pt->data = new_data;
}

void slist::list::pop(DATA_TYPE &old_data)
{
    if (head != 0) {
        old_data = head->data;
        node *pt = head;
        head = head->next;
        delete pt;
    }
}
```

We now have to modify the test program in order to gain access to the relevant members of the `slist` namespace. There are three distinct ways in which this can be done.

1. The easiest way of accessing the namespace is by means of the **using** directive, as demonstrated below, but this has the disadvantage of opening up the entire `slist` namespace.

```
// source:  my_test.cxx
// use:     Tries out singly linked list class.

#include <iostream>
#include <cstdlib>       // For EXIT_SUCCESS
#include "slist.h"
using namespace slist;
using namespace std;

int main()
{
    list s;
```

```
        DATA_TYPE j;
        for (int i = 1; i <= 5; ++i)
            s.push(10 * i);
        while (s.is_not_empty()) {
            s.pop(j);
            cout << j << "\n";
        }
        return(EXIT_SUCCESS);
    }
```

2. Another way of accessing the slist namespace is by giving the full names of all functions and data. This has been done in the following file:

```
// source:  my_test.cxx
// use:     Tries out singly linked list class.

#include <iostream>
#include <cstdlib>        // For EXIT_SUCCESS
#include "slist.h"

int main()
{
    slist::list s;
    slist::DATA_TYPE j;
    for (int i = 1; i <= 5; ++i)
        s.push(10 * i);
    while (s.is_not_empty()) {
        s.pop(j);
        std::cout << j << "\n";
    }
    return(EXIT_SUCCESS);
}
```

This technique is useful for avoiding name clashes, but can result in code that is fairly cluttered. Notice that it is only when we want to define objects, such as s and j, that we need to use the full name. When a function is invoked for an object, as in

```
s.pop(j);
```

the type of object is known and therefore the compiler can work out which function to call.

3. In general, the best way of accessing a namespace is by means of a using declaration, as shown in the following modified test program:

```
// source:  my_test.cxx
// use:     Tries out singly linked list class.
```

```
#include <iostream>
#include <cstdlib>        // For EXIT_SUCCESS
#include "slist.h"

using slist::DATA_TYPE;
using slist::list;
using std::cout;

int main()
{
    list s;
    DATA_TYPE j;
    for (int i = 1; i <= 5; ++i)
        s.push(10 * i);
    while (s.is_not_empty()) {
        s.pop(j);
        cout << j << "\n";
    }
    return(EXIT_SUCCESS);
}
```

This puts all the **namespace** information near the beginning of the file and leaves the body of the function **main()** looking uncluttered. However, in some circumstances, we would have to use the previous technique in order to avoid name clashes.

14.11 Summary

- A **namespace** is created by using the **namespace** keyword followed by declarations within a pair of braces:

```
namespace weather_station
{
    void temperature(void);
}
```

- Members of a **namespace** can be accessed by using the name of the appropriate **namespace**, together with the scope resolution operator:

```
void met_office::temperature(void)
{
    cout << "Temperature (K) is:\n";
    // etc.
}
```

- A namespace is "open". New members can be declared by using the namespace syntax:

```
namespace met_office
{
    void humidity(void);
}
```

The new members do not need to be in the same translation unit as the original namespace.

- Namespaces can be nested:

```
namespace environment
{
    namespace met_office
    {
        void temperature(void);
    }
    namespace weather_station
    {
        void temperature(void);
    }
}
```

In order to access members of a nested namespace, use the scope resolution operator together with all of the namespace names in the order in which they appear in the namespace declaration:

```
environment::met_office::temperature();
```

- A namespace alias is declared as in:

```
namespace M = met_office;
```

and enables us to use:

```
M::temperature();
```

in place of:

```
met_office::temperature();
```

- The using directive is:

```
using namespace met_office;
```

This statement makes all the names in met_office visible and accessible without qualification with the namespace name.

- The using declaration:

```
using met_office::temperature;
```

makes the `temperature()` function in the namespace `met_office` visible within the file. We can access the `temperature()` function without using the namespace name.

- All C++ header files for the Standard Library add all of their function names etc. to a namespace called `std`.

- If the name for a namespace is omitted then we have an unnamed namespace, as in:

```
namespace {
    int flag;
}
```

Within the translation unit, we can refer directly to members of the namespace. Members of an *unnamed* namespace are *not* accessible outside of the translation unit.

14.12 Exercises

1. For each of the following projects, modify the header file so that the classes are declared within a namespace. Also modify the other files associated with the project so that they have access to the namespace members. (You should *not* achieve this by means of `using` directives.)

   (a) The doubly linked list of Section 10.3.2.

   (b) The bit array class, described in Section 11.4.1.

2. Repeat the above exercise for the following projects:

   (a) The menu class, described in Section 12.7.2.

   (b) The controller classes of Section 13.5.

   For the derived classes, you should experiment with making them members of the base class namespace and with defining separate namespaces.

Chapter 15

Exception Handling

Almost any non-trivial program will contain errors, but one of the aims of a good programmer is to minimize the number and severity of errors, and to provide some method of recovery when errors do occur. Writing code to trap errors can be a difficult task and this chapter doesn't claim to be any more than an introduction.[1] Also, it should be emphasized that many of the examples given in this chapter are very small programs that are only for demonstration purposes. The benefit of the techniques introduced here for dealing with errors will only really become apparent for large programs, or programs using certain library facilities.

15.1 Errors

Errors can be of many types, including:

1. **Mistakes in the syntax**

   A typical example is:

   ```
   int i
   int j;
   ```

   Errors of this type should be caught by the compiler. However, don't forget that if you do something like:

   ```
   if (flag = 5) {
       do_something();
   }
   ```

   when you really mean:

   ```
   if (flag == 5) {
       do_something();
   }
   ```

---

[1]Further details are given in Chapters 8 and 14 of [10].

then the syntax is valid, even if it isn't what you intend.

2. **Logic errors**

   For this type of error, the program runs but doesn't perform as you expect. For example, you may have written some code to invert a matrix, but you haven't allowed for the fact that your particular matrix is close to being singular. So you get a result, but it isn't mathematically correct. The big danger here is that your program may run correctly most of the time and that you won't notice when it does give wrong results.

3. **Exceptions due to predictable problems**

   Examples of this type of error are when a program attempts to open a file that doesn't exist or attempts to dynamically allocate memory that isn't available. This is the type of error with which we are concerned in this chapter.

Whenever possible, errors should be handled locally, where they occur. For example, suppose a program asks the user to input the maximum number of iterations. We could try something like the following:

```
int iterations;
cout << "Input the number of iterations: ";
cin >> iterations;
```

This may be fine for a simple program that is only going to be used by the programmer. However, there are several things that could go wrong and it is worth anticipating these possibilities and trapping them. For instance, the user could:

- enter a negative number (or zero);

- enter a number that is so large that the program would take years to run to completion;

- enter some nonsensical answer like "half a dozen".

We can trap errors like these where they occur. Consequently, in such cases we don't need to introduce the exception handling procedures that we will meet later in this chapter. Something like the following would suffice:

```
int iterations = -1;
const int max_iterations = 10000;
while (true) {
    cout << "Input the number of iterations.\n" <<
        "This must greater than 0 and less than " <<
        max_iterations + 1 << ".\n";
    cin >> iterations;
    if (iterations > 0 && iterations <= max_iterations)
        break;
    cin.clear();
    cin.ignore(1024, '\n');
}
```

Here we have used the functions `clear()` and `ignore()` from the `streams` library. (See Chapter 18.) The `clear()` function is needed to reset the condition of `cin` after the read failure. The `ignore()` function throws away up to the specified number of characters in the stream unless it gets to the termination character, in which case it throws away all characters up to and including the termination character. In this code, the termination character is `'\n'`.

Exercise

Make the above code fragment into a program that outputs the value of `iterations`. Try to find some invalid input that the program does not handle correctly. (For example, you could try entering "half a dozen".) Make appropriate changes to the code in order to make it as robust as you can.

15.2 Introducing `throw`, `try` and `catch`

The way in which errors are handled for large programs needs special consideration. This is because the C++ language encourages the reuse of code and, in particular, the use of libraries. Consequently, an error can occur in a library due to an error in the code using the library. In general, the programmer who wrote the library won't have any knowledge of the program that uses it. Conversely, the programmer using the library may not have access to the library code. As an example, a library may supply a bounds checked array class. Suppose the user code has an error that causes an array to go out of bounds. It is the library code that detects the error, but the user code must decide what to do about it. This is a difficult situation since the way the error must be dealt with is non-local.

In order to deal with errors of this type, the C++ language provides an *exception handling* mechanism, with the keywords `throw`, `try` and `catch`. The basic idea is to put code that is likely to cause an error within a try-block, which is immediately followed by a catch-block. A test for an error condition is made within the try-block (usually within a function that is called from the try-block). If an error is found, then an object is thrown and caught by the following catch-block. This is what is meant by saying that an *exception is thrown*. The syntax should become clearer if you examine the following example in detail:

```cpp
#include <iostream>
#include <cstdlib>        // For EXIT_SUCCESS

using std::cout;
using std::cin;
using std::exit;

class dimension_error {
public:
    dimension_error() { }
};
```

```
void set_array_dimension(void)
{
    const int max_elements = 10;
    int elements;
    cout << "Enter number of elements: ";
    cin >> elements;
    if (elements > max_elements)
        throw dimension_error();
    cout << "Array dimension entered O.K.\n";
    // Do something more here.
}

int main()
{
    try {
        set_array_dimension();
    }
    catch(dimension_error) {
        cout << "The array dimension is too big.\n";
        exit(EXIT_FAILURE);
    }
    cout << "No errors so far.\n";
    // Do something more here.
    return(EXIT_SUCCESS);
}
```

Here we define a `dimension_error` class so that we can create `dimension_error` objects, even though they don't store any data. We also define a `set_array_dimension()` function. This returns the number entered by the user provided this number is less than or equal to `max_elements`. If the number entered is greater than `max_elements` then instead of the function giving the message "Array dimension entered O.K.", a `dimension_error` object is thrown. This is caught by the catch-block, which then issues a suitable error message. Notice that if an exception is thrown and successfully handled by the exception handler, then the code following the catch-block is executed, and *not* the code following the `throw`.

Exercise

Compile and run the above code. See what happens when you input different numbers. What happens if you do the following (one at a time):

(a) delete:

```
        if (elements > max_elements)
            throw dimension_error;
```

    from the code;

(b) leave out the `try` keyword;

(c) leave out the `catch` keyword;

(d) insert some code between the try-block and the catch-block.

The above example is deliberately kept simple since it is only meant to demonstrate the ideas involved in throwing an exception. In a more realistic program, the scene may be set for an error in one part of a program, which then calls a function, which calls another function, which calls another function, and so on. This can result in the function that detects an error, and hence throws an exception, being many nested function calls away from the code that actually caused the error. The function calls are stored on a stack and, when an exception is thrown, the stack is unwound until the initial call is reached. This is known as *stack unwinding* and is a one-way process; once the stack is unwound it is gone. As the stack is unwound, the destructors for local objects are invoked and the objects are destroyed.

At the other extreme, it is possible for the throw to occur directly in a try-block, rather than in a function called from within a try-block, as in the following example:

```cpp
#include <iostream>
#include <cstdlib>        // For exit()
using std::cout;
using std::cin;
using std::exit;

class dimension_error {
public:
    dimension_error() { }
};

int main()
{
    try {
        const int max_elements = 10;
        int elements;
        cout << "Enter number of elements: ";
        cin >> elements;
        if (elements > max_elements)
            throw dimension_error();
        cout << "Array dimension entered O.K.\n";
        // Do something more here.
    }
    catch(dimension_error) {
        cout << "The array dimension is too big.\n";
        exit(EXIT_FAILURE);
    }
    cout << "No errors so far.\n";
    // Do something more here.
    return(EXIT_SUCCESS);
}
```

Using throw and catch in this way is very unusual since the code would be simpler if we replaced the throw by the error message. It is better to reserve the exception handling techniques of C++ for the more complicated non-local situations outlined at the start of this section.

15.3   Throwing a Fundamental Type

It isn't necessary to define a class in order to throw an object; it is possible to throw an instance of a fundamental type. In the following example the object thrown is of type const char*:

```cpp
#include <iostream>
#include <cstdlib>        // For exit()
using std::cout;
using std::cin;
using std::exit;

void set_array_dimension(void)
{
    const int max_elements = 10;
    int elements;
    cout << "Enter number of elements: ";
    cin >> elements;
    if (elements > max_elements)
        throw "You have too many elements.\n";
    cout << "Array dimension entered O.K.\n";
    // Do something more here.
}

int main()
{
    try {
        set_array_dimension();
    }
    catch(const char* pt) {
        cout << pt;
        exit(EXIT_FAILURE);
    }
    cout << "No errors so far.\n";
    // Do something more here.
    return(EXIT_SUCCESS);
}
```

Although this is a straightforward way of catching an error message, any object of type const char* will be caught. For this reason it is more usual to throw an object of a specifically defined error class.

Exercise

Modify the root-finding program, given in Section 5.9.2, so that if an error occurs in the find_root() function, an exception consisting of an error message is thrown. Suitable exception handling should be introduced in the function main() so that this string is caught and the message is sent to the output stream.

15.4 Extracting Information from a `catch`

In many circumstances it would be useful if the object that got thrown stored data to help us diagnose the cause of the error. To do this we need to define a constructor for our error class such that the constructor takes one or more arguments. This is illustrated in the following program where if the value entered for `element` is greater than `max_elements`, then a `dimension_error` object gets thrown from the try-block.

```cpp
#include <iostream>
#include <cstdlib>        // For exit()
using std::cout;
using std::cin;
using std::exit;

class dimension_error {
public:
    dimension_error(int number_of_elements) {
        elements = number_of_elements;
    }
    int elements;
};

void set_array_dimension(void)
{
    const int max_elements = 10;
    int elements;
    cout << "Enter number of elements: ";
    cin >> elements;
    if (elements > max_elements)
        throw dimension_error(elements);
    cout << "Array dimension entered O.K.\n";
    // Do something more here.
}

int main()
{
    try {
        set_array_dimension();
    }
    catch(dimension_error x) {
        cout << "The array dimension is too big.\nYou have " <<
            x.elements << " elements.\n";
        exit(EXIT_FAILURE);
    }
    cout << "No errors so far.\n";
    // Do something more here.
    return(EXIT_SUCCESS);
}
```

Notice that if we want to refer to the `dimension_error` object when it gets caught then we must give the object a name in the argument of the catch-block. This is exactly the same syntax as for a function argument. In this case the object that gets caught is given the name `x`. It is worth emphasizing that `x` is the object, not the value stored in the object. Consequently, we need to use `x.elements` in order to access the value stored by `x`.

Although only one object gets thrown, the object is not limited to storing one item of data, as the following example shows:

```cpp
#include <iostream>
#include <cstdlib>        // For exit()
using std::cout;
using std::cin;
using std::exit;

class dimension_error {
public:
    dimension_error(int number_of_rows,
        int number_of_columns) {
        r = number_of_rows;
        c = number_of_columns;
    }
    int r, c;
};

void set_array_dimensions(void)
{
    const int max_rows = 10;
    const int max_columns = 10;
    int rows, columns;
    cout << "Enter rows: ";
    cin >> rows;
    cout << "Enter columns: ";
    cin >> columns;
    if ((rows > max_rows) || (columns > max_columns))
        throw dimension_error(rows, columns);
    cout << "Array dimensions entered O.K.\n";
    // Do something more here.
}

int main()
{
    try {
        set_array_dimensions();
    }
    catch(dimension_error x) {
        cout << "One or both of the array dimensions are too " <<
            "big.\nYou have " << x.r << " rows and " << x.c <<
```

```
               " columns\n";
             exit(EXIT_FAILURE);
        }
        cout << "No errors so far.\n";
        // Do something more here.
        return(EXIT_SUCCESS);
    }
```

In this program, the dimension_error object that is thrown from the try-block has two data members. The name of the object that is caught is x, and the data members are accessed as x.r and x.c.

Since there are two different types of error in the above example, an alternative technique is to have two different objects. Which object gets thrown could then depend on the type of error and this approach leads to the code given below.

```
#include <iostream>
#include <cstdlib>        // For exit()
using std::cout;
using std::cin;
using std::exit;

class row_error {
public:
    row_error(int number_of_rows) {
        r = number_of_rows;
    }
    int r;
};

class column_error {
public:
    column_error(int number_of_columns) {
        c = number_of_columns;
    }
    int c;
};

void set_array_dimensions(void)
{
    const int max_rows = 10;
    const int max_columns = 10;
    int rows, columns;
    cout << "Enter rows: ";
    cin >> rows;
    if (rows > max_rows)
        throw row_error(rows);
    cout << "Enter columns: ";
    cin >> columns;
```

```
        if (columns > max_columns)
            throw column_error(columns);
        cout << "Array dimensions entered O.K.\n";
        // Do something more here.
    }

    int main()
    {
        try {
            set_array_dimensions();
        }
        catch(row_error x) {
            cout << "You have " << x.r << " rows.\n" <<
                "This is too many.\n";
            exit(EXIT_FAILURE);
        }
        catch(column_error x) {
            cout << "You have " << x.c << " columns." <<
                "\nThis is too many.\n";
            exit(EXIT_FAILURE);
        }
        cout << "No errors so far.\n";
        // Do something more here.
        return(EXIT_SUCCESS);
    }
```

Here we have two separate error classes, `row_error` and `column_error`. When the `set_array_dimension()` function detects an error, it throws an instance of one of these error classes from the try-block. The object is then caught by one the catch-blocks. There can be any number of catch-blocks following the try, and in this case there are two. The succession of catch-blocks is superficially like a switch statement; the object that is thrown drops through the catch-blocks until it is caught by a catch with an argument of the appropriate type. If there isn't an appropriate catch, then the object is not caught.

Exercise

Write a function to find the real roots of a quadratic equation. Implement error classes with constructor functions taking arguments so that if complex roots or division by zero occur then appropriate exceptions are thrown. The exception handing in the function `main()` should send the data contained in the different error objects that can be caught to the output stream.

15.5   Catching Everything

A catch of the form `catch(...)` traps any throw that hasn't already been caught, but it does not provide access to any error data. This form of the `catch` is demonstrated by the following program:

```cpp
#include <iostream>
#include <cstdlib>        // For exit()
using std::cout;
using std::cin;
using std::exit;

class row_error {
public:
    row_error(int number_of_rows) {
        r = number_of_rows;
    }
    int r;
};

class column_error {
public:
    column_error(int number_of_columns) {
        c = number_of_columns;
    }
    int c;
};

class negative_error {
public:
    negative_error() { }
};

void set_array_dimensions(void)
{
    const int max_rows = 10;
    const int max_columns = 10;
    int rows, columns;
    cout << "Enter rows: ";
    cin >> rows;
    if (rows > max_rows)
        throw row_error(rows);
    if (rows < 0)
        throw negative_error();
    cout << "Enter columns: ";
    cin >> columns;
    if (columns > max_columns)
        throw column_error(columns);
    if (columns < 0)
        throw negative_error();
    cout << "Array dimensions entered O.K.\n";
    // Do something more here.
}
```

```
int main()
{
    try {
        set_array_dimensions();
    }
    catch(row_error x) {
        cout << "You have " << x.r <<
            " rows.\nThis is too many.\n";
        exit(EXIT_FAILURE);
    }
    catch(column_error x) {
        cout << "You have " << x.c <<
            " columns.\nThis is too many.\n";
        exit(EXIT_FAILURE);
    }
    catch(...) {
        cout << "Some error has occurred.\n";
        exit(EXIT_FAILURE);
    }
    cout << "No errors so far.\n";
    // Do something more here.
    return(EXIT_SUCCESS);
}
```

In this example, if the number of rows specified is greater than the maximum allowed, then a `row_error` is thrown and caught. Similarly, if the number of columns specified is greater than the maximum allowed, then a `column_error` is thrown and caught. However, if a negative number of rows or columns is specified, then a `negative_error` is thrown, for which there is no specific catch. Consequently, a `negative_error` is caught by the `catch(...)`. Notice that the order of the catch-blocks is important since the catches are attempted in the order in which they occur. This means that a `catch(...)` should always be the final catch-block, since any following catch-blocks could never be executed.

Exercise

Modify your program for the quadratic equation exercise given on page 430 so that all exceptions are caught by a `catch(...)`.

15.6   Derived Error Classes

It is quite common for there to be a hierarchy of exceptions so that it may be worth using inheritance to construct a hierarchy of exception classes. This is illustrated by the following program:

```
#include <iostream>
#include <cstdlib>        // For exit()
```

```cpp
using std::cout;
using std::cin;
using std::exit;

class dimension_error {
public:
    dimension_error() { }
    virtual void print_error(void) { }
};

class too_small : public dimension_error {
public:
    too_small() { }
    void print_error(void)
    {
        cout << "Number of elements is negative.\n";
    }
};

class too_big : public dimension_error {
public:
    too_big() { }
    void print_error(void)
    {
        cout << "There are too many elements.\n";
    }
};

void set_array_dimension(void)
{
    const int max_elements = 10;
    int elements;
    cout << "Enter number of elements: ";
    cin >> elements;
    if (elements > max_elements)
        throw too_big();
    else if (elements < 0)
        throw too_small();
    cout << "Array dimension O.K.\n";
    // Do something more here.
}

int main()
{
    try {
        set_array_dimension();
    }
```

```
    catch(dimension_error &x) {
        x.print_error();
        exit(EXIT_FAILURE);
    }
    cout << "No fatal errors so far.\n";
    // Do something more here.
    return(EXIT_SUCCESS);
}
```

In this example, `dimension_error` is a virtual base class, from which the `too_small` and `too_big` classes are derived. Notice how a reference to a `dimension_error` object is caught and that this allows the correct error message to be given in the two cases, even though there is only a single `catch`. The exception handling in this program therefore provides another demonstration of polymorphism.

Exercise

Modify the program you wrote to find the real roots of a quadratic equation (in answer to the exercise on page 430) so that it uses a `virtual` base class, together with derived classes, to implement exception handling. There should only be a single `catch` block and polymorphism should be used to distinguish the different types of error.

15.7   Exception Specifications

It is possible to have what is known as an *exception specification* as part of a function declaration, in which case the same specification must be made in the function definition.[2] The advantage of this is that only the throws specified in the function declaration can be caught. The idea of an exception specification can be illustrated by modifying our previous program so that an array with zero elements is an error that throws an exception.

```
#include <iostream>
#include <cstdlib>        // For exit()
using std::cout;
using std::cin;
using std::exit;

class dimension_error {
public:
    dimension_error() { }
    virtual void print_error(void) { }
};

class too_small : public dimension_error {
public:
```

---

[2]There are subtleties for exception specifications involving overloaded functions. See [10] and [1] for more details.

```
        too_small() { }
        void print_error(void)
        {
            cout << "Number of elements is negative.\n";
        }
};

class too_big : public dimension_error {
public:
    too_big() { }
    void print_error(void)
    {
        cout << "There are too many elements.\n";
    }
};

// This is how an exception specification is declared:
void set_array_dimension(void) throw(too_big, too_small);

// This is a definition of a function with an exception
// specification:
void set_array_dimension(void) throw(too_big, too_small)
{
    const int max_elements = 10;
    int elements;
    cout << "Enter number of elements: ";
    cin >> elements;
    if (elements > max_elements)
        throw too_big();
    else if (elements < 0)
        throw too_small();
    else if (elements == 0)
        throw "There are zero elements.\n";
    cout << "Array dimension O.K.\n";
    // Do something more here.
}

int main()
{
    try {
        set_array_dimension();
    }
    catch(dimension_error &x) {
        x.print_error();
        exit(EXIT_FAILURE);
    }
    // The following catch-block is never entered.
```

```
    catch(const char* pt) {
        cout << pt;
        exit(EXIT_FAILURE);
    }
    cout << "No fatal errors so far.\n";
    // Do something more here.
    return(EXIT_SUCCESS);
}
```

In this case, if we specify the array dimension to be zero then a string is thrown, which we attempt to catch with the catch(const char* pt) error handler. Although this technique worked for the example on page 426, it fails here because this throw is not part of the function declaration. Consequently, for zero elements the attempt to throw and catch an exception results in a call of std::unexpected().

15.8　Uncaught Exceptions

If an exception is not caught, then the std::unexpected() library function is called, which in turn calls abort(). For a small program that is only going to be used by the programmer this may well be an appropriate course of action. However, in general it is better to avoid calls to unexpected(). If this can't be done, then the programmer can still catch the calls by providing an unexpected() function. This is illustrated by the exercise given below.

Exercise

(a) Compile and run the program in the previous section. Try various array sizes and, in particular, note what error message you get for an array size of zero.

(b) Insert the following function definition in the previous program:

```
    void unexpected(void)
    {
        cout << "An unexpected error has occurred.\n";
        // Perhaps do some clearing up here.
        exit(EXIT_FAILURE);
    }
```

By observing the error message generated for the case of zero elements, demonstrate that it is the above function that is called and not std::unexpected().

15.9　Dynamic Memory Allocation

In Section 7.6.1 we introduced the idea of dynamically allocating memory by means of the **new** operator. However, the resources of any system are finite and so there is always the possibility of a memory allocation failing. It is now part of the ANSI

C++ Standard that if the **new** operator fails to dynamically allocate memory, then a
bad_alloc exception is thrown. If this isn't caught by an exception handler in the code,
then it is caught by the default **unexpected()** function and the program terminates.
Consequently, if we want to provide a different solution to the resource failure, such
as deallocating some memory, then we must catch the **bad_alloc** exception. The
following program illustrates how this is achieved. Notice that the **<new>** include file
is needed.

```
#include <iostream>
#include <cstdlib>       // For EXIT_SUCCESS
#include <new>
using std::cout;
using std::bad_alloc;

int main()
{
    const int max_arrays = 100000;
    const int elements = 10000;
    double *pt[max_arrays];
    int i;
    try {
        for (i = 0; i < max_arrays; ++i)
            pt[i] = new double[elements];
    }
    catch(bad_alloc) {
        cout << "Memory exhausted.\n" << i <<
            " arrays allocated successfully.\n";
        delete pt[i - 1];
        cout << "We have given up one array.\n";
    }
    cout << "Now we can carry on and do some more work.\n";
    return(EXIT_SUCCESS);
}
```

In this program, we attempt to allocate memory for a large number of large arrays.
The amount of memory available will vary from system to system, so you may have to
adjust the number or size of the arrays. In particular, the available memory is likely
to be very large for systems with virtual memory.[3] Assuming the dynamic memory
allocation fails, a **bad_alloc** exception is thrown and this is caught by our own error
handler. The error handler deletes one of the arrays and we are then able to carry on
and do some useful work. This scenario should be contrasted with what happens if
we don't provide our own exception handler. The program would terminate without
giving us the opportunity to continue.

Exercise

Suppose a programmer attempts to dynamically allocate memory for an
array of type **double** by means of the statement:

---

[3]You should exercise some care in running this program since it could result in a lot of disc activity
on a virtual memory system.

```
    pt = new double[elements];
```

but by mistake has assigned -10 to `elements`. Provide exception handling
to catch the `bad_alloc` exception that is thrown and send the value of the
pointer `pt` to the output stream. What is significant about this value? (See
the discussion on page 173.)

15.10   Using Exception Handling

In Section 12.7.1 we implemented a bounds checked array class. Any attempt to access
a member of this class for an index outside of its valid range gave an error message
and terminated the program. In some circumstances this action would be too drastic
and in any case is certainly too inflexible. Although it is the `checked_array` class that
detects an out-of-bounds error, the choice of how the error is handled should be up
to the user of this code. This can be achieved by using the C++ exception handling
mechanism.

As in Section 12.7.1, we have the header file `array.h`, and source files `array.cxx`
and `my_test.cxx`, together with a `makefile`. We can introduce exception handling by
changing the `array.h` file to the following:

```
// source:   array.h
// use:      Defines array class.

#ifndef ARRAY_H
#define ARRAY_H

#include <iostream>
#include <cstdlib>        // For EXIT_SUCCESS
#include <cstring>        // For memcpy()
using std::cout;
using std::memcpy;

class array {
public:
    array(int size);
    array(const array &x);
    virtual ~array() { delete pt; }
    array &operator=(const array &x);
    virtual double &operator[](int index);
    int get_size(void);
protected:
    int n;
    double *pt;
};

class checked_array : public array {
public:
```

```
        checked_array(int size) : array(size) { }
        double &operator[](int index);
private:
        void check_bounds(int index);
};

class index_error {
public:
        index_error(int index) { i = index; }
        virtual void print_error(void) { }
protected:
        int i;
};

class index_too_high : public index_error {
public:
        index_too_high(int index) : index_error(index) { }
        virtual ~index_too_high() { }
        void print_error(void)
        {
            cout << "The index " << i << " is too high.\n";
        }
};

class index_too_low : public index_error {
public:
        index_too_low(int index) : index_error(index) { }
        virtual ~index_too_low() { }
        void print_error(void)
        {
            cout << "The index " << i << " is less than 1.\n";
        }
};

// inline array class implementations:

inline int array::get_size(void)
{
    return n;
}

inline double &array::operator[](int index)
{
    return pt[index - 1];
}

// inline checked_array class implementations:
```

```
inline double &checked_array::operator[](int index)
{
    check_bounds(index);
    return array::operator[](index);
}

inline void checked_array::check_bounds(int index)
{
    if (index < 1)
        throw index_too_low(index);
    if (index > n)
        throw index_too_high(index);
}

#endif   // ARRAY_H
```

The index_too_high and index_too_low classes both have index_array as a virtual
base class. If the index is out of range in the check_bounds() function, then an error
is thrown. The use of inheritance in the error classes, together with a virtual base
class, allows us to leave the correct choice of error message to polymorphism, as in
Section 15.6.

The modified array.cxx file is shown below.

```
// source:  array.cxx
// use:     Implements array class.

#include <iostream>
#include <cstdlib>        // For exit()
#include <new>            // For bad_alloc
#include "array.h"
using std::cout;
using std::exit;
using std::bad_alloc;
using std::memcpy;

array::array(int size)
{
    n = size;
    try {
        pt = new double[n];
    }
    catch(bad_alloc) {
        cout << "Failed to allocate array.\n";
        exit(EXIT_FAILURE);
    }
}
```

```
array::array(const array &x)
{
    n = x.n;
    try {
        pt = new double[n];
    }
    catch(bad_alloc) {
        cout << "Failed to allocate array.\n";
        exit(EXIT_FAILURE);
    }
    memcpy(pt, x.pt, n * sizeof(double));
}


array &array::operator=(const array &x)
{
    delete pt;
    n = x.n;
    try {
        pt = new double[n];
    }
    catch(bad_alloc) {
        cout << "Failed to allocate array.\n";
        exit(EXIT_FAILURE);
    }
    memcpy(pt, x.pt, n * sizeof(double));
    return *this;
}
```

The only change to this file is that we have used the C++ exception handling mechanism to catch the exhaustion of dynamically allocated memory.

Finally, the my_test.cxx file is modified to:

```
// source:  my_test.cxx
// use:     Tests array class.

#include <iostream>
#include <cstdlib>        // For EXIT_SUCCESS
#include "array.h"
using std::cout;

double sum(array &a, int first_index, int last_index)
{
    double result = a[first_index];
    for (int i = first_index + 1; i <= last_index; ++i)
        result += a[i];
    return result;
}
```

```cpp
int main()
{
    const int n = 10;

    // Define an object:
    checked_array x(n);

    // Access the array size:
    cout << "The array x has " << x.get_size() << " elements.\n";

    // Store some data. Also find the total directly so it can be
    // used for comparison:
    double total = 0.0;
    for (int i = 1; i <= n; ++i) {
        x[i] = i * 25.0;
        total += i * 25.0;
    }

    // Retrieve some data:
    cout << "The data stored in x are:\n";
    for (int i = 1; i <= n; ++i)
        cout << x[i] << "\n";
    cout << "\n";

    // Define another object using copy constructor:
    checked_array y = x;

    // Check the array size:
    cout << "The array y has " << y.get_size() << " elements.\n" <<
        "y was created by the copy constructor.\n\n";

    // Check that the copy is identical:
    int errors = 0;
    for (int i = 1; i <= n; ++i)
        if (x[i] != y[i])
            cout << "x[" << i << "] != y[" << i << "]\n";
    if (errors)
        cout << errors << " elements of the arrays x and y " <<
            "differ in their data.\n";
    else
        cout << "Arrays x and y have identical data.\n\n";

    // Define an object with half the size but no bounds checking:
    array z(n/2);

    // Check the array size:
    cout << "The array z has " << z.get_size() << " elements.\n\n";
```

```
    // Try out the assignment operator:
    z = x;

    // Check the array size:
    cout << "After assignment of x to z, the array z has " <<
        z.get_size() << " elements.\n\n";

    errors = 0;
    // Check that the copy is identical:
    for (int i = 1; i <= n; ++i)
        if (z[i] != x[i])
            cout << "z[" << i << "] != x[" << i << "]\n\n";
    if (errors)
        cout << errors << " elements of the arrays x and z " <<
            "differ in their data.\n";
    else
        cout << "Arrays x and z have identical data.\n\n";

    // Find sum for z[i], going out of bounds. If you get the
    // correct result, try changing the third argument in sum()
    // to some other value greater than n:
    cout << "The sum of the data in z is " << sum(z, 1, 2 * n) <<
        ".\nThe sum should be " << total << ".\n\n";

    // Find sum for x[i], going out of bounds (too high):
    try {
        cout << "The sum of the data in x is " <<
            sum(x, 1, 2 * n) << ".\nThe sum should be " <<
            total << ".\n\n";
    }
    catch(index_error &error_object) {
        error_object.print_error();
    }

    // Find sum for x[i], going out of bounds (too low):
    try {
        cout << "The sum of the data in x is " <<
            sum(x, -10, n) << ".\nThe sum should be " <<
            total << ".\n\n";
    }
    catch(index_error &error_object) {
        error_object.print_error();
    }

    return(EXIT_SUCCESS);
}
```

The changes here are the introduction of `try` and `catch` blocks, together with an example of attempting to access an array with a negative index.

The makefile remains unchanged as:

```
my_test: my_test.cxx array.o
    g++ my_test.cxx array.o -o my_test
array.o: array.cxx array.h
    g++ -c array.cxx
```

Exercise

(a) Compile and run `my_test`.

(b) Change the initialization of `n` in `main()` from 10 to -10. Explain what happens when you run this modified `my_test`.

15.11   Summary

- If `dimension_error()` is the constructor for a `dimension_error` class, then an exception is thrown by:

```
if (elements > max_elements)
    throw dimension_error();
```

The exception is handled by:

```
try {
    set_array_dimension();
}
catch(dimension_error) {
    cout << "The array dimension is too big.\n";
    exit(EXIT_FAILURE);
}
```

where `set_array_dimension()` is the function from which the error is thrown. Any `catch` blocks must immediately follow the `try` block.

- It is possible to throw a fundamental type:

```
void set_array_dimension(void)
{
    // Some code goes in here.
    if (elements > max_elements)
        throw "You have too many elements.\n";
    // Do something more here.
}
```

This could be handled by:

```
try {
    set_array_dimension();
}
catch(const char* pt) {
    cout << pt;
    exit(EXIT_FAILURE);
}
```

- If the object thrown has a constructor that takes an argument, as in:

```
class dimension_error {
public:
    dimension_error(int number_of_elements) {
        elements = number_of_elements;
    }
    int elements;
};
```

then the object thrown can contain data, perhaps giving the cause of the error:

```
void set_array_dimension(void)
{
    // Code goes in here.
    if (elements > max_elements)
        throw dimension_error(elements);
    // Do something more here.
}
```

The catch can extract information on the cause of the exception:

```
try {
    set_array_dimension();
}
catch(dimension_error x) {
    cout << "You have " << x.elements << "elements.\n";
    exit(EXIT_FAILURE);
}
```

- There may be a number of different objects that can be thrown:

```
void set_array_dimensions(void)
{
    // Code goes in here.
    if (rows > max_rows)
        throw row_error(rows);
    if (columns > max_columns)
        throw column_error(columns);
    // Do something more here.
}
```

and these can be caught by a number of `catch` blocks following a single `try` block:

```
try {
    set_array_dimensions();
}
catch(row_error x) {
    cout << "You have " << x.r << " rows.\n";
    exit(EXIT_FAILURE);
}
catch(column_error x) {
    cout << "You have " << x.c << " columns.\n";
    exit(EXIT_FAILURE);
}
```

- A catch of the form `catch(...)` traps any throw that hasn't already been caught, but it does not provide access to any error data. A `catch(...)` should always be the final catch-block:

```
try {
    set_array_dimensions();
}
catch(row_error x) {
    cout << "You have " << x.r << " rows.\n";
    exit(EXIT_FAILURE);
}
catch(column_error x) {
    cout << "You have " << x.c << " columns.\n";
    exit(EXIT_FAILURE);
}
catch(...) {
    cout << "Some error has occurred.\n";
    exit(EXIT_FAILURE);
}
```

- If there is a hierarchy of exceptions then it may be worth using inheritance to construct a hierarchy of exception classes:

```
class dimension_error {
public:
    dimension_error() { }
    virtual void print_error(void) { }
};

class too_small : public dimension_error {
public:
    too_small() { }
    void print_error(void)
    {
```

```
                    cout << "Number of elements is negative.\n";
            }
    };

    class too_big : public dimension_error {
    public:
        too_big() { }
        void print_error(void)
        {
            cout << "There are too many elements.\n";
        }
    };
```

Different derived class objects can be thrown depending on the error:

```
    void set_array_dimension(void)
    {
        const int max_elements = 10;
        int elements;
        cout << "Enter number of elements: ";
        cin >> elements;
        if (elements > max_elements)
            throw too_big();
        else if (elements < 0)
            throw too_small();
        cout << "Array dimension O.K.\n";
        // Do something more here.
    }
```

If a **virtual** base class is used then only a single **catch** statement is needed and polymorphism results in the appropriate object being caught.

- A function declaration can have an exception specification, in which case the same specification must be made in the function definition:

```
    void set_dimension(void) throw(too_big, too_small);

    void set_dimension(void) throw(too_big, too_small)
    {
        // Code goes in here.
        if (elements > max_elements)
            throw too_big();
        else if (elements < 0)
            throw too_small();
        // Do something more here.
    }
```

Only throws specified in the function declaration can be caught.

- If an exception is not caught, then the `std::unexpected()` library function is called, which in turn calls `abort()`. Such calls can also be caught if the programmer provides an `unexpected()` function.

- If the `new` operator does not succeed in dynamically allocating memory, then a `bad_alloc` exception is thrown. The default is that this exception is caught by the `std::unexpected()` function and the program terminates. Alternatively, the exception can be caught by an exception handler provided by the programmer:

```
try {
    for (i = 0; i < max_arrays; ++i)
        pt[i] = new double[elements];
}
catch(std::bad_alloc) {
    cout << "Memory exhausted.\n";
    delete pt[i - 1];
    cout << "We have given up one array.\n";
}
```

The `<new>` include file is needed with this approach.

15.12   Exercises

1. Modify and test the `menu` class, described in Section 12.7.2, so that it uses the techniques introduced in this chapter to deal with errors. You should try modifying both the original class and also the result of Exercise 2a on page 419.

2. Modify the bit array class, described in Section 11.4.1, so that the errors are dealt with by using the C++ exception handling techniques. You should define an error class with constructors that take arguments so that error data can be accessed in the function `main()` of the `my_test.cxx` file.

Chapter 16

Templates

Templates (also known as *parameterized types*) are an important part of the C++ language. They provide a tool for constructing generic classes and functions that are valid for different data types. Templates are intrinsic to the *Standard Template Library* (also known as the STL), which contains the templates for many useful classes. (See Chapter 17.) Templates can be provided for both functions and classes.

16.1 Function Templates

In order to introduce the idea of templates, we start with a very simple example. Consider the following function that returns the maximum of two integers.

```
int max(int a, int b)
{
    return(a > b ? a : b);
}
```

A straightforward substitution of `double` for `int` would produce a function suitable for arguments of type `double`. Similar substitutions could be performed for any other fundamental or derived type for which a greater than operator is defined. It would be tedious to implement a large number of such closely related functions; what is needed is a function for a generic type and this can be achieved by using a template.

The general syntax for a *template definition* is:

```
template<template_argument_list> definition;
```

where `definition` may refer to either a function or a class. It is important to realize that throughout this chapter the angle bracket pair, $<\ \ >$, is an essential part of the template syntax. As an example, a *function template* returning the maximum of two values is:

```
template<class T> T max(T a, T b)
{
    return(a > b ? a : b);
}
```

In this definition, the parameterized type T represents any fundamental or user-defined type for which the greater than operator is defined. The scope of T is limited to the template definition. Any valid identifier could be used to represent the parameterized type; we are not restricted to using T. It is important to realize that although the notation is "class T", T is not restricted to being a class. A straightforward illustration of using the max() function template is provided by the types int and double in the following program:[1]

```
#include <iostream>
#include <cstdlib>        // For EXIT_SUCCESS
using std::cout;

template<class T> T max(T a, T b)
{
    return(a > b ? a : b);
}

int main()
{
    int i = 2, j = 4;
    double x = 6.7, y = 3.4;
    cout << "The maximum of " << i << " and " << j << " is " <<
        max(i, j) << "\n";
    cout << "The maximum of " << x << " and " << y << " is " <<
        max(x, y) << "\n";
    return(EXIT_SUCCESS);
}
```

The template doesn't actually define a function, rather it provides a recipe for constructing functions. For instance, invoking the max(i, j) function causes the function appropriate for arguments of type int to be constructed. This process is known as *template instantiation* (or simply, *instantiation*). The individual function is known as an *instantiation* of the function template.

As with ordinary functions, we may want to declare a function template and define it somewhere else, in which case we simply omit the function body. The general syntax for a *template declaration* is:

```
template<template_argument_list> declaration;
```

where declaration may refer to either a function or a class.

As an example, the declaration for the max() function defined above, would be:

```
template<class T> T max(T a, T b);
```

There is no need to use the same identifier to represent the parameterized type in the template definition and declaration. However, it would be a bit perverse not to do so and would make the code more difficult to understand.

There must be an exact match between argument types for the functions invoked and the functions that can be generated from the template, so that the program:

---

[1]In order to keep things simple we have placed the function template in the same file as main(). Generally we wouldn't do this for a more realistic program.

```
#include <iostream>
#include <cstdlib>        // For EXIT_SUCCESS
using std::cout;

template<class T> T max(T a, T b)
{
    return(a > b ? a : b);
}

int main()
{
    int i = 2;
    double x = 6.7;
    cout << "The maximum of " << x << " and " << i << " is " <<
        max(i, x) << "\n";
    return(EXIT_SUCCESS);
}
```

fails to compile. This is very reasonable since there is no way for the compiler to know whether T is supposed to be of type int or double. Consequently, the return type for the max() function is ambiguous.

One way of overcoming this problem is to use *specialization*, that is we specify which class we want to replace the parameterized type. The way in which this is done is similar to the syntax for a function template declaration. For example, if we want to invoke the max() function with T replaced by the type double, then we need a statement of the form:

```
max<double>(x, j);
```

Hence a correct version of the previous program is as follows:

```
#include <iostream>
#include <cstdlib>        // For EXIT_SUCCESS
using std::cout;

template <class T> T max(T a, T b)
{
    return(a > b ? a : b);
}

int main()
{
    int i = 2;
    double x = 6.7;
    cout << "The maximum of " << x << " and " << i << " is " <<
        max<double>(i, x) << "\n";
    return(EXIT_SUCCESS);
}
```

Although the second argument of the template specialization is of type `int`, the compiler performs an implicit cast to `double`.

The `max()` function is very simple. However, a more complicated example of the use of function templates is provided by the bubble sort that we implemented in Section 7.8.2. The code can easily be modified so that it applies to any data type for which the greater than operator is meaningful. This has been done for the following program:

```cpp
#include <iostream>
#include <cstdlib>        // For EXIT_SUCCESS
using std::cout;
using std::cin;

template<class T> void bubble_sort(T *data, int elements)
{
    int n = elements - 1;
    for (int i = 0; i < n; ++i)
        for (int j = n; j > i; --j)
            if (data[j - 1] > data[j]) {
                T temp = data[j - 1];
                data[j - 1] = data[j];
                data[j] = temp;
            }
}

int main()
{
    int elements;

    // Perform bubble sort on integers:
    cout << "How many integers do you want to enter: ";
    cin >> elements;
    int *integer_data = new int[elements];
    cout << "Enter " << elements << " integers.\n";
    for (int i = 0; i < elements; ++i)
        cin >> integer_data[i];
    bubble_sort(integer_data, elements);
    cout << "The ordered list of your input is:\n";
    for (int i = 0; i < elements; ++i)
        cout << integer_data[i] << " ";
    cout << "\n";

    // Perform bubble sort on floats:
    cout << "How many floating point numbers do you want to " <<
        "enter: ";
    cin >> elements;
    double *double_data = new double[elements];
    cout << "Enter " << elements << " floating point numbers.\n";
```

```
    for (int i = 0; i < elements; ++i)
        cin >> double_data[i];
    bubble_sort(double_data, elements);
    cout << "The ordered list of your input is:\n";
    for (int i = 0; i < elements; ++i)
        cout << double_data[i] << " ";
    cout << "\n";

    return(EXIT_SUCCESS);
}
```

In this program there is one function template for a bubble sort, but two different data types. In each case, the compiler constructs the appropriate function according to the data type of the function arguments.

Exercise

(a) Try the above program with your own data.

(b) Modify the alphabetic sort program, given in Section 7.8.2, so that it uses the function template version of a bubble sort.

16.2 Class Templates

Now consider the following declaration for a one-dimensional array class:

```
class array {
public:
    array(int size);
    array(const array &a);
    virtual ~array() { delete pt; }
    double &operator[](int index);
    array &operator=(const array &a);
private:
    int n;
    double *pt;
};
```

This class is suitable for storing elements of type `double`. However, simple changes would make the class equally suitable for storing many other types, such as `int`, `complex`, `list`, `matrix`, etc. Rather than declaring (and implementing) a separate class for each of these cases, a *class template* can be used to specify how an `array` class is to be constructed from a template that has elements of a parameterized type. The syntax is similar to the function template, so that the `array` class template becomes:

```
template<class T> class array {
public:
    array(int size);
    array(const array &x);
```

```
        virtual ~array() { delete pt; }
        array &operator=(const array &x);
        T &operator[](int index);
    private:
        int n;
        T *pt;
};
```

As in the case of the function template, any valid identifier could be used for the parameterized type, T. Furthermore, T does not have to be the name of a user-defined class; it can also be the name of a fundamental type. Notice that a class template does not actually declare a class; a compiler can only generate a class declaration if the template is used with the parameter, T, replaced by a known type, as in:

```
array<double> x(20);
array<complex> z(10);
```

In this case the compiler would generate two classes with the names array<double> and array<complex>. This process is again known as *template instantiation* (or simply, *instantiation*). The individual class (such as array<double>) is known as an *instantiation* of the class template. Examples of using this array class template are given in the following program:

```
#include <iostream>
#include <cstdlib>        // For EXIT_SUCCESS
#include <cstring>        // For memcpy()
using std::cout;
using std::memcpy;

template<class T> class array {
public:
    array(int size);
    array(const array &x);
    virtual ~array() { delete pt; }
    array &operator=(const array &x);
    T &operator[](int index);
private:
    int n;
    T *pt;
};

template<class T> array<T>::array(int size)
{
    n = size;
    pt = new T[n];
}

template<class T> array<T>::array(const array &x)
{
```

```
        n = x.n;
        pt = new T[n];
        memcpy(pt, x.pt, n * sizeof(T));
}

template<class T> array<T> &array<T>::operator=(const array &x)
{
        delete pt;
        n = x.n;
        pt = new T[n];
        memcpy(pt, x.pt, n * sizeof(T));
        return *this;
}

template<class T> T &array<T>::operator[](int index)
{
        return pt[index - 1];
}

int main()
{
        const int n = 10;
        array<double> x(n);
        array<double> y(n);
        for (int i = 1; i <= n; ++i)
            x[i] = 2 * i + 1;
        y = x;
        for (int i = 1; i <= n; ++i)
            cout << "y[" << i << "] = " << y[i] << "\n";
        return(EXIT_SUCCESS);
}
```

Notice that within the template declarations and definitions, the class can be referred to without qualification (in this case as array rather than array<T>). However, if a member function is defined outside of the template declaration, then we need to prefix the definition by template<class T> so that T is known to be a template. Also, we need to use the parameterized type in the function name. (See, for instance, the definition of the copy constructor.) In the function main(), the objects x and y are instantiations of the array<T> template. Once these objects have been defined, they are referred to simply by their names, x and y.

Exercise

Use the array<T> template given in this section, together with the complex class (described in Section 9.6.1 and extended by Exercise 3 on page 286), to implement a complex array class.

16.3   Static Members

A class template can have static data members, in which case each instantiation has
its own static data. For example, we might want to know how many arrays have
been created for our array<T> class template. This could be achieved by means of
a static int total data member of the template. Since each instantiation has its
own data, we can keep separate tallies of the number of integer and double arrays
that are created. The member function that returns the total number of array<T>
objects (for each parameterized type T) should also be declared static so that the
function doesn't need to be invoked for a particular object. This is demonstrated in
the following program:

```
#include <iostream>
#include <cstdlib>      // For EXIT_SUCCESS
#include <cstring>      // For memcpy()
using std::cout;
using std::memcpy;

template<class T> class array {
public:
    array(int size);
    array(const array &x);
    virtual ~array() { delete pt; }
    array &operator=(const array &x);
    T &operator[](int index);
    static int number_of_arrays(void) { return total; }
private:
    int n;
    T *pt;
    static int total;
};

template<class T> int array<T>::total;

template<class T> array<T>::array(int size)
{
    n = size;
    pt = new T[n];
    ++total;
}

template<class T> array<T>::array(const array &x)
{
    n = x.n;
    pt = new T[n];
    memcpy(pt, x.pt, n * sizeof(T));
    ++total;
}
```

```
template<class T> array<T> &array<T>::operator=(const array &x)
{
    delete pt;
    n = x.n;
    pt = new T[n];
    memcpy(pt, x.pt, n * sizeof(T));
    return *this;
}

template<class T> T &array<T>::operator[](int index)
{
    return pt[index - 1];
}

int main()
{
    const int n = 10;
    cout << "There are " << array<double>::number_of_arrays() <<
        " arrays of type double.\n";
    cout << "There are " << array<int>::number_of_arrays() <<
        " arrays of type int.\n";
    array<double> x(n);
    array<double> y(n);
    cout << "There are " << array<double>::number_of_arrays() <<
        " arrays of type double.\n";
    array<int> a(n);
    cout << "There are " << array<int>::number_of_arrays() <<
        " arrays of type int.\n";
    return(EXIT_SUCCESS);
}
```

Notice that although `total` is declared in the template, it must be separately defined by means of the statement:

```
template<class T> int array<T>::total;
```

This is consistent with the situation for `static` members of non-template classes.

Exercise

The destructor for the above program doesn't decrement `total` when an `array` object is destroyed. Remedy this defect and test your modified class template.

16.4   Class Templates and Functions

It isn't possible to pass a template as a function argument. If, for instance, we try to declare a function `f()` that takes our `array<T>` template as an argument, as in:

```
void f(array<T> x);        // WRONG.
```

then we get a compiler error since there cannot be an **array<T>** object. The declaration:

```
void f(array x);           // WRONG.
```

also fails to compile since in this context **array** is not a class. However, it is possible to pass an object that is an instantiation of a class template as a function argument. The declaration:

```
void f(array<int> x);      // O.K.
```

is valid since in this case **x** is an instance of a genuine class, **array<int>**.

   If we want to declare a function with an argument that is a class template, then we need to use a function template, as in:

```
template<class T> void f(array<T> x);
```

This idea is illustrated in the program given below.

```
#include <iostream>
#include <cstdlib>         // For EXIT_SUCCESS
using std::cout;

template<class T> class X {
public:
    T t;
};

template<class T> T sum(X<T> x1, X<T> x2)
{
    return (x1.t + x2.t);
}

int main()
{
    X<int> a, b;
    a.t = 12;
    b.t = 15;
    cout << a.t << " + " << b.t << " = " << sum(a, b) << "\n";
    return(EXIT_SUCCESS);
}
```

The class **X** has a single public data member. An instantiation of the function template **sum()** returns an object that is the sum of the data members of its two arguments. Because we want **sum()** to apply to any instantiation of the **X** class template, we are forced to make **sum()** a function template.

Exercise

Modify the above program so that it also defines two **X<double>** objects. Assign constants of type **double** to these objects and verify that the correct sum is returned.

16.5   Function Template Arguments

A function template may take any number of arguments, and the arguments may be a mixture of class templates, instantiations of templates, and fundamental types. This is illustrated by the following program:

```
#include <iostream>
#include <cstdlib>        // For EXIT_SUCCESS
using std::cout;

template<class T> class X {
public:
    T t;
};

template<class T> class Y {
public:
    T t;
};

template<class T> T f(X<T> x1, Y<int> i1, int i2)
{
    return (x1.t / i1.t + i2);
}

int main()
{
    X<double> a;
    Y<int> b;
    a.t = 2.468;
    b.t = 2;
    int i = 3;
    cout << a.t << " / " << b.t << " + " << i << " = " <<
        f(a, b, i) << "\n";
    return(EXIT_SUCCESS);
}
```

Notice how the function template, f(), takes three arguments of type: class template, class template specialization, and int.

16.6   Template Parameters

The parameters in a class template can be of any type and there can be more than one parameter. These parameters could be a mixture of user-defined and fundamental types. The following program demonstrates these features:

```
#include <iostream>
#include <cstdlib>        // For EXIT_SUCCESS
```

```
using std::cout;

template<class T1, class T2, int i> class array_pair {
public:
    T1 x[i];
    T2 y[i];
};

int main()
{
    const int dimension = 10;
    array_pair<double, int, dimension> data;
    for (int j = 0; j < dimension; ++j) {
        data.x[j] = 1.1 * j;
        data.y[j] = 2 * j + 1;
    }
    for (int j = 0; j < dimension; ++j)
        cout << "data.x[" << j << "] = " << data.x[j] <<
            "\tdata.y[" << j << "] = " << data.y[j] << "\n";
    return(EXIT_SUCCESS);
}
```

This class template may give the impression that memory for the x and y arrays is dynamically allocated. However, this impression is not correct since although the value of i is not known in the class template, it *is* known when the template is instantiated.

Exercise

What is the consequence of leaving out the const qualifier from the function main() in the above program?

16.7   Templates and Friends

Class templates can declare friends and, in general, there are three possible ways in which a class template can declare classes and functions to be friends. A friend of a class template can be:

- a function or class template;

- a specialization of a function or class template;

- an ordinary (non-template) function or class.

In the following program, the equality operator provides an example of a friend function in a class template:

```
#include <iostream>
#include <cstdlib>      // For EXIT_SUCCESS
#include <cstring>      // For memcpy()
```

```
using std::cout;
using std::memcpy;

template<class T> class array {
public:
    array(int size);
    array(const array &x);
    virtual ~array() { delete pt; }
    array &operator=(const array &x);
    T &operator[](int index);
    friend bool operator==<>(const array &x1, const array &x2);
private:
    int n;
    T *pt;
};

template<class T> array<T>::array(int size)
{
    n = size;
    pt = new T[n];
}

template<class T> array<T>::array(const array &x)
{
    n = x.n;
    pt = new T[n];
    memcpy(pt, x.pt, n * sizeof(T));
}

template<class T> array<T> &array<T>::operator=(const array &x)
{
    delete pt;
    n = x.n;
    pt = new T[n];
    memcpy(pt, x.pt, n * sizeof(T));
    return *this;
}

template<class T> T &array<T>::operator[](int index)
{
    return pt[index - 1];
}

template<class T> bool operator==(const array<T> &x1,
    const array<T> &x2)
{
    for (int i = 0; i < x1.n; ++i)
```

```
        if (x1.pt[i] != x2.pt[i])
            return false;
    return true;
}

int main()
{
    const int n = 10;
    array<double> u(n), v(n);

    for (int i = 1; i <= n; ++i) {
        u[i] = 2.2 * i + 1.1;
        v[i] = 2.2 * i + 1.1;
    }
    if (u == v)
        cout << "u == v\n";
    else
        cout << "u != v\n";
    u[5] = 3.7;
    if (u == v)
        cout << "u == v\n";
    else
        cout << "u != v\n";

    return(EXIT_SUCCESS);
}
```

Here we have added the function:

```
    friend bool operator==<>(const array &x1, const array &x2);
```

to the array<T> class declaration.[2] This function returns true if the arrays x1 and x2 are equal; otherwise it returns false. Both properties are tested in the function main(), where we define two arrays of type double. These arrays are filled with the same numbers and tested for equality. Element u[5] is then changed and the test for equality is repeated.

Exercise

Modify the program given above so that it defines two arrays of type array<complex> and assigns the same complex numbers to both arrays. Test the equality operator in a similar way to the original program.

16.8   Specialized Templates

If a template is defined for a parameterized type T, then a template instantiation simply replaces every occurrence of T by the specified type. However, in some situations there

---

[2]Notice that the <> appearing in the function declaration is required by some compilers.

may be a requirement for one instantiation of a template to behave differently to most other instantiations of the template. For example, we might want an `array<int>` class to initialize all the class data to one.[3] Such an initialization would not be appropriate in general so we need to give alternative classes. We could do this by using a different name for the integer array, but C++ provides a better technique. If the general array is declared by:

```
template<class T> class array {
public:
    array(int size);
    virtual ~array() { delete pt; }
    T &operator[](int index);
private:
    int n;
    T *pt;
};
```

then we can declare a *specialization* by:

```
template<> class array<int> {
public:
    array(int size);
    virtual ~array() { delete pt; }
    int &operator[](int index);
private:
    int n;
    int *pt;
};
```

The `template<>` prefix denotes that what follows is a specialization. A complete program demonstrating template specialization in action is given below.

```
#include <iostream>
#include <cstdlib>        // For EXIT_SUCCESS

using std::cout;

template<class T> class array {
public:
    array(int size);
    virtual ~array() { delete pt; }
    T &operator[](int index);
private:
    int n;
    T *pt;
};
```

---

[3]In practice, it is more likely that we would want to initialize the elements of an integer array to zero, but a non-zero value makes it easier to be certain that initialization has indeed taken place.

```
template<class T> array<T>::array(int size)
{
    n = size;
    pt = new T[n];
}

template<class T> T &array<T>::operator[](int index)
{
    return pt[index - 1];
}

template<> class array<int> {
public:
    array(int size);
    virtual ~array() { delete pt; }
    int &operator[](int index);
private:
    int n;
    int *pt;
};

array<int>::array(int size)
{
    n = size;
    pt = new int[n];
    for (int i = 0; i < n; ++i)
        pt[i] = 1;
}

int &array<int>::operator[](int index)
{
    return pt[index - 1];
}

int main()
{
    const int n = 10;
    array<int> x(n);

    cout << "Integer array initialized to:\n";
    for (int i = 1; i <= n; ++i)
        cout << x[i] << "\n";

    return(EXIT_SUCCESS);
}
```

Notice that the way in which the array<int> function definitions are given is different from the template definitions. For example, instead of:

```
template<class T> array<T>::array(int size)
{
    // Body goes here.
}
```

we have:

```
array<int>::array(int size)
{
    // Body goes here.
}
```

Exercise

Since the body of the specialization:

```
int &array<int>::operator[](int index)
```

is the same as for the template version of the function, it might be thought that this specialization could be omitted. What happens if you do this?

16.9   Member Function Specialization

In the example of specializing the `array` template to `array<int>`, we provided specific integer versions of all the class member functions even though only the constructor functions differed. In this type of situation it is easier to use a *member function specialization*. This is achieved by prefixing `template<>` to the function definition that provides the specialization. The modified program using this technique is given below.

```
#include <iostream>
#include <cstdlib>        // For EXIT_SUCCESS
using std::cout;

template<class T> class array {
public:
    array(int size);
    virtual ~array() { delete pt; }
     T &operator[](int index);
private:
    int n;
    T *pt;
};

template<class T> array<T>::array(int size)
{
    n = size;
    pt = new T[n];
}
```

```
template<class T> T &array<T>::operator[](int index)
{
    return pt[index - 1];
}

// Member function specialization:
template<> array<int>::array(int size)
{
    n = size;
    pt = new int[n];
    for (int i = 0; i < n; ++i)
        pt[i] = 1;
}

int main()
{
    const int n = 10;
    array<int> x(n);
    cout << "Integer array initialized to:\n";
    for (int i = 1; i <= n; ++i)
        cout << x[i] << "\n";
    return(EXIT_SUCCESS);
}
```

Notice that there are two versions of the constructor, but only one version of all the other functions. This demonstrates that in many cases using member function specialization is much simpler than the corresponding class template specialization.

Exercise

Implement a member function that calculates the magnitude of each element of an array<T> class template and returns the element with the maximum magnitude. Provide a specialization for the case when T is the complex class.

16.10   Program Structure

In the examples given in this chapter, the template declarations and definitions have all been in the same file as the function main(). This contrasts with our previous emphasis on using individual header files and source files for each related set of classes, and keeping code that uses the classes separate from these files. However, having all of the code in one file has enabled us to keep this introduction to templates as simple as possible. Since a template only provides a recipe for constructing a class (or function) and is not actually a class (or function), you might wonder whether a template should go in a .h or a .cxx file. The simplest approach at this stage is to put templates in .h files. An alternative approach is given in Section 13.7 of [10], but involves using the export directive, which currently isn't implemented by all C++ compilers.

16.11   Using Templates

An obvious candidate for using templates is the linked list class introduced in Section 10.3.1. Instead of using a typedef to change the type of data stored, we can use class templates. The modified header file is given below.

```
// source:  slist.h
// use:     Singly linked list and node class templates.

template<class T> class list;

template<class T> class node {
    friend class list<T>;
public:
    T data;
private:
    node *next;
};

template<class T> class list {
public:
    list(void);
    virtual ~list();
    void push(T new_data);
    void pop(T &old_data);
    bool is_not_empty(void);
private:
    node<T> *head;
};

template<class T> bool list<T>::is_not_empty(void)
{
    if (head == 0)
        return false;
    else
        return true;
}

template<class T> list<T>::list(void)
{
    head = 0;
}

template<class T> list<T>::~list()
{
    while (head != 0) {
        node<T> *pt = head;
        head = head->next;
```

```
            delete pt;
        }
    }

    template<class T> void list<T>::push(T new_data)
    {
        node<T> *pt = new node<T>;
        pt->next = head;
        head = pt;
        pt->data = new_data;
    }

    template<class T> void list<T>::pop(T &old_data)
    {
        if (head != 0) {
            old_data = head->data;
            node<T> *pt = head;
            head = head->next;
            delete pt;
        }
    }
```

One important change from Section 10.3.1 is that we need to make a forward declaration of the list class with the statement:

```
    template<class T> class list;
```

This is so that the compiler realizes that list in the friend class declaration is in fact a class template.

The modified test program, which can be compiled with:

```
    g++ my_test.cxx -o my_test
```

is as follows:

```
    // source:  my_test.cxx
    // use:     Tests singly linked list class template.

    #include <iostream>
    #include <cstdlib>        // For EXIT_SUCCESS
    #include "slist.h"
    using std::cout;

    int main()
    {
        list<int> s;
        int j;
        for (int i = 1; i <= 5; ++i)
            s.push(10 * i);
        while (s.is_not_empty()) {
```

```
            s.pop(j);
            cout << j << "\n";
        }
        return(EXIT_SUCCESS);
    }
```

Notice that it is not entirely trivial to convert non-template classes to templates, but it is certainly much easier if fully tested implementations of the non-template classes already exist. It is easy to make errors if you try to implement a template from scratch, so the best approach is to first design and test a class without using templates, and then turn it into a class template. Even the experts do this!

16.12 Summary

- The general syntax for declaring a template is:

    ```
    template<template_argument_list> declaration;
    ```

- A function template is declared by:

    ```
    template<class T> T max(T a, T b);
    ```

    where the parameterized type T is any fundamental or user-defined type.

- A function template is defined as in:

    ```
    template<class T> T max(T a, T b)
    {
        return(a > b ? a : b);
    }
    ```

- A function template definition does not actually define a function. The template is instantiated (i.e. an instantiation of the template is generated) when the function is invoked.

- A function specialization is obtained by using the < > syntax:

    ```
    max<double>(x, j);
    ```

- A class template is declared by:

    ```
    template<class T> class array {
    public:
        array(const array &x);
        T &operator[](int index);
        // etc.
    private:
        T *pt;
    };
    ```

- Class template member function definitions made outside of the template declaration need to have the template prefix:

```
template<class T>array<T> &array<T>::operator=(const array &x)
{
    // Template body goes here.
}
```

- A class template is instantiated by using the < > syntax:

```
array<double> x(20);
```

- Each instantiation of a class template has its own copy of any static data members.

- An ordinary function cannot have a class template as an argument. However, an instantiation of a class template can be in a function argument list:

```
void f(array<int> x);
```

- A function template can have a template in its argument list, as in:

```
template<class T> void f(array<T> x);
```

The parameterized type in the function argument must also appear in the template argument.

- A template can take any number of arguments and the arguments may be a mixture of class templates, instantiations of templates, and fundamental types, as in:

```
template<class T> T f(X<T> x1, Y<int> i1, int i2);
```

and:

```
template<class T1, class T2, int i> class array_pair {
// etc.
};
```

- A class template can declare friends and these may be either functions or classes:

```
template<class T> class array {
public:
    friend bool operator==<>(const array &x1,
        const array & x2);
// etc.
};
```

- A specialization of a class template can be declared by means of the template<> prefix:

```
template<> class array<int> {
public:
    array(const array &x);
    int &operator[](int index);
    // etc.
private:
    int *pt;
};
```

- A class template member function specialization can be declared by using the `template<>` prefix:

```
template<> array<int>::array(int size)
{
    n = size;
    pt = new int[n];
    for (int i = 0; i < n; ++i)
        pt[i] = 1;
}
```

- When designing a class template, the standard approach is to first design and test a non-template version of the class.

16.13 Exercises

1. Modify the doubly linked list class of Section 10.3.2 so that it is a class template with the data being a parameterized type instead of a `typedef`. Test the template for data of type `int` and for the `string` class of Section 9.6.2.

2. Modify the array class of Section 12.7.1 so that it is a class template with the data being a parameterized type. Test the template for data of type `double` and the `complex` class of Section 9.6.1.

Chapter 17

Standard Library

17.1   Introduction

In addition to the C++ language itself, the ANSI Standard defines an extensive *Standard Library* (which we will often abbreviate to "the Library"). It is worth emphasizing the distinction between the language and the Library. The language defines such things as the meaning of operations on the fundamental types. For example, the language defines what is meant by:

```
int i = 4;
```

and there is no way in which we can change the language so that this statement means something different. The Library defines things like useful classes and functions. For example, the `float sqrt(float)` function, declared in the header file `<cmath>`, returns the square root of a number of type `float`. If we really want to, we can omit to include the header file and define our own `sqrt()` function.

The Library makes considerable use of the techniques introduced in previous chapters. In particular, a significant part of the Library is based on the use of templates and is sometimes known as the *Standard Template Library* or STL. In fact, there are books devoted entirely to the STL. (See, for instance, [8].) However, it is the facilities offered by the Library that are important, rather than the techniques on which they are based. Consequently, the STL is not singled out for special treatment in this chapter (nor in the ANSI Standard).

All of the facilities offered by the Library require header files and they all add names to the `std` namespace. The entire `std` namespace can be made visible by means of the statement:

```
using namespace std;
```

and in small projects this may be the most convenient approach. However, in order to avoid potential name clashes it is preferable to have a using declaration, as in:

```
using std::sqrt;
```

or to use the full name:

```
std::cout << "Square root of 2 is: " << std::sqrt(2.0) << "\n";
```

The ANSI Standard divides the Library into ten different categories and we follow the same division in this chapter. The categories are given in Table 17.1. Each of these categories requires a number of header files and each of these header files declares various functions, classes, templates etc. The only part of the Library that we consider in any detail is Input and Output. This topic is so important that Chapter 18 is devoted to a description of input and output streams. To describe the remainder in similar detail would take another book. Consequently, we don't attempt to do any more than outline what is available. This will at least ensure you don't reinvent what is already in the Library. When you do need to use part of the Library in earnest, then you should consult more detailed documentation. Your compiler may have specific documentation on the Standard Library. Other sources of information include various books ([8], [10]), together with the ANSI Standard ([1]). You can also learn a lot by looking at the header file for the part of the Library that you wish to use.

Within a category, the facilities offered by the Library can generally be labelled by the header files that are required. As mentioned previously, some of these are derived from C header files and are easily distinguished from the newer C++ header files. If a C header file is included by `<X.h>`, then the C++ version is included by `<cX>`. For example, the C header file `<stddef.h>` becomes `<cstddef>`. Some of these files are needed for more than one category of the Library. This is because different function declarations from one of the original C header files may fit naturally into different categories of the Library.

You should be careful when using the Library since it is here that compilers show the greatest divergence from the ANSI Standard. What we present in this chapter (and the next) attempts to follow the Standard rather than any particular compiler.

Category	Section or Chapter
Language Support	Section 17.2.1
Diagnostics	Section 17.2.2
General Utilities	Section 17.2.3
Strings	Section 17.2.4
Localization	Section 17.2.5
Containers	Section 17.2.6
Iterators	Section 17.2.7
Algorithms	Section 17.2.8
Numerics	Section 17.2.9
Input and Output	Chapter 18

Table 17.1: Library categories.

17.2 Library Categories

In each of the following subsections, we give an outline of one of the categories listed in Table 17.1. Input and Output are covered in Chapter 18.

17.2.1 Language Support

The Library facilities described here are designed to support the C++ language in some way. The header files associated with these facilities are given in Table 17.2 and, as you can see, many of them are derived from C header files. The `<cstddef>` header defines `size_t` and `ptrdiff_t`. These are the types returned by `sizeof()` and the result of a pointer subtraction, respectively. The `<climits>` and `<cfloat>` files give the limits on the integral and floating point types. For example, `INT_MAX` is the largest number of type `int` that can be represented on a particular computer. The `<limits>` header is concerned with the same kind of information, but in terms of classes, templates and specializations. The `<cstdlib>` header defines values for `EXIT_SUCCESS` and `EXIT_FAILURE`, together with declarations for the `abort()`, `atexit()` and `exit()` functions. The `<new>` header supports dynamic memory allocation; for example, it is needed if you want to catch an exception thrown by a failed memory allocation. The `<exception>` header defines classes and functions for handling exceptions. In particular, these facilities are used by the Library. The `<typeinfo>`, `<cstdarg>`, `<csetjmp>` and `<csignal>` header files are all connected with facilities that you are unlikely to need, except in advanced applications.

Header	Purpose
`<cstddef>`	`ptrdiff_t` and `size_t`
`<limits>`	templates for limits of integral and floating types
`<climits>`	C-style limits for integral types
`<cfloat>`	C-style limits for floating types
`<new>`	dynamic memory allocation
`<typeinfo>`	run-time type identification
`<exception>`	exception handling
`<cstdarg>`	variable-length function argument lists
`<csetjmp>`	stack unwinding
`<csignal>`	signal handling
`<cstdlib>`	program termination

Table 17.2: Language support header files.

17.2.2 Diagnostics

The facilities described here can be used to detect and report error conditions. The associated header files are given in Table 17.3. The `<stdexcept>` header declares a comprehensive range of classes for reporting different kinds of error. By contrast, the `<cassert>` and `<cerrno>` headers contain a few macros, derived from the C headers.

Header	Purpose
`<stdexcept>`	exceptions
`<cassert>`	assert macro
`<cerrno>`	C-style error handling

Table 17.3: Diagnostics header files.

17.2.3   General Utilities

The facilities described here are general utilities that are either required by other components of the Library or are of general use in C++ programs. The relevant headers are given in Table 17.4. The facilities associated with the `<utility>` header are basic template functions and classes that are used by other components of the Library. The `<functional>` header deals with *function* objects, that is objects with an `operator()` defined. These are important for the algorithms component of the Library, outlined in Section 17.2.8. The `<memory>` header is concerned with what are known as *allocators*. An allocator is used by other parts of the Library to allocate and deallocate memory. The `<cstring>` header declares some functions that are useful for copying, setting and comparing memory. For example, in this book we frequently use the `memcpy()` function to copy the bit pattern from one area of memory to another. (See, for instance, Sections 7.8.2, 9.6.2, 10.1.1 and 12.7.1.) The `<ctime>` header declares many useful functions connected with time. For example, the `ctime()` function converts calendar time to local time.

Header	Purpose
`<utility>`	operators used by other parts of Library
`<functional>`	function objects
`<memory>`	allocators for containers
`<cstring>`	operations on raw memory
`<ctime>`	C-style time and date

Table 17.4: General utilities header files.

17.2.4   Strings

The facilities described here are all connected with handling strings of characters. The `<string>` header defines a string class template. This should be used in preference to the `string` class that we developed in Section 9.6.2 since the Library version is far more sophisticated. In general, if satisfactory software already exists, there is no point in reinventing it. The reuse of software is one of the main benefits of object-oriented languages, such as C++.

As can be seen from Table 17.5, the remainder of the facilities are derived from the equivalent C header files. The headers `<cwctype>` and `<cwchar>` are to do with what are known as *wide characters*. These are characters (such as Unicode) that are outside

Header	Purpose
`<string>`	string template
`<cctype>`	character classification
`<cwctype>`	wide-character classification
`<cstring>`	C-style string functions
`<cwchar>`	C-style wide-string functions
`<cstdlib>`	C-style string functions

Table 17.5: String header files.

those normally stored by the type `char` and are only relevant for rather specialized applications. The `<cctype>` header declares functions for deciding whether or not a particular character is a digit, upper case etc. It also has functions to convert between upper and lower case. We have already met the `<cstring>` header in the previous section, where we were concerned with functions that manipulated raw memory, without regard to the significance of the stored data. However, the header also declares functions for handling null-terminated C-style strings. For instance, there are functions to compare, concatenate and copy such strings. However, in many applications it is better to use a string class. The `<cstdlib>` header declares more functions that act on null-terminated C-style strings. For example, the `atoi()` function converts the string pointed to by its argument to an integer.

17.2.5 Localization

The headers connected with this part of the Library are given in Table 17.6. They are both concerned with encapsulating cultural and language differences, such as money and date or time formatting. Whereas the `<locale>` header contains an extensive collection of class templates for dealing with such things, `<clocale>` is derived from the C header and contains a few functions and macros.

Header	Purpose
`<locale>`	cultural and language differences
`<clocale>`	C-style cultural and language differences

Table 17.6: Localization header files.

17.2.6 Containers

This part of the Library is concerned with what are known as *containers*; these are objects that store other objects. As can be seen from Table 17.7, none of the headers are derived from equivalent C header files. In fact, this part of the Library makes extensive use of the purely C++ concepts of classes and templates. The first five headers of Table 17.7 are concerned with *sequences*; these are containers that organize, in a linear fashion, a finite set of objects of the same type.

Header	Purpose
`<deque>`	double-ended queue
`<list>`	doubly-linked list
`<queue>`	queue
`<stack>`	stack
`<vector>`	one-dimensional array
`<map>`	associative array
`<set>`	set
`<bitset>`	set of Booleans

Table 17.7: Container header files.

A `list` is a doubly linked list, similar in idea to the `dlist` class that we implemented in Section 10.3.2. Now that our `dlist` class has served its purpose, you should abandon it in favour of the more sophisticated Library facility given in `<list>`. Lists are useful if you want to do large numbers of insertions and deletions at any point in the sequence.

The `<vector>` header does not provide what a mathematician would think of as a vector. For instance, there is no concept of a scalar product. The `vector` class template can be used to provide a container in which objects are accessed by an index. This implies that a `vector` object is like a sophisticated array.

A `queue` is a container that allows objects to be inserted at one end and extracted from the other. Consequently, objects in a queue emerge in the order in which they are inserted. A `deque` is a double-ended queue; that is operations can be carried out at either end.

A `stack` is a sequence container for which objects can only be inserted or deleted at one end. We have already met the idea of a `stack` in Section 10.3.1.

The `map` and `set` class templates deal with *associative containers*; these are containers that are concerned with pairs of values. Given one value, known as a *key*, we can access another value, known as the *mapped value*. A *map* is a sequence of pairs, with at most one value for each key. A *set* is a map for which only the keys are relevant.

The `<bitset>` header contains the class template for an array of single bits, much like our `bit_array` class of Section 11.4.1.

17.2.7   Iterators

The facilities of this part of the Library are associated with the single header file, `<iterator>`. An *iterator* is a generalization of a pointer and is used to move between elements of a container. An iterator points to a current element and has the ability to change so that it points to the next element. The ANSI Standard distinguishes five different types of iterator: *input, output, forward, bidirectional* and *random*. Iterators are an important way of moving between objects in containers. (See, for instance, Chapter 19 of [10] and Chapter 2 of [8].)

17.2.8   Algorithms

The two header files associated with this part of the Library are given in Table 17.8. The `<cstdlib>` is derived from the `<stdlib.h>` C header file and declares two functions: `bsearch()` and `qsort()`. The `bsearch()` function performs a binary search of a sorted array. The `qsort()` function is an implementation of the quick-sort algorithm.

The `<algorithm>` header defines an extensive collection of operations on containers. The operations include counting, copying, swapping, shuffling, sorting, searching etc.

Header	Purpose
`<algorithm>`	algorithms on sequences
`<cstdlib>`	`bsearch()` and `qsort()`

Table 17.8: Algorithm header files.

17.2.9 Numerics

The header files associated with this part of the Library are given in Table 17.9. The `<cmath>` header is derived from `<math.h>` and has declarations for many common mathematical functions, such as `cos()`, `sin()`, `log()`, `exp()`, `sqrt()` etc. The functions are overloaded so that there are `float`, `double`, and `long double` versions. The `<cstdlib>` header is derived from `<stdlib.h>`. It is concerned with the generation of random numbers, and finding the quotient and remainder that results from integer division.

Header	Purpose
`<complex>`	complex arithmetic
`<valarray>`	efficient operations on one-dimensional numerical arrays
`<numeric>`	numerical operations
`<cmath>`	standard mathematical functions
`<cstdlib>`	random numbers, quotient, remainder

Table 17.9: Numerics header files.

The `<complex>` header declares a class template for complex numbers. This should be used in preference to our own `complex` class that was introduced in Section 9.6.1. The template has all the usual complex arithmetic operations and comes with specializations in which the ordered pair of real numbers that constitute a complex number are of type `float`, `double`, or `long double`.

The core of many numerical calculations consists of relatively simple operations on one-dimensional arrays. Such calculations are important in many areas of science and engineering, and consequently some machine architectures are optimized for operations on one-dimensional numerical arrays. The `<valarray>` header declares class templates for operations on one-dimensional numerical arrays. It was designed for efficiency rather than ease of use, but is likely to be important to many scientists and engineers.

The `<numeric>` part of the Library implements a few numerical algorithms, analogous to `<algorithm>`. The available operations are:

- accumulating the results of an operation on a sequence;

- accumulating the results of an operation on two sequences;

- generating a sequence by an operation on two sequences;

- generating a sequence by an operation on one sequence.

Calculating the inner product of two vectors is an example of what can be achieved here.

17.3   Using the Standard Library

17.3.1   Complex Arithmetic

It is difficult to give a very substantial program demonstrating the use of complex arithmetic without getting fairly involved in a specific application. However, the following

program should give some idea of what is available through using the class templates
in <complex>:

```
#include <iostream>
#include <complex>
#include <cstdlib>        // For EXIT_SUCCESS
using std::cout;
using std::cin;
using std::complex;
using std::real;
using std::imag;
using std::conj;
using std::abs;
using std::sqrt;
using std::polar;
using std::pow;
using std::cos;
using std::sin;

int main()
{
    // Define z = 3 + 4 i:
    complex<double> z(3, 4);
    cout << "z = " << z << "\n";

    // Give real and imaginary parts:
    cout << "real part of z = " << real(z) << "\n";
    cout << "imaginary part of z = " << imag(z) << "\n";

    // Initialize with complex conjugate:
    complex<double> z_conj = conj(z);
    cout << "complex conjugate of z = " << z_conj << "\n";

    // Find magnitude:
    cout << "magnitude of z = " << abs(z) << "\n";

    // Define i:
    const complex<double> i(0.0, 1.0);

    // Make assignment using polar form:
    double theta = 0.25 * M_PI;
    z = sqrt(2.0) * polar(1.0, theta);

    // Check de Moivre's theorem:
    cout << "\nCheck de Moivres's theorem for z = " << z << "\n";
    complex<double> z1, z2;
    for (int m = 0; m < 10; ++m) {
        z1 = pow(z, m);
```

```
    z2 = pow(sqrt(2.0), m) * (cos(m * theta) + i *
        sin(m * theta));
    cout << "z1 = " << z1 << "\nz2 = " << z2 << "\n";
}

// Do an impedance type calculation:
z = (2.0 + i) / (1.0 - i);
complex<double> impedance = z + 1.0 / z;
cout << "\nimpedance = " << impedance << "\n";

// Check we can input a number:
cout << "\nEnter a complex number as (x,y): ";
cin >> z;
cout << "z = " << z << "\n";

return(EXIT_SUCCESS);
}
```

It its worth highlighting following features:

- The program initializes z to $3 + 4i$ by using the **complex** constructor function. Notice that we must specify the type of the ordered pair of real numbers that constitute a complex number. In this case the type is **double**.

- The real and imaginary parts of z are given by **real(z)** and **imag(z)** respectively. Notice how a complex number is inserted in the output stream by the **<<** operator. The way in which this is achieved is explained in Chapter 18.

- The complex conjugate of z (written mathematically as $\bar{z}$ or z^*) is given by the **conj(z)** function, and the result is used to initialize **z_conj**.

- The magnitude of z (written mathematically as $|z|$) is given by **abs(z)**.

- The imaginary constant i is initialized by means of the complex constructor. Engineers may want to replace i by j.

- The **sqrt()** and **polar()** functions are used to assign $\sqrt{2}e^{i\pi/4}$ to z. We then evaluate:

$$z_1 = z^m$$

and:

$$z_2 = (\sqrt{2})^m \left(\cos(m\theta) + i\sin(m\theta) \right)$$

According to de Moivre's theorem, we should get $z_1 = z_2$. When you run this program you will probably find slight discrepancies due to the finite precision of floating point arithmetic on a real computer.

- We calculate:

$$z + \frac{1}{z}$$

where:

$$z = \frac{(2+i)}{(1-i)}$$

This is typical of an impedance calculation in an AC circuit, and also demonstrates complex division. (The exact result for this calculation is $(7 + 9i)/10$.)

- Finally, we demonstrate that it is possible to read complex numbers from the input stream by means of the >> operator.

17.3.2   Boolean Arrays

In this section we revisit the bitwise operations of Sections 11.4.1 and 11.4.2. Instead of using our own `bit_array` class, we make use of the `vector<bool>` specialization from the Standard Library. The modified `my_test.cxx` program is given below.

```
// source:     my_test.cxx
// use:        Tries out the vector<bool>
//             template specialization.

#include <iostream>
#include <cstdlib>          // For EXIT_SUCCESS, atoi()
#include <vector>
using std::cin;
using std::cout;
using std::get;
using std::exit;
using std::vector;
using std::atoi;

// Function declarations:
void wait(void);
void display_bits(vector<bool> &b);
void set_bits(vector<bool> &b);
void clear_bits(vector<bool> &b);

void wait(void)
{
    cout << "\nHit <Enter> to continue." << "\n";
    cin.get();
}

void display_bits(vector<bool> &b)
{
    int bits = b.size();
    const int display_width = 80;
    for (int i = 0; i < bits; ++i) {
        if (!(i % display_width))
            cout <<"\n";
        if (b[i])
            cout << "x";
        else
```

```
            cout << ".";
        }
        cout << "\n";
}

void set_bits(vector<bool> &b)
{
        int bits = b.size();
        for (int i = 0; i < bits; ++i) {
                b[i] = true;
        }
}

void clear_bits(vector<bool> &b)
{
        int bits = b.size();
        for (int i = 0; i < bits; ++i) {
                b[i] = false;
        }
}

int main(int argc, char *argv[])
{
        if (argc != 2) {
                cout << "Usage:  my_test <length>\n";
                exit(EXIT_FAILURE);
        }
        int bits = atoi(argv[1]);
        vector<bool> b(bits);
        cout << "max_size = " << b.max_size() << "\nsize = " <<
                b.size() << "\ncapacity = " << b.capacity() << "\n";
        for (int i = 0; i < bits; ++i)
                if (i % 2)
                        b[i] = true;
                else
                        b[i] = false;
        display_bits(b);
        wait();
        clear_bits(b);
        display_bits(b);
        wait();
        set_bits(b);
        display_bits(b);
        wait();
        return(EXIT_SUCCESS);
}
```

As can be seen, it is a lot easier to use a pre-existing class than to develop our own. Since the `vector<bool>` template specialization is part of the Standard Library, the following command is all that is needed to compile and link the program, producing an executable file called `my_test`:

```
g++ my_test.cxx -o my_test
```

In this program we make use of three `vector<bool>` member functions to provide some interesting information about the object b. The `max_size()` function gives the largest Boolean array that can be created. This is likely to be very large indeed, and only limited by the amount of memory available. The `size()` function gives the number of elements in the b object, and for this program the number should be what is typed in at the system prompt. The `capacity()` function gives the total number of elements that are available without resizing the array. This is a multiple of the natural size of memory blocks on the computer being used. For instance, if I enter `my_test 19` on my computer, then the program gives the capacity as 32.

The remainder of the program follows the code given in Section 11.4.1. Notice that `display_bits()`, `clear_bits()`, and `set_bits()` are not member functions, but instead have a `vector<bool>` argument. It is particularly important that these are reference arguments since we don't want to do unnecessary copying of potentially large arrays.

Having demonstrated the `bit_array` class in Section 11.4.1, we used it in Section 11.4.2 to generate a list of prime numbers by implementing the Sieve of Eratosthenes. We can now modify this program to use the `vector<bool>` template specialization. The modified program is given below.

```
// source:    primes.cxx
// use:       Implements the Sieve of Eratosthenes using
//            the vector<bool> template specialization.

#include <iostream>
#include <cstdlib>        // For EXIT_SUCCESS, atoi()
#include <cmath>          // For sqrt()
#include <vector>

using std::cin;
using std::cout;
using std::flush;
using std::exit;
using std::vector;
using std::atoi;
using std::sqrt;

// function declarations:
void wait(void);
void list_primes(vector<bool> &b);
void display_bits(vector<bool> &b);
void set_bits(vector<bool> &b);
```

```cpp
void wait(void)
{
    cout << "\nHit <Enter> to continue.\n";
    cin.get();
}

void list_primes(vector<bool> &b)
{
    const int display_height = 24;
    int primes_displayed = 1;
    int bits = b.size();
    for (int i = 0; i < bits; ++i) {
        if (b[i]) {
            cout << 2 * i + 3 << "\n";
            if (primes_displayed % display_height)
                ++primes_displayed;
            else {
                primes_displayed = 1;
                cout.flush();
                wait();
            }
        }
    }
}

void display_bits(vector<bool> &b)
{
    int bits = b.size();
    const int display_width = 80;
    for (int i = 0; i < bits; ++i) {
        if (!(i % display_width))
            cout <<"\n";
        if (b[i])
            cout << "x";
        else
            cout << ".";
    }
    cout << "\n";
}

void set_bits(vector<bool> &b)
{
    int bits = b.size();
    for (int i = 0; i < bits; ++i) {
        b[i] = true;
    }
}
```

```
int main(int argc, char *argv[])
{
    if (argc != 2) {
        cout << "Usage:  primes <largest integer>\n";
        exit(EXIT_FAILURE);
    }
    int max_int = atoi(argv[1]);

    // Only store odd numbers:
    int bits = (max_int - 1) / 2;
    vector<bool> b(bits);

    // Set all bits:
    set_bits(b);
    int max_m = static_cast<unsigned>(sqrt(max_int));

    // These 2 loops are the Sieve of Eratosthenes:
    for (int m = 3; m <= max_m; m += 2) {
        int max_n = max_int / m;
        for (int n = m; n <= max_n; n += 2)
            b[m * n / 2 - 1] = false;
    }

    cout << "Bit array set for primes:\n";
    display_bits(b);
    wait();
    cout << "Primes between 3 and " << max_int << ":\n";
    list_primes(b);
    wait();
    return(EXIT_SUCCESS);
}
```

Again notice how much easier it is to use the template specialization provided in the Standard Library than to implement our own class.

17.4   Summary

This chapter is effectively summarized by its tables. The library facilities are divided into categories, as given in Table 17.1. The header files for each of these categories are then given in Tables 17.2 to 17.9, and Table 18.1.

17.5   Exercises

1. (a) Implement the Sieve of Eratosthenes, described in Section 11.4.2, by using the bitset class template outlined in Section 17.2.6.

(b) You now have three different implementations of the Sieve of Eratosthenes. Modify each program so that there are two command line parameters. These parameters should specify the lower and upper limits of the primes that are to be displayed. Time all three programs when they are used to find a range of very large primes.

2. Use the Standard Library to implement a program that sorts a list of strings into alphabetical order.

3. Instantiations of the `vector` class template in the Standard Library would be called arrays by mathematicians. Use the Standard Library to implement a true vector class. Your class should define such operations as the addition, dot product, and cross product of two vectors.

4. (a) Suppose you have a list of data pairs; the first element of the pair is of type `double` and the second element is a string (as in `<string>`). For instance, the data could represent the distance of a star from the Earth and the name of the star. Use the Standard Library to produce a list that is sorted by increasing distance.

   (b) Suppose you have two sorted lists of data pairs. Use the merge sort function provided by the Standard Library to produce a merged and sorted list of both sets of data.

Chapter 18

Input and Output

18.1 Introduction

Input and output (I/O) are not part of the C++ language, but are part of the ANSI C++ Standard Library. The I/O library provides a strongly typed and very safe system for performing input and output operations; errors either get trapped at compile-time or they set error flags associated with I/O library objects. This library uses many of the more advanced techniques of C++, such as classes, templates, inheritance and exception handling. It is clearly better to have some understanding of such techniques before using the I/O library. However, even without this background, it should be possible to use this chapter to perform such tasks as formatting your I/O or writing to and reading from files.

The basic idea behind the I/O library is that information flows to or from a program as a "stream" of data. For example, data may stream from a program to a computer screen, or from the keyboard to a program.

It isn't possible to cover every detail of the I/O library here, but additional sources of information are [1] and [10], together with the documentation for your compiler. It is also worth looking at the relevant header files on your system. As is the case for those parts of the Standard Library presented in Chapter 17, you may find that your compiler is not entirely compatible with the ANSI Standard. In this chapter, we follow the Standard rather than any particular compiler.

18.2 Input and Output Classes

The class hierarchy is shown in Figure 18.1, but you should be able to use a lot of the features of the I/O library without knowing details of the class structure. Apart from `ios_base`, all the classes shown are actually templates and they have a `basic_` prefix. The `basic_ios` class is a virtual base class of the `basic_istream` and `basic_ostream` classes. Fortunately, `typedefs` are provided for most of the template specializations; for example, instead of `basic_ofstream<char>` we can use `ofstream` when opening a file for output. There are ten header files that are associated with this library and these are shown in Table 18.1. The headers contain declarations for various classes, templates, functions, `typedefs` etc. associated with the I/O library. However, due to

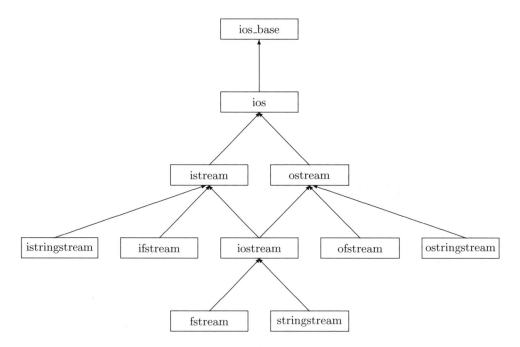

Figure 18.1: Hierarchy diagram for the I/O library classes.

the hierarchical nature of the library, it is usually only necessary to include one or two of these headers in a file, since any other relevant files are automatically included. For instance, so far we have only needed to include the `<iostream>` header.

The `<iosfwd>` header contains **typedefs** and declarations of stream I/O classes and templates so that they can be referred to but cannot be used to perform operations. You are unlikely to need to include this header.

The `<iostream>` header declares the following important objects:

- `cin` is an instance of the `istream` class and is the standard input stream (typically the keyboard).

Header	Purpose
`<iosfwd>`	forward declarations
`<iostream>`	standard iostream objects
`<ios>`	iostreams base classes
`<streambuf>`	stream buffers
`<istream>`	input stream template
`<ostream>`	output stream template
`<iomanip>`	formatting and manipulators
`<sstream>`	string streams
`<fstream>`	file streams
`<cstdio>`	C-style I/O

Table 18.1: I/O library header files.

- `cout` is an instance of the `ostream` class and is the standard output stream (typically the screen).

- `cerr` is an instance of the `ostream` class. It is the standard error stream (typically the screen) and is effectively unbuffered.

- A buffered version of `cerr`, called `clog`, is also defined. Although this error stream is also typically the screen, the output is often redirected to a file. One way to achieve this is to use stream assignment, as described in Section 18.7.

The header file also defines wide-character versions of these four streams.

The `<ios>` header declares the I/O base classes. It contains important declarations of manipulators (such as `io_base::hex`) and flags (such as `ios_base::badbit`). Manipulators and flags are described in Section 18.5.

The `<streambuf>` header declares a `basic_streambuf` class template that is used to provide various stream buffers. The class template does not appear in Figure 18.1 because it is not a base class for any of the classes shown.

The `<istream>` header declares the basic buffered input stream template class, `basic_istream`. The `<ostream>` header declares the basic buffered output stream template class, `basic_ostream`.

The `<iomanip>` header declares functions that enable inserters and extractors to be used to alter the state of an I/O class. For example, the `setw()` manipulator can be used to set the width for the next output operation. This header is only needed if you want to use manipulators that take arguments.

The `<sstream>` header defines template classes that allow us to associate stream buffers with the string class in the Standard Library. Include this header if you want to insert and extract C++ strings (rather than C strings).

The `<fstream>` header defines template classes that associate stream buffers with a file. This enables us to read from and write to files.

The `<cstdio>` header contains macros and function declarations for C-style I/O. It does not use the concept of streams and should be avoided if at all possible. C-style I/O is not described in this chapter, but if you do need to use it then [5] is a good source of information.

All of the facilities offered by the I/O library add names to the `std` namespace. As has already been pointed out, the entire `std` namespace can be made visible by means of the statement:

```
using namespace std;
```

In small projects this may be the most convenient approach and we generally assume that this statement is included in any programs containing the code fragments given in this chapter. Alternatively, potential name clashes can be avoided by a **using** declaration, as in:

```
using std::cin;
```

or by giving the full name:

```
std::cout << "Welcome to the world of C++!\n";
```

18.3   Output

Data is inserted in an output stream by using the binary *insertion operator*,<< (also known as an *inserter*). We have used this technique throughout this book. For instance, the following sends a string and the value of x to the output stream:

```
cout << "The value of x is " << x;
```

This operator is overloaded for all of the fundamental types; as a simple example we can list the addresses of elements of an array:

```
double x[10];
for (int i = 0; i < 10; ++i)
    cout << "Address of element " << i << " is " << (x + i) <<
        "\n";
```

Notice that insertions can be concatenated. This is because the expression:

```
cout << "Address of element "
```

returns an `ostream` object (as does `cout << i` etc.). An alternative for long strings is to use the fact that adjacent string constants are concatenated, as in:

```
cout << "This long sentence demonstrates how to continue "
    "long strings by using concatenation";
```

However, repeating the insertion operator will probably make your intention more obvious, especially if you put the operator at the end of the line that is to be continued.

The `ostream` object, `cout`, provides buffered output. This is usually more efficient than unbuffered output, but may sometimes be inconvenient. For instance, if we are trying to debug a program, we might attempt to send the values of some crucial variables to `cout`. However, if the messages are still stuck in a buffer when disaster strikes, then we may not get any output. The `flush()` function overcomes this problem, by *flushing* data from the appropriate stream buffer. This is demonstrated by the following code fragment:

```
double max_error;
// Suppose we forget to set max_error.
// Put in a debug statement:
cout << "max_error: " << max_error;
cout.flush();
```

The `flush()` function is unnecessary (but harmless) in the following situation:

```
cout << "Input the start time: ";
cout.flush();
cin >> t;
```

This is because `cout` is *tied* to `cin`; that is the `cout` buffer is automatically flushed before any attempt is made to extract data from `cin`.

The object, `cerr`, is similar to `cout`, except that it is the standard error stream (usually the screen) and this is flushed after each insertion. The flushing has the advantage that critical error messages cannot get stuck in a buffer:

```
if (n != m)
    cerr << "Cannot invert matrix: matrix is not square\n";
```

There are a few features of the insertion operator that could initially cause some confusion. Consider the code fragment:

```
char array[] = "A message.";
char *string = "Another message.";
cout << "Base address of char array is: " << array <<
    "\nValue of char pointer is: " << string << "\n";
```

The result is:

```
Base address of char array is: A message.
Value of char pointer is: Another message.
```

In contrast to the other fundamental types, if the right operand of an insertion operator has the type `char*`, then a string, rather than the value of an address, is inserted. A cast to `void*` produces the address stored by the pointer and:

```
char array[] = "A message.";
char *string = "Another message.";
cout << "Base address of char array is: " <<
    static_cast<void *>(array) << "\nValue of char pointer is: " <<
    static_cast<void *>(string) << "\n";
```

typically gives the (hexadecimal) result:

```
OXbffff7f0
0X80487cb
```

Another feature of the insertion operator is that the parentheses in the following code fragment are *not* redundant since without them the code fails to compile:

```
bool b1 = true, b2 = false;
cout << "b1 || b2 is: " << (b1 || b2) << "\n";
```

This is because the insertion operator has a higher precedence than the logical OR operator. (See Chapter 11 and Appendix B.) Insertion expressions involving other low-precedence operators also need parentheses and analogous remarks apply to the extraction operator.

18.3.1 Unformatted Output

Single characters can be inserted in an output stream by means of the `put()` library function:

```
cout.put('C');
```

The `put()` function can be concatenated, as can any function that returns a reference to an appropriate class object:

```
cout.put('C').put('+').put('+').put('\n');
```

but statements of this form are cumbersome and usually unnecessary.

There is also a `write()` function that takes two arguments. The first argument is a pointer to a buffer and the second argument is the maximum number of characters to be written to the stream. The following code fragment illustrates both the action of this function and what can go wrong if it isn't used carefully:

```
char buffer[3] = {'C', '+', '+'};
for (int i = 1; i < 8; ++i) {
    cout.write(buffer, i);
    cout << "\n";
}
```

The output on my computer is:

```
C
C+
C++
C++¿
C++¿8
C++¿8ø
C++¿8øÿ
```

Notice that the first three lines of output correspond to the contents of `buffer`, but then increasing amounts of garbage are appended to C++ as the `write()` function accesses memory beyond that assigned to `buffer`. Even if there is a newline character in `buffer`, the `write()` function will attempt to insert the specified number of characters in the stream, as running the following code will demonstrate:

```
char buffer[4] = {'C', '+', '+', '\n'};
for (int i = 1; i < 8; ++i) {
    cout.write(buffer, i);
    cout << "\n";
}
```

So the message is that if you really must use a function that accesses raw memory, such as `write()`, then you need to pay special attention to the number of characters stored.

Exercise

Make the above code fragment into a program and try running it on your computer. Try increasing the maximum number of characters sent to the output stream.

18.4 Input

Data is removed from an input stream by means of the binary *extraction operator*, >>, also known as an *extractor*. For instance:

```
cin >> i;
```

extracts the value of i from the input stream, cin. Extractions can be concatenated; the expression:

```
cin >> i >> j >> k;
```

extracts i, then j, then k, from cin. This is because the expression cin >> i returns an istream object and this is then used as the first argument of the next inserter function.[1] Since white space is ignored, typing:

```
1   2   3
```

or:

```
1
2   3
```

assigns the same values to i, j and k. Notice that statements such as:

```
cin >> i, j, k;      // WRONG!
```

superficially look alright and even compile, but do not have the desired effect since only i is assigned to.

The extraction operator is overloaded for all of the fundamental types. For example, it is straightforward to extract a string, as illustrated by the following code fragment:

```
char buffer[128];
cout << "Input a string:\n";
cin >> buffer;
cout << "You input: " << buffer << "\n";
```

When extracting a string, the "end of string" character, '\0', is automatically appended so it is necessary to make certain that the buffer is large enough to hold this additional character. Failure to do so may result in a run-time error. Note that characters are extracted from the cin stream until white space is encountered, so that input of the form:

```
This is a message.
```

results in the output:

```
This
```

It is straightforward to overcome this problem by using techniques associated with unformatted input.

18.4.1   Unformatted Input

We can use the functions get() and getline() to extract data from an input stream without regard as to what the data represents. There are three different get() functions.

The expression cin.get(), with no arguments, returns the value of the character extracted as an integer. Consequently, the following code fragment outputs the ASCII code of whatever character is input. (Try it!)

---

[1]Don't forget that all binary operators, apart from assignment, are left-associative.

```
cout << "Input one character:\n";
int i = cin.get();
cout << "You input: " << i << "\n";
```

In order to output the actual character, we need to do a cast to a **char**. This occurs implicitly in the following code fragment:

```
cout << "Input a string:\n";
char c = cin.get();
if (c != '\n') {
    cout << "Your string contains the characters:\n" << c  << "\n";
    while ((c=cin.get()) != '\n')
        cout << c << "\n";
}
```

Notice that it is not possible to use the extraction operator with this form of the **get()** function, so expressions such as **cin.get() >> i** do not compile.

The second form of the **get()** function takes a reference **char** argument and again reads a single character. Since this function does return an **istream** object, concatenation is possible, as in:

```
char a, b, c;
cout << "Input a string of three characters:\n";
cin.get(a).get(b).get(c);
cout << "You input: " << a << b << c << "\n";
```

It is also possible to use the extraction operator, as in:

```
char a, b, c;
cout << "Input a string of three characters:\n";
cin.get(a) >> b >> c;       // Note use of >>.
cout << "You input: " << a << b << c << "\n";
```

However, don't forget that whereas **get(a)** extracts any character, the insertion operator ignores white space. Consequently, the behaviour of the previous two fragments of code will differ if you input white space.

The third form of the **get()** function takes three arguments. The first argument is a pointer to a **char** buffer, the second is the maximum number of characters to be extracted, and the third is the character that terminates the extraction. If this termination character (also known as a delimiter) is omitted, then it defaults to the newline character. Suppose we have the statement:

```
cin.get(buffer, length, '#');
```

This **get()** function reads **length** - 1 characters into **buffer** and then appends the end-of-string character. If the termination character, **#**, is reached before **length** - 1 characters are extracted, then the extraction stops and the end-of-string character is appended to the characters that have been placed in **buffer**. There are two features that may trap the unwary:

1. It is **length** - 1 characters that are extracted rather than **length**. This is to allow one position in the buffer for the end-of-string character.

2. The delimiter character is *not* extracted.

A code fragment demonstrating the use of this form of the get() function with 'x' as the delimiter, together with an illustration of what can go wrong, is given below.

```
const int buffer_length = 128;
char buffer[buffer_length];
cout << "Input a string including the character x: ";
cin.get(buffer, buffer_length, 'x');
cout << "You input: " << buffer << "\n" <<
    "Input a string including the character x: ";
cin.get(buffer, buffer_length, 'x');
cout << "You input: " << buffer << "\n";
```

The result might typically be as follows:

```
Input a string including the character 'x': abcxyz
You input: abc
Input a string including the character 'x': You input:
```

The first three characters entered at the keyboard are extracted, stored in buffer, and the end-of-string character is appended. However, since the delimiter, 'x', is not extracted, we are never given the opportunity to enter a second set of characters.

Exercise

Make the above code fragment into a program and try running it with various inputs. Include examples where several lines are input before there is a line containing the 'x' character.

The moral to be learned from this exercise is that you should never call this three-argument form of the get() function twice without extracting the delimiter. This can be achieved as follows:

```
const int buffer_length = 128;
char buffer[buffer_length];
cout << "Input a string including the character x: ";
cin.get(buffer, buffer_length, 'x');
cout << "You input: " << buffer << "\n";
cin.get();
cout << "Input a string including the character x: ";
cin.get(buffer, buffer_length, 'x');
cout << "You input: " << buffer << "\n";
```

This discussion assumes that the first extraction from the input stream terminates because the delimiter is reached, rather than because the specified number of characters have been extracted. More robust code would test for this possibility and take appropriate action.

Another useful function is getline(). This function takes the same parameters as the three-argument version of get(). It also behaves in a similar way to the three-argument version of get(), except that it *does* extract the termination character, as the following code fragment demonstrates:

```
const int buffer_length = 128;
char buffer[buffer_length];
cout << "Input a string including the character x: ";
cin.getline(buffer, buffer_length, 'x');
cout << "You input: " << buffer << "\n" <<
    "Input a string including the character x: ";
cin.getline(buffer, buffer_length, 'x');
cout << "You input: " << buffer << "\n";
```

Exercise

Make the above code fragment into a program and try running it with
various inputs. Explain the output that you obtain.

Notice that although the delimiter *is* extracted, it is not put in the buffer; it is simply
discarded.

Another useful function is gcount(). This function takes no arguments, and returns
the number of characters that have been extracted by any of the unformatted input
functions. The following code fragment illustrates the use of gcount():

```
const int buffer_length = 128;
char buffer[buffer_length];
cout << "Input a string:\n";
cin.getline(buffer, buffer_length);
cout << "You input:\n" << buffer << "\n";
int chars_input = cin.gcount();
cout << chars_input << " characters were extracted.\n" <<
    "Input another string:\n";
cin.get(buffer, buffer_length);
cout << "You input:\n" << buffer << "\n";
chars_input = cin.gcount();
cout << chars_input << " characters were extracted.\n";
```

Exercise

Try out the above code. In particular, try inputting the same string twice.

As you will have observed in this exercise, gcount() includes the delimiter when
getline() is invoked because in this case the delimiter *is* extracted.

The read() function (which is the counterpart of the ostream function, write())
takes two arguments. The first argument is the address of the buffer where the extracted
characters are to be stored, and the second is the number of characters to be extracted.
This function does not attempt to find a delimiter, nor does it append an end-of-
string character to the buffer. If necessary, we have to explicitly add the end-of-string
character to the buffer, as the following program illustrates:

```
const int buffer_length = 128;
char buffer[buffer_length];
cout << "Input a string of at least 20 characters.\n" <<
    "You should include some white space:\n";
```

```
cin.read(buffer, 20);
buffer[20] = '\0';
cout << "You input:\n" << buffer << "\n";
```

Sometimes we may want to simply remove characters from the input stream and then discard them. The `ignore()` function serves this purpose. It takes two arguments; the first is the maximum number of characters to be extracted (which defaults to one) and the second is a delimiter (which defaults to end-of-file).[2] As with `getline()`, the delimiter *is* extracted (assuming it is reached).

As an example of unformatted input, a common requirement is for a program to prompt the user to answer a question, such as:

```
char c;
do {
    cout << "Do you want to continue? (Enter Y or N) ";
    cin >> c;
} while (c != 'Y' && c != 'N');
```

Unfortunately, incorrect replies of the form, "perhaps", send us spinning many times round the `do while` loop. Our program should remove all of the incorrect characters for each attempted reply and the following code achieves this:

```
char c;
do {
    cout << "Do you want to continue? (Enter Y or N) ";
    cin >> c;
    cin.ignore(1280, '\n');
} while (c != 'Y' && c != 'N');
```

The value of 1280 in the first argument of `ignore()` is just an arbitrarily large integer that a user is unlikely to exceed by inputting incorrect characters. An alternative approach is:

```
const int length = 128;
char buff[length];
do {
    cout << "Do you want to continue? (Enter Y or N) ";
    cin.getline(buff, length, '\n');
} while (buff[0] != 'Y' && buff[0] != 'N');
```

Exercise

Both code fragments given above are not completely satisfactory since they also accept input, such as "Newton-Raphson" or "Y-chromosome", as correct. Modify the second version so that only sensible inputs, such as "YES", "Y", "Yes", "NO", "N" and "No" are accepted.

---

[2]Notice that the delimiter is end-of-file, *not* end-of-line.

18.5   Flags, Manipulators and Formatting

A stream has an associated state that is held as a set of *flags*. These flags (and hence the state of the stream) can be manipulated directly by means of functions. For instance, we have already used the `flush()` function to empty the output stream buffer, as in:

```
cout.flush();
```

This statement effectively changes the state of the buffer associated with `cout` from not empty to empty. A large number of flags hold the state of an I/O stream and these are members of the `ios_base` class. For example, the `ios_base::showpos` flag is used to signal whether or not an output stream should show a + sign in front of non-negative numbers.[3]

A flag can be set by using the `setf()` function. The following program demonstrates how to set the output to show a + sign in front of a non-negative number. It also demonstrates how to ensure that scientific notation is used for the output of floating point numbers.[4]

```
#include <iostream>
#include <cstdlib>        // For EXIT_SUCCESS
using namespace std;

int main()
{
    const double deg = 0.01745;
    cout << deg << "\t\t(default floating point format)\n";
    cout.setf(ios_base::showpos);
    cout << deg << "\t(show + for non-negative number)\n";
    cout.setf(ios_base::scientific, ios_base::floatfield);
    cout << deg << "\t(also use scientific notation)\n";
    return(EXIT_SUCCESS);
}
```

The formatting flags are implemented as bit-masks and the effect of the first invocation of `setf()` is to take the bitwise OR of `ios_base::showpos` with the stream state. This only works when a single bit controls a feature. Consequently, the second invocation of the `setf()` function has two arguments. This ensures that there are no side effects and that the only change is to the way in which floating point numbers are printed.

Changing the formatting state by using a function notation to set and unset bits is not very convenient and is also very error-prone. Fortunately, there is a better way of achieving the same objective. The insertion operator is overloaded so that we can use it to change the state of an output stream. For instance, to flush the stream we can use the following statement:

```
cout << flush;
```

---

[3]A complete list of the many I/O flags is not given here since it is generally better to use manipulators, which are introduced shortly. If you do want to use flags, then [1] and [10] are good sources of information.

[4]If your compiler does not conform to the ANSI Standard then you may have to change `ios_base` to `ios`.

Manipulator	Effect
`boolalpha`	insert and extract `bool` in alphabetic form (e.g. `true`)
`noboolalpha`	insert and extract `bool` in default form (e.g. `1`)
`showbase`	prefix integer output to show base (i.e. `0` for octal and `0x` for hexidecimal)
`noshowbase`	do no prefix integer output to show base
`showpoint`	add trailing zeros in floating point output as required by precision (e.g. `2.400`)
`noshowpoint`	do not add trailing zeros in floating point output
`showpos`	show a `+` sign for non-negative numeric output
`noshowpos`	do not show a `+` sign for non-negative output
`skipws`	ignore leading white before certain input operations
`noskipws`	do not ignore leading white before input operations
`uppercase`	use `X` and `E` rather than `x` and `e`
`nouppercase`	use `x` and `e` rather than `X` and `E`
`unitbuf`	flush after each output operation
`nounitbuf`	do not flush after each output operation
`internal`	add fill characters between sign and value
`left`	add fill characters to right of output
`right`	add fill characters to left of output
`dec`	I/O in decimal form
`hex`	I/O in hexadecimal base (e.g. `1f3ab`)
`oct`	I/O in octal base (e.g. `377`)
`fixed`	output in fixed-point notation (e.g. `12.5`)
`scientific`	output in scientific notation (e.g. `1.25e1`)
`ws`	eat white space in input stream
`endl`	insert `'\n'` and flush
`ends`	insert `'\0'` and flush
`flush`	flush stream

Table 18.2: Manipulators not taking arguments.

In this context, `flush` is known as a *manipulator* since it manipulates an I/O state flag. The statement implicitly invokes the `flush()` function and has the advantage that we can concatenate insertions, as in:

```
cout << i << flush << j << flush;
```

This is an important reason for preferring manipulators to functions. Many manipulators can be used with input as well as output streams, although a few of them are only appropriate for one type of stream or the other. A list of I/O manipulators that don't take arguments is given in Table 18.2.

Some manipulators take arguments, and for such manipulators we must include the `<iomanip>` header file. For instance:

```
cout << setprecision(4) << M_PI;
```

prints the value of `M_PI` with four digits after the decimal point. Notice that manipulators that do not take arguments must *not* be given with empty parentheses. For example, the following does not compile:

```
cout << i << flush();   // WRONG: no parentheses.
```

There are also manipulators that insert data in addition to changing the state of a stream. For instance, `endl` inserts a new line and flushes the buffer, as in:

```
cout << "The value of t is " << t << endl;
```

A list of I/O manipulators taking arguments (and therefore requiring the header `<iomanip>`) is given in Table 18.3. Most manipulators set the state of a stream until it is reset in some way. The notable exception is `setw(n)` since this sets the field width for the next operation only.

Manipulator	Effect
setfill(i)	sets fill to character represented by i
setprecision(n)	use n digits after the decimal point
setw(n)	field width for *next* operation is n

Table 18.3: Formatting manipulators that take arguments.

Before leaving this section, it is worth pointing out that we have introduced a number of different ideas concerning the format state of an I/O stream and we should be careful to distinguish between:

- flags, such as `ios_base::scientific`;

- functions that change flags, as in `cout.flush()`;

- manipulators, as in `cout << flush`;

- manipulators taking arguments, as in `cout << setprecision(10)`.

Exercise

Write a program that demonstrates the effect of each manipulator in Tables 18.2 and 18.3.

18.6   File Input and Output

This section describes how to perform input and output on a file. Such files are typically stored on a disc and are often crucial to numerical applications that process very large amounts of experimental data or produce complicated simulations.

The I/O library class used for output from a program to a file is `ofstream`, so to open a file for output only, we must define an `ofstream` object, as in:

```
ofstream my_file("example.dat");
```

In the above statement, `example.dat` is the name of the file. If the complete path name isn't given then the name refers to a file in the directory in which the program is run. The `<fstream>` header file (which in turn includes `<iostream>`) must be included for all operators and functions concerned with file input and output.

You should be aware of a potential source of confusion with file I/O. Opening a file for output means output from the application program; that is *input* to the file! The compiler will reject any attempt to perform an extraction from a file only "open for output".

Before attempting to use a file, it is always worth testing whether or not the file has been opened successfully. The way this is done is illustrated in the following code fragment:

```
if (!my_file) {
    cerr << "Failed to open my_file.\n";
    exit(EXIT_FAILURE);
}
```

Having opened a file for output, data can subsequently be inserted in the stream by using the insertion operator, as in:

```
my_file << "This is an example file.\n" <<
    "The approximate value of pi is " << 3.142 << "\n";
```

When the `ofstream` object goes out of scope, a destructor flushes the buffer and closes the file. The file can also be closed explicitly by the statement:

```
my_file.close();
```

in which case any necessary flushing is done automatically.

To open a file for input to the program, we need to create an `ifstream` object. The following program demonstrates how to open a file for output, write some data, close the file, open the file for input, and finally read the data:

```
#include <fstream>
#include <cstdlib>        // For exit()
using namespace std;

int main()
{
    ofstream my_file("example.dat");
    if (!my_file) {
        cerr << "Failed to open my_file.\n";
        exit(EXIT_FAILURE);
    }
    my_file << "This is an example file.\n" <<
        "The approximate value of pi is " << 3.142 << "\n";
    my_file.close();
    ifstream file("example.dat");
    if (!file) {
        cerr << "Failed to open file.\n";
```

```
        exit(EXIT_FAILURE);
    }
    const int length = 128;
    char buffer[length];
    file.getline(buffer, length, '\n');
    cout << "The first string is:\n\t" << buffer << "\n";
    file.getline(buffer, length, '\n');
    cout << "The second string is:\n\t" << buffer << "\n";
    return(EXIT_SUCCESS);
}
```

Running the above code gives the result:

```
The first string is:
    This is an example file.

The second string is:
    The approximate value of pi is 3.142
```

Exercise

Check that the above output is produced by your system. Also check that
the file `example.dat` is created on your system and use an editor to look
at its contents.

The `ifstream` and `ofstream` constructors each have a second argument and this
controls the mode in which the stream is to be opened. Default values are provided for
this second argument. For instance, in the case of the `ofstream` constructor the value
defaults to:

```
ios_base::out|ios_base::trunc
```

This indicates that the stream is to be opened for output and that if the file already
exists then the contents should be discarded (i.e. truncated). Notice that the second
argument is obtained by taking the bitwise OR of certain constants. These constants
are members of `ios_base` and a complete list is given in Table 18.4.[5] The options are
selected with the bitwise OR operator because they are implemented as bit-masks.

Option	Effect if set
app	seek to the end before each write (i.e. append)
ate	open and seek to end
binary	perform I/O in binary mode
in	open for input
out	open for output
trunc	truncate an existing stream on opening

Table 18.4: Open mode options.

---

[5]The effect of the `binary` option will depend on your system. In UNIX and Linux, all files are
treated as binary and it is up to the programmer to know how to handle the data.

Creating an `fstream` object, opens a file for both input and output. The following program writes data to a file and then reads from the file. Notice that no `flush()` is needed before reading from the file.

```
#include <fstream>
#include <cstdlib>        // For exit()
using namespace std;

int main()
{
    fstream file("temp.dat", ios_base::in|ios_base::out);
    if (!file) {
        cerr << "Failed to open file.\n";
        exit(EXIT_FAILURE);
    }
    const int items = 10;
    for (int i = 0; i < items; ++i)
        file << (100.0 * i) << " ";
    // Perhaps do some work here.
    // Go to start of file:
    file.seekg(0);
    double x;
    for (int i = 0; i < items; ++i) {
        file >> x;
        // Check what was stored:
        cout << x << " ";
    }
    cout << endl;
    return(EXIT_SUCCESS);
}
```

The second argument of the `fstream` constructor in the above code is the bitwise OR of (`ios_base::in`) and (`ios_base::out`), indicating that the stream is to be opened for both input and output. There is no default for the second argument of `fstream()` so we have to specify the mode of opening the file whenever we use this constructor. This is the first time that we have come across the `seekg()` function and it requires some explanation. The file has an associated pointer that points to the position from which the next byte can be obtained. The relative position of the pointer can be changed by means of the `seekg()` function and, as in this example, `seekg(0)` positions the pointer at the start of the file. The g in `seekg` stands for get. A corresponding `seekp()` function exists for insertion operations, where the p stands for put.

The previous examples use formatted input and output. (Try examining the files, `example.dat` and `temp.dat`, with an editor.) However, if data only needs to be read by a subsequent program (rather than displayed on a terminal for instance) it is much more efficient to use unformatted file I/O; the file stream classes are the same, but we use different functions, namely the `read()` and `write()` that we have already met. For example, we could use unformatted I/O for a file that is open for both input and output:

```
#include <fstream>
#include <cstdlib>        // For exit()
using namespace std;

int main()
{
    fstream file("scratch.dat",ios_base::in|ios_base::out);
    if (!file) {
        cerr <<"Failed to open file.\n";
        exit(EXIT_FAILURE);
    }
    const int items = 20;
    double buffer_1[items];
    double buffer_2[items];
    // Create some data:
    for (int i = 0; i < items; ++i)
        buffer_1[i] = 100.0 * i;
    file.write(buffer_1, items * sizeof(double));
    // Perhaps do some work here.
    // Go to start of file:
    file.seekg(0);
    file.read(buffer_2, items * sizeof(double));
    // Check what was stored:
    for (int i = 0; i < items; ++i) {
        cout << buffer_2[i] << "\n";
    }
    return(EXIT_SUCCESS);
}
```

Now try examining the file, scratch.dat, with an editor.

18.7   Assigning Streams

Although data inserted in the cout, cerr and clog streams is often sent to the screen
by default, it may be more convenient to send one or other of the streams to a separate
disc file. Fortunately it is possible to make assignments to the cout, cin, cerr and
clog streams. For example, the following program sends all subsequent error messages
to the error.dat file:

```
#include <fstream>
#include <cstdlib>        // For exit()
using namespace std;

int main()
{
    ofstream error_file("error.dat");
    if (!error_file) {
        cerr << "Failed to open error_file.\n";
```

```
        exit(EXIT_FAILURE);
    }
    cerr = error_file;
    cerr << "This is an example error file.\n";
    return(EXIT_SUCCESS);
}
```

Exercise

Try running the above program and use an editor to check what has been written to the file `error.dat`.

18.8 Stream Condition

Each I/O stream has an associated condition state, indicating whether or not an error has occurred. There are four possible states:

good No error has occurred.

eof No more data could be inserted or extracted because the end-of-file has been reached.

fail An error has occurred, but the stream is probably usable if the error state is cleared.

bad A serious error has occurred and the stream is probably not usable.

There are also four functions that test for these states:

good() Returns `true` if no errors have occurred.

eof() Returns `true` if an end-of-file was encountered.

fail() Returns `true` if an operation has failed. This includes an `eof` or `bad` error.

bad() Returns `true` if there has been a serious failure.

The logical NOT operator is overloaded so that, if we have a file called `my_file`, then `!my_file` gives the same result as `my_file.fail()`. We have already used this notation to test for successful opening of a file.

The error state can be reset to `good` by invoking the `clear()` function, as in:

```
my_file.clear();
```

but this would probably not be useful if the condition was `bad`.

Exercise

In order to demonstrate the various possible condition states, edit a file called, `input.dat`, so that it contains the single floating point number, `1e1`. Now compile and run the following program:

```
#include <fstream>
#include <cstdlib>      // For exit()
using namespace std;
```

```
int main()
{
    ifstream file("input.dat");
    if (!file) {
        cerr << "Failed to open file.\n";
        exit(EXIT_FAILURE);
    }
    double x;

    file.seekg(1024);
    cout << "\nThe condition states are:\n" <<
        "\tgood is " << file.good() <<
        "\teof is " << file.eof() <<
        "\tfail is " << file.fail() <<
        "\tbad is " << file.bad() << "\n\n";
    file >> x;
    cout << "Data read was " << x <<
        "\nThe condition states are:\n" <<
        "\tgood is " << file.good() <<
        "\teof is " << file.eof() <<
        "\tfail is " << file.fail() <<
        "\tbad is " << file.bad() << "\n\n";

    file.clear();

    file >> x;
    cout << "Data read was " << x <<
        "\nThe condition states are:\n" <<
        "\tgood is " << file.good() <<
        "\teof is " << file.eof() <<
        "\tfail is " << file.fail() <<
        "\tbad is " << file.bad() << "\n\n";

    return(EXIT_SUCCESS);
}
```

Notice that attempting to read beyond the end of the file sets the eof and fail conditions to true and that any attempt to clear the file stream is futile.

Now delete the file.seekg(1024) statement, which attempts to position the get pointer past the end-of-file marker, and try running the program with different data in input.dat. Suggestions are: 1e1, 1e500, 1e1f, 2.4, Z, etc. You could also try changing x from double to other types.

18.9   User-defined Types

As an example of input and output with a user-defined type, consider the `complex` class, defined in Section 9.6.1. Overloaded insertion and extraction operators can be defined by:

```
ostream &operator<<(ostream &os, const complex &z)
{
    os << '(' << z.re << ',' << z.im << ')';
    return os;
}


istream &operator>>(istream &is, complex &z)
{
    char c = 0;
    is >> c;
    if (c == '(') {
        is >> z.re;
        is >> c;
        if (c == ',') {
            is >> z.im;
            is >> c;
        }
        else
            z.im = 0.0;
    }
    else {
        is.putback(c);
        is >> z.re;
        z.im = 0.0;
    }
    return is;
}
```

Notice that both operators must be declared friends of the `complex` class and that the operators also return the appropriate class object, making operator concatenation possible. More sophisticated implementations would carry out error checking and ensure that formatting states of the streams are appropriately set. Our implementation of the extraction operator uses the `putback()` function. This puts the specified character back into the input stream and allows us to handle the extraction of a real number, such as `3.142`.

Exercise

Try out our insertion and extraction operators for the `complex` class by means of a short program that requests a complex number and then outputs the number entered. You should test for the three valid forms of input, such as: `3.142`, `(3.142)` and `(3.142, 2.718)`. Also find out what happens if you input other types of data, such as integers and characters.

18.10   Using Input and Output

There are a lot of features in the I/O part of the Standard Library, so we don't intend
to cover all of them in this section. We present three programs in order to simulate
the role of I/O in processing data. The first program creates some data by writing
two columns of numbers to a file. The numbers are formatted and the columns are
separated by a # character. The second program reads the formatted data from the
file, and outputs each column as a separate unformatted file. The third program reads
the two files of unformatted data and sends them to the screen as two columns.

The first program is given below.

```
// source:  program1.cxx
// use:     Writes initial data to file.

#include <fstream>
#include <cstdlib>        // For exit()
using std::cout;
using std::cerr;
using std::exit;
using std::ofstream;
using std::endl;

int main(int argc, char *argv[])
{
    if (argc != 2) {
        cout << "Usage: program1 <file name>\n";
        exit(EXIT_FAILURE);
    }
    ofstream output_file(argv[1]);
    if (!output_file) {
        cerr << "Failed to open file\n";
        exit(EXIT_FAILURE);
    }
    cout << "Number of lines of data required: ";
    int lines;
    cin >> lines;
    for (int i = 0; i < lines; ++i) {
        output_file << 2.222 * i << " # ";
        if (i % 2)
            output_file << 6.66 * i << endl;
        else
            output_file << -6.66 * i << endl;
    }
    return(EXIT_SUCCESS);
}
```

This program takes a single argument, which is a user-supplied file name. For instance,
for an output file f1.dat, the user would type:

```
program1 f1.dat
```

The file name is given by the string `argv[1]` and this is used as the argument to the `ofstream` constructor. The program then generates two lots of data and this is sent to the output stream. There is no need to explicitly close the output file, since this will be done by the `ofstream` destructor when the program terminates. Since the data is formatted, you can look at the output file with an editor. The motivation behind the program is that the two columns of data simulate what you might have entered as the result of an experiment.

The second program is shown below.

```cpp
// source:  program2.cxx
// use:     Processes initial data.

#include <fstream>
#include <cstdlib>       // For exit()
using std::cout;
using std::cerr;
using std::exit;
using std::ifstream;
using std::ignore;
using std::clear;
using std::seekg;
using std::open;
using std::close;
using std::ofstream;
using std::write;

int main(int argc, char *argv[])
{
    if (argc != 4) {
        cout << "Usage: program2 <input file> " <<
            "<output file 1> <output file 2>\n";
        exit(EXIT_FAILURE);
    }
    ifstream input_file(argv[1]);
    if (!input_file) {
        cerr << "Failed to open input file: " << argv[1] << endl;
        exit(EXIT_FAILURE);
    }
    int lines = 0;
    while (input_file.ignore(1280, '\n'))
        ++lines;
    input_file.clear();
    input_file.seekg(0);
    cout << "There are " << lines << " lines of data.\n";
    double *data_1 = new double[lines];
    double *data_2 = new double[lines];
```

```
        for (int i = 0; i < lines; ++i) {
            input_file >> data_1[i];
            input_file.ignore(80, '#');
            input_file >> data_2[i];
        }
        input_file.close();
        ofstream output_file(argv[2]);
        if (!output_file) {
            cerr << "Failed to open output file: " << argv[2] << endl;
            exit(EXIT_FAILURE);
        }
        output_file.write(&lines, sizeof(lines));
        output_file.write(data_1, lines * sizeof (data_1[0]));
        output_file.close();
        output_file.open(argv[3]);
        if (!output_file) {
            cerr << "Failed to open output file: " << argv[3] << endl;
            exit(EXIT_FAILURE);
        }
        output_file.write(&lines, sizeof(lines));
        output_file.write(data_2, lines * sizeof (data_2[0]));
        output_file.close();
        return(EXIT_SUCCESS);
    }
```

This program takes three command line arguments, and these are the names of one
input and two output files. For instance, the user might enter:

```
program2 f1.dat f2.dat f3.dat
```

The input file name is given by the string `argv[1]` and this is taken as the argument
to the `ifstream` constructor. The `ignore()` function is used to find out the number of
lines of data in the input file. The `seekg()` function puts the get pointer at the start
of the file, but notice that we must first use the `clear()` function. This is because
the `while` loop containing the `ignore()` function only terminates when the end-of-
file is reached. Knowing the number of data points in the input file, we dynamically
allocate memory for the arrays `data_1[]` and `data_2[]`, and then read in the data.
The `ignore()` function is used to discard the `#` symbol. When we have finished with
the input file, we close it by means of the `close()` function. The total number of files
that can be open at any one time is limited (although usually fairly large) so it is a
good idea to close files that are no longer needed. The `write()` function is used to
send unformatted output to the two output files. In each case, the first piece of data is
the number of data points. This is then followed by the data itself. Notice that instead
of using the `ofstream` constructor twice, we close the first output file and then use the
`open()` function to open the second output file with the same `ofstream` object.

The third program is as follows:

```
// source:  program3.cxx
// use:     Displays data files.
```

```cpp
#include <fstream>
#include <cstdlib>        // For exit()
#include <iomanip>        // For setw()
using std::cout;
using std::cerr;
using std::exit;
using std::ifstream;
using std::read;
using std::open;
using std::close;
using std::setw;
using std::endl;

int main(int argc, char *argv[])
{
    if (argc != 3) {
        cout << "Usage: program3 <input file 1> " <<
            "<input file 2>\n";
        exit(EXIT_FAILURE);
    }
    ifstream input_file(argv[1]);
    if (!input_file) {
        cerr << "Failed to open input file: " << argv[1] << endl;
        exit(EXIT_FAILURE);
    }
    int data_points_1;
    input_file.read(&data_points_1, sizeof(data_points_1));
    cout << "There are " << data_points_1 <<
        " data points in file: " << argv[1] << endl;
    double *data_1 = new double[data_points_1];
    input_file.read(data_1, data_points_1 * sizeof (data_1[0]));
    input_file.close();
    input_file.open(argv[2]);
    if (!input_file) {
        cerr << "Failed to open input file: " << argv[2] << endl;
        exit(EXIT_FAILURE);
    }
    int data_points_2;
    input_file.read(&data_points_2, sizeof(data_points_2));
    cout << "There are " << data_points_2 <<
        " data points in file: " << argv[2] << endl;
    if (data_points_1 != data_points_2) {
        cout << "Inconsistent data in files: " << argv[1] <<
            " and " << argv[2] << endl;
        exit(EXIT_FAILURE);
    }
```

```
        double *data_2 = new double[data_points_2];
        input_file.read(data_2, data_points_2 * sizeof (data_2[0]));
        input_file.close();
        cout << "  data file 1    data file 2\n";
        for (int i = 0; i < data_points_1; ++i)
            cout << setw(10) << data_1[i] << "     " << setw(10) <<
                data_2[i] << endl;
        return(EXIT_SUCCESS);
}
```

This program takes two command line arguments. For example, the user might enter:

```
program3 f2.dat f3.dat
```

The program extracts the unformatted data from the input files by means of the `read()` function. The data is then sent to the screen as two columns. Notice the use of `setw(10)` in an attempt to line up the columns.

Exercise

Compile and run the above programs, and then make the following improvements:

(a) The three programs do not carry out a lot of checking. For instance, the first program simply counts the number of lines of input and doesn't allow for the possibility of the final input line being blank. Modify the program so that it recovers gracefully from such an error.

(b) The third program doesn't line up the data properly in two columns, due to the varying number of digits in a number. Modify the code so that the decimal points are lined up in each column.

18.11   Summary

- Input and output streams are declared in the header file, `<iostream>`.

- The standard input stream, `cin`, is a predefined `istream` object.

- The standard output, `cout`, and standard error, `cerr`, streams are predefined `ostream` objects. The `cerr` stream is effectively unbuffered. There is also a standard buffered error stream, `clog`, which is an `ostream` object.

- The inserter or insertion operator, `<<`, is used to insert data into an output stream.

- The extractor or extraction operator, `>>`, is used to extract data from an input stream.

- The `flush()` function is used for flushing a buffered output stream, such as `cout`:

```
        cout.flush();
```

- The functions, `put()` and `write()`, are used for unformatted insertion into an output stream:

  ```
  cout.put('C');
  cout.write(buffer, 3);
  ```

- The functions, `get()` and `getline()`, are used for unformatted extraction from an input stream. There are three different versions of the `get()` function:

  ```
  int a = cin.get();
  cin.get(b);
  cin.get(buffer, length, '#');
  ```

  The default delimiter for the three-argument version of `get()` is the newline character. Never invoke the three-argument version of `get()` twice, unless the delimiter is removed. A better alternative is to use `getline()`, which does remove the delimiter:

  ```
  cin.getline(buffer, length, '#');
  ```

- The `gcount()` function returns the number of characters extracted by an unformatted extraction:

  ```
  int chars_extracted = cin.gcount();
  ```

- The `read()` function extracts characters from a stream without regard for their meaning:

  ```
  cin.read(buffer, 20);
  ```

- The `ignore()` function discards characters in an input stream:

  ```
  cin.ignore(20, '#');
  ```

- Use manipulators to change the state of a stream:

  ```
  cout << flush;
  ```

  Some manipulators take arguments, but don't forget to include the `<iomanip>` header file:

  ```
  cout << setprecision(4);
  ```

  Manipulators are listed in Tables 18.2 and 18.3, and are particularly important for formatting.

- File I/O uses the classes declared in `<fstream>`. To open a file for output *from* a program, use the `ofstream` constructor:

```
ofstream output_file("file.dat");
```

To open a file for input *to* a program, use the `ifstream` constructor:

```
ifstream input_file("file.dat");
```

To open a file for both input and output use the `fstream` constructor:

```
fstream file("file.dat", ios_base::in|ios_base::out);
```

All three constructors have second arguments that determine the mode in which the file is opened. (See Table 18.4.) The `ofstream` and `ifstream` constructors have defaults for these arguments.

- Always check that a file has been successfully opened:

```
if (!output_file)
    cerr << "Failed to open file.\n";
```

- Use `close()` to explicitly close a file:

```
output_file.close();
```

- The function, `seekg()`, changes the position of the `get` pointer, associated with a file open for input:

```
input_file.seekg(0);
```

- The function, `seekp()`, changes the position of the `put` pointer, associated with a file open for output:

```
output_file.seekp(0);
```

- Streams can be assigned:

```
ofstream error_file("error.dat");
cerr = error_file;
```

In this example, insertions are sent to the `error.dat` file instead of the screen.

- The condition of a stream can be good, `eof`, `fail` or bad.

- The error state of a stream is tested by the `good()`, `eof()`, `fail()` and `bad()` functions:

```
if (my_file.good())
    cout << "Condition state is good.\n";
```

- The error state can be reset to good by using the `clear()` function:

```
my_file.clear();
```

- The insertion operator, <<, can be overloaded for a user-defined class, X, by a friend function with the declaration:

  ```
  ostream &operator<<(ostream &os, const X &obj);
  ```

- The extraction operator, >>, can be overloaded for a user-defined class, X, by a friend function with the declaration:

  ```
  istream &operator>>(istream &is, X &obj);
  ```

- For details of the I/O Library that are not discussed in this chapter consult [1] and [10], together with the reference manual for your particular system.

18.12 Exercises

1. Explain what the result of the following code would be:

   ```
   double x = 128.0;
   cout.write(&x, sizeof(x));
   ```

2. Using the techniques learned in this chapter, modify the activate() function for the menu class discussed in Section 12.7.2 so that the function handles incorrect user-supplied values for choice, such as "which one?".

3. By assigning cin and cout, modify the alphabetic sort program, developed in Section 7.8.2, so that it reads the data from a file and writes the results to a different file.

4. Implement insertion and extraction operators for the string class, described in Section 9.6.2.

5. For the list class (developed in Section 10.3.1) replace the print() function by an inserter.

6. Implement an inserter for a two-dimensional self-describing array class. (See Chapter 8.) For small arrays, your operator should produce output in the style:

$$\begin{bmatrix} 1 & 4 & 7 \\ 2 & 3 & 0 \\ 9 & 1 & 5 \end{bmatrix}$$

For large arrays the required style is:

```
row 0:
            1  3  2  7  8  9  ...

row 1:
            0  9  2  1  4  6  ...
```

Ensure that the output is appropriately formatted, irrespective of the initial format setting for the output stream.

7. Design and implement a class so that objects can store an arbitrary length, signed integer. Implement an appropriate inserter and extractor.

8. Use the I/O library to output a *neatly* formatted calendar of the form:

		January						February		
Sun		5	12	19	26		2	9	16	23
Mon		6	13	20	27		3	10	17	24
Tue		7	14	21	28		4	11	18	25
Wed	1	8	15	22	29		5	12	19	26
Thu	2	9	16	23	30		6	13	20	27
Fri	3	10	17	24	31		7	14	21	28
Sat	4	11	18	25		1	8	15	22	29

Appendix A

The ASCII Character Codes

Decimal	Octal	Hex.	Meaning
0	0	0	null
1	1	1	SOH
2	2	2	STX
3	3	3	ETX
4	4	4	EOT
5	5	5	ENQ
6	6	6	ACK
7	7	7	bell
8	10	8	backspace
9	11	9	horizontal tab
10	12	a	new line
11	13	b	vertical tab
12	14	c	form feed
13	15	d	carriage return
14	16	e	SO
15	17	f	SI
16	20	10	DLE
17	21	11	DC1
18	22	12	DC2
19	23	13	DC3
20	24	14	DC4
21	25	15	NAK
22	26	16	SYN
23	27	17	ETB
24	30	18	CAN
25	31	19	EM

Decimal	Octal	Hex.	Meaning
26	32	1a	SUB
27	33	1b	escape
28	34	1c	FS
29	35	1d	GS
30	36	1e	RS
31	37	1f	US
32	40	20	space
33	41	21	!
34	42	22	"
35	43	23	#
36	44	24	$
37	45	25	%
38	46	26	&
39	47	27	'
40	50	28	(
41	51	29	)
42	52	2a	*
43	53	2b	+
44	54	2c	,
45	55	2d	-
46	56	2e	.
47	57	2f	/
48	60	30	0
49	61	31	1
50	62	32	2
51	63	33	3

Table A.1: ASCII character codes.

Decimal	Octal	Hex.	Meaning	Decimal	Octal	Hex.	Meaning
52	64	34	4	90	132	5a	Z
53	65	35	5	91	133	5b	[
54	66	36	6	92	134	5c	\
55	67	37	7	93	135	5d	]
56	70	38	8	94	136	5e	^
57	71	39	9	95	137	5f	_
58	72	3a	:	96	140	60	`
59	73	3b	;	97	141	61	a
60	74	3c	<	98	142	62	b
61	75	3d	=	99	143	63	c
62	76	3e	>	100	144	64	d
63	77	3f	?	101	145	65	e
64	100	40	@	102	146	66	f
65	101	41	A	103	147	67	g
66	102	42	B	104	150	68	h
67	103	43	C	105	151	69	i
68	104	44	D	106	152	6a	j
69	105	45	E	107	153	6b	k
70	106	46	F	108	154	6c	l
71	107	47	G	109	155	6d	m
72	110	48	H	110	156	6e	n
73	111	49	I	111	157	6f	o
74	112	4a	J	112	160	70	p
75	113	4b	K	113	161	71	q
76	114	4c	L	114	162	72	r
77	115	4d	M	115	163	73	s
78	116	4e	N	116	164	74	t
79	117	4f	O	117	165	75	u
80	120	50	P	118	166	76	v
81	121	51	Q	119	167	77	w
82	122	52	R	120	170	78	x
83	123	53	S	121	171	79	y
84	124	54	T	122	172	7a	z
85	125	55	U	123	173	7b	{
86	126	56	V	124	174	7c		
87	127	57	W	125	175	7d	}
88	130	58	X	126	176	7e	~
89	131	59	Y	127	177	7f	delete

Table A.2: Table A.1: ASCII character codes—*continued*.

Appendix B

Operators

Table B.1 lists the available C++ operators. This table also gives the sections where the operators are first introduced since this should allow you to easily look up what an operator does. (If there is no section number, the operator is not used in this book.) All of the operators within a group in this table have the same precedence. A group of operators higher up in the table has a higher precedence than a group further down the table. All binary operators, except for the assignment operators, are left associative. The assignment operators and unary operators are right associative. Note that it is *not* true that all operators in the same group have the same associativity.

Within an expression, operators with the higher precedence are evaluated first. If the resulting expression contains operators with the same precedence, then these are evaluated right to left, or left to right, according to their associativity. For instance, in the expression:

```
6 + 3 * 4 / 3 + 2
```

the * and / operators have the same highest precedence. Since they are not unary or assignment operators, they also have the same left to right associativity. Therefore 3 * 4 is evaluated first, giving 4 for the expression 3 * 4 / 3. The whole expression then reduces to:

```
6 + 4 + 2
```

which is evaluated, left to right, giving the final result of 12.

It is not worth learning the precedence of every operator; for the less well used operators either consult Table B.1 or else use parentheses. In any case, if you are uncertain about how an expression is evaluated according to operator precedence and associativity, then using parentheses will probably make it much more readable.

Operator		Section
`::`	scope resolution	8.5, 14.3
`::`	global scope resolution	5.4
`.`	member selection	8.3
`->`	member selection	8.9
`[]`	subscripting	6.2
`()`	function call	5.1
`++`	postfix increment	3.1.5
`--`	postfix decrement	3.1.5
`typeid`	type identification	
`typeid`	run-time type identification	
`dynamic_cast<>()`	run-time conversion	3.3.2
`static_cast<>()`	compile-time conversion	3.3.2
`reinterpret_cast<>()`	unchecked conversion	3.3.2
`const_cast<>()`	const conversion	3.3.2
`sizeof`	size of object	3.4.1
`sizeof`	size of type	3.4.1
`++`	prefix increment	3.1.5
`--`	prefix decrement	3.1.5
`-`	unary minus	3.1.6
`()`	C-style cast	Appendix C
`!`	logical negation	4.2
`&`	address-of	6.1.1
`*`	dereference	6.1.2
`new`	allocate	7.6.1
`new []`	allocate	7.6.1
`delete`	deallocate	7.6.2
`delete[]`	deallocate	7.6.2
`~`	one's complement	11.1.1
`->*`	member selection	8.10
`.*`	member selection	8.10
`*`	multiplication	3.1.2
`/`	division	3.1.3
`%`	modulo (remainder)	3.1.4
`<<`	left shift	11.1.5
`>>`	right shift	11.1.5
`>`	greater than	4.1
`>=`	greater than or equal to	4.1
`<`	less than	4.1
`<=`	less than or equal to	4.1

Table B.1: Operator precedence.

Operator		Section
+	addition	3.1
−	subtraction	3.1
==	equal	4.3
!=	not equal	4.3
&	bitwise AND	11.1.2
^	bitwise XOR	11.1.3
|	bitwise OR	11.1.4
&&	logical AND	4.2
||	logical OR	4.2
=	assignment	3.1
+=	add and assign	3.4.3
-=	subtract and assign	3.4.3
*=	multiply and assign	3.4.3
/=	divide and assign	3.4.3
%=	modulo and assign	3.4.3
<<=	left shift and assign	11.1.6
>>=	right shift and assign	11.1.6
&=	AND and assign	11.1.6
^=	exclusive OR and assign	11.1.6
|=	inclusive OR and assign	11.1.6
?:	conditional operator	4.11
throw	throw exception	15.2
,	comma	4.9

Table B.2: Table B.1: Operator precedence—*continued.*

Appendix C

Differences between C and C++

The basic premiss underlying this book is that C++ can be learned from scratch, rather than as an adjunct to C. However, you may have to reuse existing code written in C.[1] The aim of this Appendix is to highlight some of the differences between the two languages. There are also some differences between the original, 1978, edition of [5] and the ANSI C Standard; these differences are discussed in many references, including [5]. In fact, many improvements to the C language are features that were originally part of C++. However, a further complication is that the ANSI C++ Standard differs from earlier versions of C++. Here we concentrate on the major differences between ANSI C and C++. Since the assumption is that you will mainly program in C++ rather than C, this Appendix only highlights the major differences. A more complete discussion is given in Appendix B of [10].

Features only available in C++

Many features that are available in C++ are not available in C. The following is a brief summary of features that are only available in C++:

- Support is provided for object-oriented programming. Classes can be declared and these can have data hiding, member functions, `virtual` functions, overloaded operators, derived classes, `friend` functions and classes, constructors, destructors etc. (See Chapters 8 to 18 for numerous examples.)

- Function and class templates can be declared.

- Function names can be overloaded:

      int square(int i);
      double square(double x);

- The `inline` specifier can be used to suggest that the compiler directly substitutes the function body:

---

[1] If you do need information specifically on C, rather than C++, then consult [4], [5] and [6].

```
inline int square(int i) { return i * i; }
```

- The `new` and `delete` operators perform dynamic memory management:

```
double *pt = new double[1000];
// ...
delete pt;
```

- Objects can be declared anywhere within a block:

```
int length;
length = 1000;
vector v(length);
```

- Declarations can occur in `for` statement initializers and conditions:

```
for (int i = 0; i < i_max; ++i)
    sum += f(i);
```

This is a significant departure from earlier versions of C++.

- An explicit type conversion can be made by a casts of the form:

```
dynamic_cast<>()       static_cast<>()
const_cast<>()         reinterpret_cast<>()
```

These replace the C-style casts:

```
x = double(i);      // C-style cast.
y = (double)j;      // Older C-style cast.
```

which are still available in C++, but should be avoided.

- There is a distinct Boolean type, `bool`, together with the defined keywords `true` and `false`.

- Comments can be denoted by `//`:

```
// This is a comment.
```

- The standard `iostream` library overloads operators to perform type-safe I/O:

```
cout << "Enter dimension: " << flush;
cin >> dim;
```

- Errors can be trapped by means of exception handling, which introduces the C++ keywords, `throw` and `catch`.

Unemphasized features common to C and C++

There are some features, common to C and C++, which have not been described in
detail because they have been superseded by features available in C++:

- The #define preprocessor directive is often used in C programs to implement
  complicated macros and examples occur in many of the header files for the standard C libraries. The C++ technique of declaring inline functions provides a
  safer alternative.

- Structures are widely used in C and are like *very* restricted classes that can only
  have public data members:

```
struct complex {
    double re, im;
};

struct complex z;
z.re = 11.11;
z.im = -2.13;
```

  Structures have not been introduced since classes give a uniform notation.

- Dynamic memory management is performed in C by library functions (such as
  malloc()) rather than the new and delete operators. The C++ operators are
  safer and more convenient.

- Input and output can be carried out by library functions, such as scanf() and
  printf(). The C++ iostream library provides a safer, more convenient and
  object-oriented technique.

- Using the ellipsis in a function declaration, as in:

```
void results(...);
```

  indicates that the function has an unknown number of arguments. Since any type
  checking is lost it is best to avoid defining such functions if it is at all possible.

Some incompatibilities between C and C++

There are some very minor differences of interpretation between statements that are
valid C and C++:

- Whereas in C++ 'x' has the type char, in C it has the type int. As a consequence, sizeof('x') is equal to sizeof(char) in C++ and sizeof(int) in
  C.

- Whereas in C++ an empty argument list in a function declaration means that
  there are no arguments (as does void), in C it indicates unknown arguments.
  Although interpreting an empty argument list in this way is now obsolescent in
  C, you may still come across it.

- A C++ structure, `struct`, is a synonym for a class in which all of the members are `public` by default. Since a C++ structure can have function members, it is a significant extension of the C construct.

- The default linkage of an identifier, preceded by the `const` specifier, is `extern` in C. This is not true in C++ and it must either be initialized or explicitly declared `extern`.

- A C++ compiler "mangles" function names (in a well-defined way) in order to provide type-safe linkage. This means that object code resulting from a C compiler will not link with code from a C++ compiler. The simplest solution may be to recompile the C source files using the same C++ compiler. The alternative is to use a linkage directive in the C++ code:

```
extern "C" double old_C_function(double x);
```

which prevents name mangling for this particular function. There is an alternative form of the linkage directive for more than one function:

```
extern "C" {
    double old_C_function(double x);
    int ancient_C_function(char *pt);
}
```

Bibliography

[1] ISO/IEC 14882:1998, *Programming languages – C++*.
This is what is commonly known as "The ANSI C++ Standard". You can download the complete Standard for a small fee from the website, `www.webstore.ansi.org`. It is certainly not a good starting point for beginners, but is a very useful source of information for more experienced C++ programmers.

[2] Grady Booch, Robert Martin and James Newkirk. *Object-Oriented Analysis and Design with Applications*. Addison Wesley Longman, 2000.
This book is about design, rather than language, and is not specifically about numerical applications. It is recommended reading if you are designing large applications using object-oriented techniques.

[3] Margaret A. Ellis and Bjarne Stroustrup. *The Annotated C++ Reference Manual*. Addison Wesley Longman, 1990.
This is very much a reference manual rather than a tutorial. It is a good, well structured, source of information on the details of why various features of the language are defined as they are. It is certainly not an easy book for the beginner nor is there any discussion of the Standard Library. Note that it pre-dates the approval of the ANSI Standard.

[4] Al Kelly and Ira Pohl. *A Book On C*. Addison Wesley Longman, 1998.
This is an excellent introduction to C and is recommended if you need to use those aspects of C++ that are included to provide backward compatibility with C.

[5] Brian W. Kernighan and Dennis M. Ritchie. *The C Programming Language*. Prentice-Hall, Englewood Cliffs, New Jersey, 1988.
Some may find the style rather terse, but this is the standard reference on the C programming language.

[6] Andrew Koenig. *C Traps and Pitfalls*. Addison Wesley Longman, 1998.
A very readable description of some of the features of C that can trap the programmer. Most of the examples are also relevant to C++; those that are not serve to demonstrate what you are missing.

[7] Bertrand Meyer. *Object-Oriented Software Construction*. Prentice-Hall, 1997.
Here is a wealth of useful ideas on object-oriented design. Example code fragments use the author's own language, Eiffel, but much of the book is relevant to C++.

[8] Robert Robson. *Using the STL: the C++ standard template library*. Springer-Verlag, New York, 1999.
A very good description of the STL, but without any emphasis on numerical applications.

[9] James T. Smith. *C++ Toolkit for Scientists and Engineers, Second Edition*. Springer-Verlag, New York, 1999.
A description of useful application packages rather than the language itself. The applications include scalar equations, vectors, polynomials and matrices.

[10] Bjarne Stroustrup. *The C++ Programming Language, Third Edition*. Addison-Wesley, Reading, Massachusetts, 1997.
This is the definitive text book on C++ by the designer of the language. There is no emphasis on numerical applications and it is certainly not for beginners. However, the book does contain lots of useful insights, and the chapters on design and development are particularly valuable.

The following references provide the background for many of the applications that appear in this book. With one exception, they are not directly relevant to C++ and some of them are quite specialized:

[11] Milton Abramowitz and Irene A. Segun. *Handbook of Mathematical Functions*. Dover Publications, 1968.
A standard source of formulas and graphs for a large variety of mathematical functions.

[12] R.B.J.T. Allenby and E.J. Redfern. *Introduction to Number Theory with Computing*. Edward Arnold, 1989.
This is a delightful introduction to number theory and a useful background to some of our programming examples.

[13] E. Atlee Jackson. *Perspectives of Nonlinear Dynamics, Volume 1*. Cambridge University Press, 1989.
Contains a useful description of the logistic map, together with many references.

[14] William H. Press, Brian P. Flannery, Saul A. Teukolsky and William T. Vetterling. *Numerical Recipes in C, The Art of Scientific Computing*. Cambridge University Press, 1993.
This is a very readable source of information on numerical techniques.

[15] Robert Sedgewick. *Algorithms in C++*. Addison Wesley Longman, 1992.
Another good source of algorithms, with more emphasis on non-numerical techniques than [14].

Index